Excel

SUCCESS ONE® HSC*

CHEMISTRY

Past HSC Questions & Answers
2001–2003 by Topic
2009–2017 by Paper

PASCAL
PRESS

ISBN 978 1 74125 668 0

Pascal Press
PO Box 250
Glebe NSW 2037
(02) 8585 4044
www.pascalpress.com.au

Publisher: Vivienne Joannou
Commissioning and project editor: Mark Dixon
Chapters 1–5 edited by Brendan Atkins, Big Box Publishing Pty Ltd and Bruce Howarth
Chapter 6 edited by Bruce Howarth
Chapters 7–9 edited by Christine Eslick
Chapter 10 edited by Leanne Poll
Chapters 11–13 edited by Karen Pearce
Chapters 14–15 edited by Rosemary Peers
Chapters 1–14 answers checked by Geoffrey Thickett
Chapter 15 answers checked by Debbie Noyes and Geoffrey Thickett
Cover design by Michael Sherman
Page layout and typesetting by ljDesign (Julianne Billington), DiZign Pty Ltd and Midland Typesetters
Cover photo by Janie Barrett, Fairfax Syndication
Printed by Vivar Printing/Green Giant Press

Foreword

Congratulations on choosing the Chemistry course for your HSC. In undertaking this course you have the opportunity to gain valuable knowledge, understanding and skills, and to extend yourself to meet new challenges.

Excel Success One HSC Chemistry is a valuable learning tool that has been developed to assist students with their HSC preparation, and this edition has been kept up to date with the inclusion of the 2017 HSC Examination paper with sample answers.

On behalf of the Science Teachers' Association of New South Wales, we hope that your year will be enjoyable and productive, and the results a just reward for your efforts.

STANSW thanks those members who continue to be involved in the production of this book.

Margaret Shepherd
STANSW President
December 2017

The Mark Maximizer Guide was prepared by Diane Alford.

HSC EXAMINATION PAPERS

SAMPLE ANSWERS

The sample answers to HSC questions contained in this publication are examples of answers which the authors believe would score full marks. However, they are not endorsed by NESA.

The sample answers to the Core section and the Shipwrecks, Corrosion and Conservation option in the 2001–2016 HSC Examination papers were written by Craig Seawright, and in the 2017 HSC Examinaton paper by Geoffrey Thickett. The Industrial Chemistry and Forensic Chemistry options in the 2011–2017 HSC Examination papers were written by Dr Jason Hoare. Other contributors to the 2001–2012 HSC Examination papers were Annu Mishra and Theo Leondios.

COMMENTS

The Science Teachers' Association of NSW (STANSW) and Pascal Press welcome constructive comments on the questions and answers for future editions of this book.

Contents

Mark Maximizer Guide

This Mark Maximizer Guide will arm you with strategies and tips so you can maximize your marks by making every minute of your HSC Examination count. And it has some great tips for your exam preparation too. Read it. Use it.

The structure of the HSC Examination paper

All Science exams consist of two sections. The following table summarises the structure, which changed in 2010.

Section	*Part*	*Compulsory/ Optional*	*Marks*	*Question type*	*Answers*	*Recommended time*
I	A	All questions are compulsory.	Each question is worth one mark – a total of 20.	Objective-response – select one alternative from A, B, C or D.	Fill in the response oval completely on the answer sheet.	About 35 min
I	B	All questions are compulsory.	55	Short-answer questions with different mark values, each indicated on the question: some questions may be in parts; some may require a more integrated approach.	Answer the questions in the space provided in the exam paper.	About 1 hour 40 min
II		Attempt only one question.	25	The question is divided into parts, with each part having the mark clearly indicated.	Answer the questions in a writing booklet.	About 45 min
Inform-ation	The paper contains general instructions on the front cover but in addition, Chemistry contains a Data Sheet and Periodic Table.					

Time allocation

The examination begins with five minutes reading time. You will not be able to write in this time. Rather than starting to read the objective-response questions at the beginning, you could:

- spend this time reading some of the short-answer questions in Part B and Section II,

  or

- select some of the easy questions to attempt earlier in the exam after completing the objective-response questions, to build your confidence for the more difficult questions later.

Time management

A planned approach to time management in the exam is important. There are several strategies you could use:

- Prepare an exam timetable
  The table above shows the suggested times for each section as recommended by NESA. If you choose to proceed through the exam in an orderly sequence of sections

or parts, you can work out an exam timetable before the exam. You will then know when you should be changing sections and check that you are on schedule. Try to be a bit ahead of schedule so that you have time to check answers, make corrections or return to more difficult questions.

- Calculate time allocated per mark
  If you do not choose to work through the exam in a fixed sequence, you may wish to calculate time allocations for questions based on the 100 marks and 180 minutes in which to record answers. Each mark should take 1 minute 48 seconds to earn. The following table gives calculations of approximately how long you should spend on questions of a particular mark value.

Question value (marks)	*Suggested time*
1	1 min 45 s
2	3 min 30 s
3	5 min 20 s
4	7 min 00 s
5	9 min 00 s
6	10 min 45 s
7	12 min 45 s
8	14 min 30 s
9	16 min 00 s
10	18 min 00 s

- Don't panic!
  Worrying about time too much will only add to your stress and waste time. There is no need to time yourself for each question, and particularly not for the low-value questions. However, if you panic or lose track of time or waste too much time on a question of relatively small mark value, you may be severely disadvantaged.

Analysing the questions

Questions in the HSC are designed to enable students to show their achievement of a range of outcomes over a range of levels. They are not trying to trick you, and the question will state clearly all that is required of you. If marks are to be awarded for giving an answer in the correct units, then the question will clearly state that you need to give the answer in the correct units.

- Highlight key terms
  Highlight or underline the key terms or things you have to do (the verbs) and the key ideas or concepts that make up the question — a useful strategy.

- Learn the lingo
  The glossary of terms that are often used in questions will help you analyse what is required of the question (see the Glossary of key words on page ix).

- Remember that *questions are planned to vary in their level of difficulty*
  Don't panic or waste excessive time on a question that you find particularly difficult. The following table may give you some ideas on how to find the easier questions that you can attempt first. Then, when you have finished everything that you can do, you can move on to spend time with the very difficult questions.

Question type	*Key words that indicate level of difficulty*
Difficult or complex questions often demanding high levels of skills to use the facts	assess, construct, critically analyse, critically evaluate, evaluate, justify, propose, recommend
Questions of medium levels of difficulty requiring some thinking and using the facts	account for, analyse, apply, assess, calculate, classify, compare, contrast, deduce, demonstrate, discuss, examine, explain, estimate, extract, extrapolate, how…, interpret, investigate, predict or recommend, sketch, why...
Simple or direct questions of low levels of difficulty based on remembering and communicating the facts	account, clarify, define, describe, identify, outline, recall, recount or summarise, state, what..., which...

  Also, there may be variation in difficulty within the topics you have studied. A relatively easy remembering-type question may become a little more difficult if it is about a particularly difficult idea.

- Look at question structure and mark value
  The structure of the question may also contribute to differences in levels of difficulty. The mark value for the question may give you some clues. For example if the question is only two lines long but is worth six marks, that is a clue that it may require extra care to ensure all aspects of the question have been covered.

  Many students find the questions that are divided into subsections easier to attempt. They feel that each small portion of the question can be 'bitten off' at a time. Often the answer to one subsection helps to provide some ideas to help with the other subsections.

  The more difficult integrated questions may seem a bit more difficult to swallow, and when you do, they may give you indigestion. Practising these questions using examples from this book will really help you develop your skills and help avoid the need for antacid tablets!

- Answer all parts
  A common error is that students often only answer one part of a two-part question. This is very easy to do if the parts are not clearly numbered but are written as sentences following each other. It is worth making a note on the paper of any multiple-part questions and make sure you return to the question to finish all sections.

Glossary of key words

For each subject you should prepare summaries and a glossary of the key definitions. You cannot answer questions if you are unfamiliar with the content of the subject.

You also need to be familiar with the language of exam questions. This involves understanding the verbs that are the keywords that tell you what you have to do to earn the marks. Below is a table of the key verbs that may be contained in questions and the definitions of those words.

Key word	*Definition*
Account for	state reasons for; report on
(Give an) account of	narrate a series of events or transactions
Analyse	Identify components and the relationship among them; draw out and relate implications
Apply	Use, utilise, employ to a particular situation
Appreciate	Make a judgment about the value of
Assess	Make a judgment of value, quality, outcomes, results or size
Calculate	Ascertain/determine from given facts, figures or information
Clarify	Make clear or plain
Classify	Arrange or include in classes/categories
Compare	Show how things are similar or different
Construct	Make; build; put together items or arguments
Contrast	Show how things are different or opposite
Critically (analyse/evaluate)	Add a degree or level of accuracy, depth, knowledge and understanding, logic, questioning, reflection and quality to
Deduce	Draw conclusions
Define	State meaning and identify essential qualities
Demonstrate	Show by example
Describe	Provide characteristics and features
Discuss	Identify issues and provide points for and/or against
Distinguish	Recognise or note/indicate as being distinct or different from; to note differences between
Evaluate	Make a judgment based on criteria; determine the value of
Examine	Inquire into
Explain	Relate cause and effect; make the relationships between things evident; provide why and/or how
Extract	Choose relevant and/or appropriate details
Extrapolate	Infer from what is known
Identify	Recognise and name

Key word	*Definition*
Interpret	Draw meaning from
Investigate	Plan, inquire into and draw conclusions about
Justify	Support an argument or conclusion
Outline	Sketch in general terms; indicate the main features of
Predict	Suggest what may happen based on available information
Propose	Put forward (for example a point of view, idea, argument, suggestion) for consideration or action
Recall	Present remembered ideas, facts or experiences
Recommend	Provide reasons in favour
Recount	Retell a series of events
Summarise	Express concisely the relevant details
Synthesise	Putting together various elements to make a whole

From Senior Science Support Document © NSW Education Standards Authority 1999

Answering the question

Objective-response questions

Many people find objective-response questions relatively easy. Recording the answer is certainly easy, but they can be deceptive in their level of difficulty. Some make up for the time saved recording the answer by requiring time to think about the answer. Some of the distracters (i.e. the wrong answer options) may be based on common mistakes or misconceptions and may lead you away from the correct answer. In one 2000 HSC Science exam only 15% of students got one particular objective-response question correct, showing how effective the distracters can be.

Careful reading and interpretation of questions is critical. Sometimes you may be left with two similar answers and you have to dig into your deeper understanding to work out the implications of the subtle difference. The more practice you get on objective-response questions, the more skilful you will become. In this book we have provided explanations for making the correct choices; you do not have to do this in an exam.

Short-answer questions

- Core modules

  The short-answer questions from Part B relate to the three core HSC modules. After you have analysed the question, link it to the specific module and the concepts of that module. If the question is in parts, look at the parts and see if they give you clues about the other sections of the question. If the question gives you the freedom to select the format of your answer, choose a format that you feel most comfortable with. For example, you may prefer to express some ideas in a table, flowchart or diagram, or in writing. This may be especially important in those questions with a high mark value that require you to integrate ideas.

  Use the mark value and space provided to give you a rough guide to the number of points in the answer and the length of the answer. If it is worth six marks, you may need to include at least six key points or one point with a set of arguments, supportive statements, reasons or counter-arguments.

- Options
  Short-answer questions from Section II all relate to the specific option you have chosen. For the integrated questions, it may be worth taking a minute to do a mental revision about the major ideas of that option and checking which sections relate to the particular question. Again, if you are given the opportunity, choose an answer format that best expresses the key ideas. Answers should be as succinct as possible.

- Read through your answers
  It is important to read through your answers to short-answer questions to check that you have clearly communicated your understanding. Check that any assumptions you have made have been stated and that your thinking is clear. If you are running out of time it is important to attempt questions in a very concise manner; even if you remain uncertain as to what the question is about, think back to the option and try to make some links to key ideas.

- Attempt all questions
  It is essential you attempt all questions. You do not lose marks for guessing or recording wrong answers. Remember, however, if you ramble on with a lengthy, confused or contradictory answer you may make a satisfactory answer meaningless and therefore worth no marks. Clear and concise answers are always preferable. If you are running out of time, writing key points, tables or diagrams may help communicate an answer or earn partial marks.

  Avoid persevering with one question to make it perfect if it means you leave out several others. The mark value of questions can serve as a guide for prioritising your attempts at questions when running out of time. You will do yourself a great disservice if you do not attempt questions worth six or more marks.

- Show your working
  In Chemistry it is essential that you show the relevant working to questions that involve calculations.

Common mistakes

Examiners provide reports after each HSC Exam that indicate areas of strengths and weaknesses in student responses. From these common mistakes, it is clear that students generally can benefit by:

- more precise and scientific use of terminology, especially that specified in the syllabus;
- more precision with graphs and labelling of diagrams;
- avoiding answers that are too long or detailed or that use ambiguous terms;
- correct interpretation of questions, eg listing instead of describing, or giving a description not an explanation;
- answering in specific terms not generalities;
- ensuring all parts of multi-part questions are attempted;

- being able to clearly describe procedures, apparatus, etc for mandatory practical activities as a first-hand experience.

Golden rules

Preparing for the exam

- Obviously the exam will reward those students who have a thorough knowledge of the whole course. Do all assignments and assessment tasks to your best ability and use them in your revision.
- Get as much practice at completing and getting feedback on exam-type questions as you possibly can. This book is perfect for this!
- Make certain you complete, know and revise the mandatory practicals from the syllabus so you can describe the procedure or evaluate its methods.
- Revise a question about an open-ended investigation you have done and be prepared to use it as an answer.
- Make sure you have learnt a few examples that show something about the history, nature and practice, applications and uses, and the implications of science as it relates to your specific subject.

In the exam

- Don't give up! If you find it difficult, so will most other students. Attempt all questions with concise, clear answers.
- Have a plan that ensures you manage time well. If the option is your particular strength, you may start with it, but do not spend more than 45 minutes on it. Alternatively start with the objective-response because the correct answer must be there, and it can help you get over your nerves.
- Know yourself and work to your strengths. If you want to experiment with different strategies for tackling the exam, do that in the trial exam and reflect on how the plan worked.
- At the end re-read the questions and the answers you have written. Check that you have really communicated what you intended to write.
- Don't panic! It can lead you to make some silly decisions. There will be some difficult questions in the exam but there will also be some easy ones.
- Even if the option you have studied looks difficult, and another one looks like it has some 'easy' questions, you are far better off answering the question from the option you have studied in class than one you have not studied.

Finally, GOOD LUCK!

CHAPTER 1

Core Topic
Production of Materials

Multiple-choice Questions

Past HSC Questions

1 Ethene may be converted into poly(ethene).

What type of reaction is this?

(A) Condensation

(B) Hydrolysis

(C) Oxidation/reduction

(D) Polymerisation

2 Which of the following is a major component of biomass?

(A) Cellulose

(B) Ethanol

(C) Natural gas

(D) Oil

3 Cellulose is a linear polymer which is a basic structural component of plant cell walls.
Which is the correct representation of part of a cellulose polymer?

(A)
CH_2OH CH_2OH CH_2OH CH_2OH
O O O O O O O O O

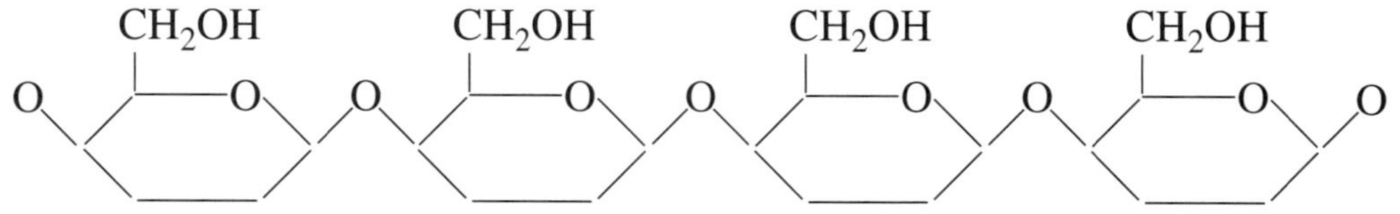

(B)
CH_2OH CH_2OH
O O O O O O O O O
CH_2OH CH_2OH

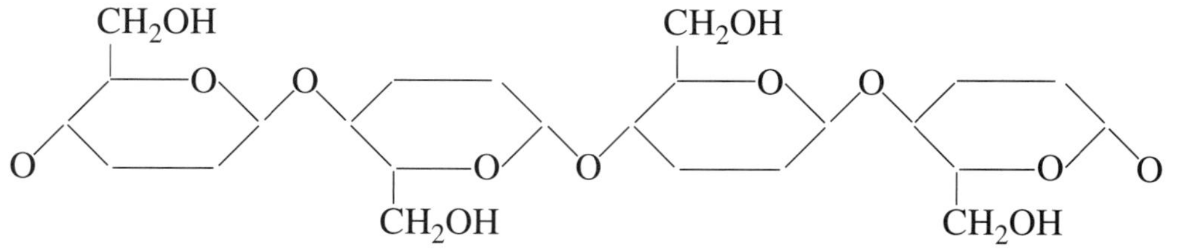

(C)
CH_2OH CH_2OH
O O O O O O O O O
CH_2OH CH_2OH

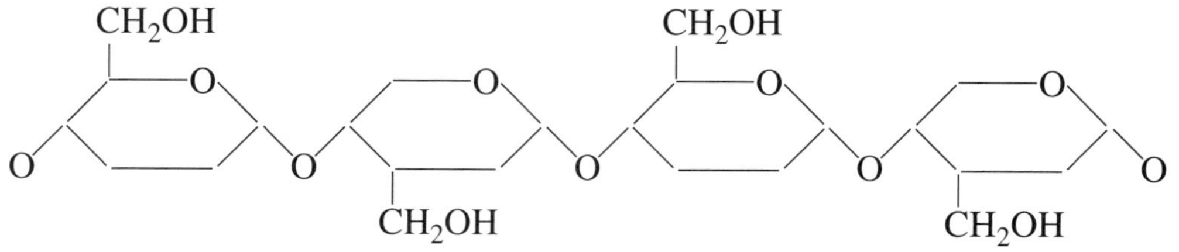

(D)
O O O O O O O O O
CH_2OH CH_2OH CH_2OH CH_2OH

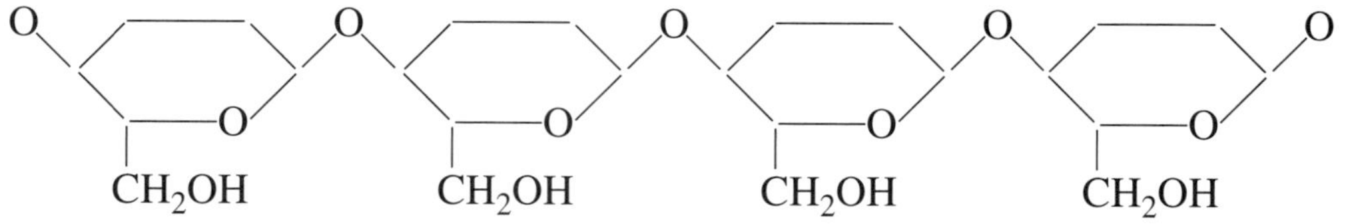

4 Which conditions would be best for the fermentation of sugars by yeast?

(A) Low oxygen concentration and a temperature between 25°C and 35°C

(B) High oxygen concentration and a temperature between 25°C and 35°C

(C) Low oxygen concentration and a temperature between 45°C and 60°C

(D) High oxygen concentration and a temperature between 45°C and 60°C

5 What is the catalyst for the conversion of ethanol to ethene?

(A) NaOH

(B) H_2SO_4

(C) HNO_3

(D) Pt

6 Which statement concerning galvanic cells is correct?

(A) Oxidation occurs at the anode.

(B) They are also known as electrolytic cells.

(C) The cathode is assigned a negative charge.

(D) An external power source must be present.

7 Which of the following is a transuranic element?

(A) Caesium

(B) Cerium

(C) Chromium

(D) Curium

8 Which instrument is used to detect radiation from radioactive isotopes?

(A) pH meter

(B) Geiger counter

(C) Ion-selective electrode

(D) Atomic absorption spectrophotometer (AAS)

9 The table gives the heat of combustion in kJ g^{-1} for a number of different fuels.

Fuel	*Heat of combustion* (kJ g^{-1})
Methanol	22.7
Ethanol	29.6
Propanol	33.6
Petrol (octane)	47.8

The heat of combustion in kJ mol^{-1} for one of the fuels was calculated as 2016 kJ mol^{-1}.

What was the fuel?

(A) Methanol

(B) Ethanol

(C) Propanol

(D) Petrol

10 A student performed three tests to investigate the relative activity of metals. In each test a metal strip was placed in a solution containing ions of a different metal. The results are shown in the diagrams.

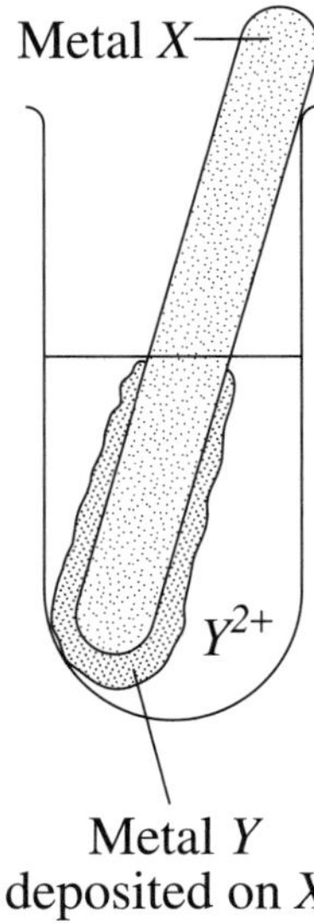

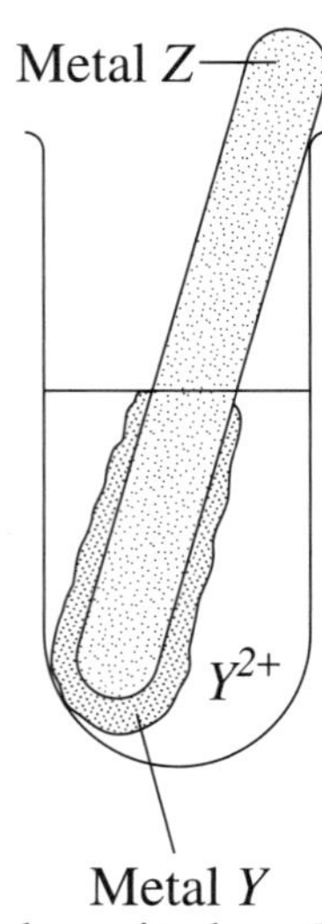

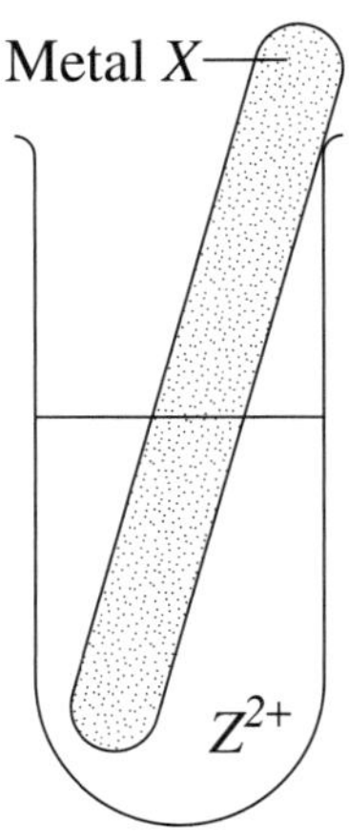

What is the order of activity of the metals, based on these results?

(A) $X > Z > Y$

(B) $Y > X > Z$

(C) $Z > Y > X$

(D) $Z > X > Y$

11 Which polymer is made by the polymerisation of methyl methacrylate?

$$H_2C{=}C(CH_3){-}COOCH_3$$

methyl methacrylate

(A) $-CH_2-C(CH_3){=}C(OCH_3)-CH_2-C(CH_3){=}C(OCH_3)-CH_2-C(CH_3){=}C(OCH_3)-$

(B) $-CH_2-C(CH_3)(COOCH_3)-CH_2-C(CH_3)(COOCH_3)-CH_2-C(CH_3)(COOCH_3)-$

(C) $-CH[HC(CH_3)COOCH_3]-CH[HC(CH_3)COOCH_3]-CH[HC(CH_3)COOCH_3]-$

(D) $-C(COOCH_3)(CH_3)-C(COOCH_3)(CH_3)-C(COOCH_3)(CH_3)-$

Free-response Questions

Past HSC Questions

Question 1 (3 marks) **Marks**

Radioisotopes are used in industry, medicine and chemical analysis. For ONE of these fields, relate the use of a named radioisotope to its properties. **3**

*(6 lines)**

Question 2 (6 marks)

Students were asked to perform a first-hand investigation to determine the molar heat of combustion of ethanol.

The following extract is from the practical report of one student.

Apparatus used:

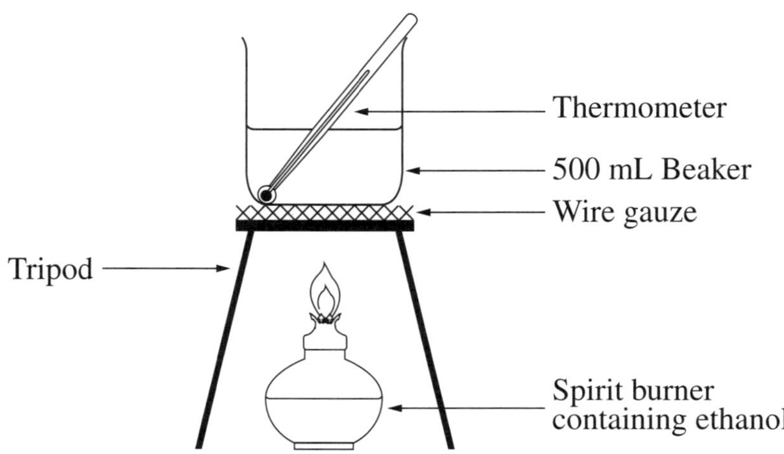

Lab data:

Mass of water	=	250.0 g
Initial mass of burner	=	221.4 g
Final mass of burner	=	219.1 g
Initial temperature of water	=	19.0°C
Final temperature of water	=	59.0°C

(a) After completing the calculations correctly, the student found that the answer did not agree with the value found in data books. Suggest ONE reason for this. **1**

(2 lines)

(b) Propose TWO adjustments that could be made to the apparatus or experimental method to improve the accuracy of the results. **2**

(4 lines)

Question 2 continues

* Shows the number of lines available in the HSC Answer Book for this question.

Question 2 (continued) **Marks**

(c) Calculate the molar heat of combustion of ethanol, using the student's data. **3**

(8 lines)

Question 3 (3 marks)

A galvanic cell was made by connecting two half-cells. One half-cell was made by putting a copper electrode in a copper (II) nitrate solution. The other half-cell was made by putting a silver electrode in a silver nitrate solution. The electrodes were connected to a voltmeter as shown in the diagram.

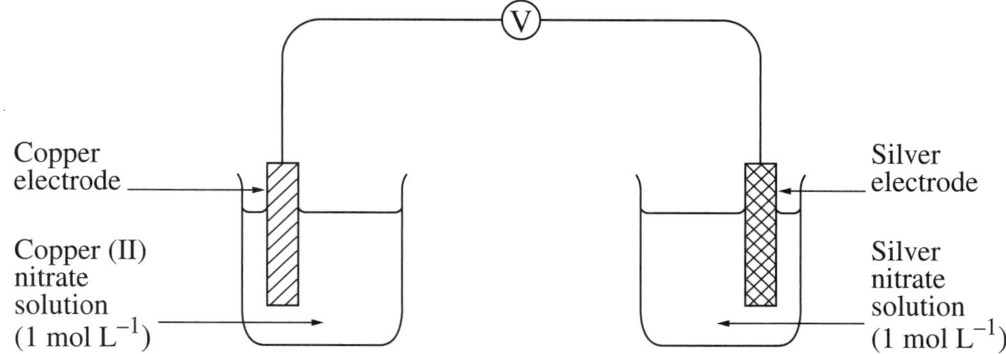

(a) Complete the above diagram by drawing a salt bridge. **1**

(b) Using the *standard potentials* table in the data sheet, calculate the theoretical voltage of this galvanic cell. **2**

(4 lines)

Question 4 (7 marks)

Name ONE type of cell, other than the dry cell or lead–acid cell, you have studied. Evaluate it in comparison with either the dry cell or lead–acid cell, in terms of chemistry and the impact on society. Include relevant chemical equations in your answer. **7**

(15 lines)

Question 5 (6 marks)

You have carried out a first-hand investigation to compare the reactivity of an alkene with its corresponding alkane.

(a) State the name of the alkene. **1**

(1 line)

(b) Outline a procedure to compare the reactivity of this alkene with its corresponding alkane. **2**

(4 lines)

Question 5 continues

Question 5 (continued) **Marks**

(c) Describe the results obtained from this first-hand investigation and include relevant chemical equations. **3**

(6 lines)

Question 6 (3 marks)

Explain why alkanes and their corresponding alkenes have similar physical properties, but very different chemical properties. **3**

(6 lines)

Question 7 (1 mark)

Name the type of polymerisation shown in the following reaction: **1**

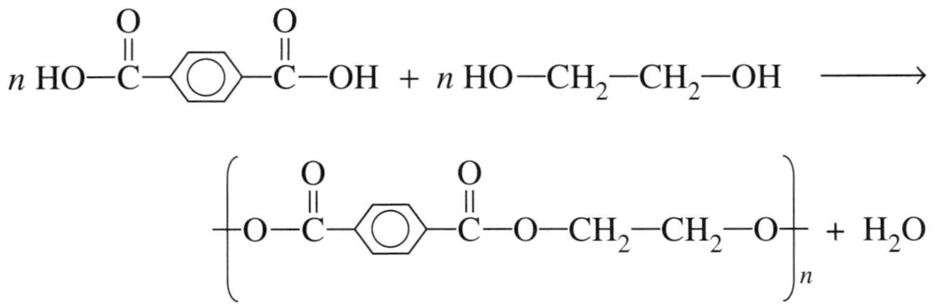

$$n\,HO{-}\overset{\overset{\displaystyle O}{\|}}{C}{-}C_6H_4{-}\overset{\overset{\displaystyle O}{\|}}{C}{-}OH + n\,HO{-}CH_2{-}CH_2{-}OH \longrightarrow \left[O{-}\overset{\overset{\displaystyle O}{\|}}{C}{-}C_6H_4{-}\overset{\overset{\displaystyle O}{\|}}{C}{-}O{-}CH_2{-}CH_2{-}O \right]_n + H_2O$$

(1 line)

Marks

Question 8 (5 marks)

(a) Describe the conditions under which a nucleus is unstable. **2**

(6 lines)

(b) The following is a flow diagram showing the sequence of products released during the decay of uranium. **3**

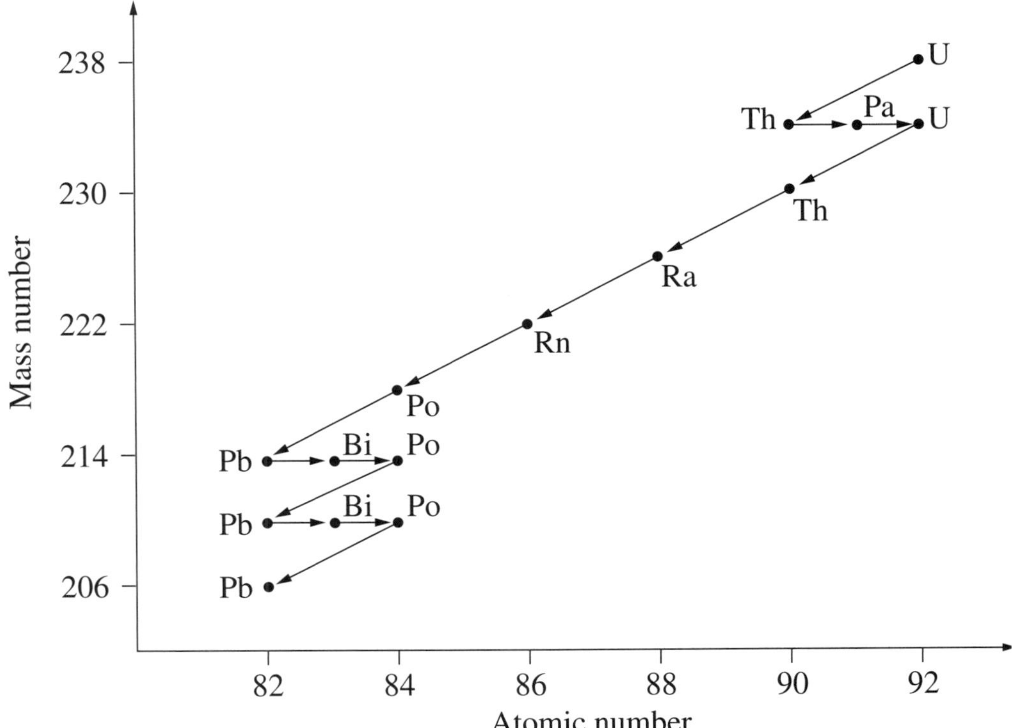

Use examples from the flow diagram to describe processes by which an unstable isotope undergoes radioactive decay.

(4 lines)

Question 9 (3 marks)

You performed a first-hand investigation that monitored mass changes during the fermentation of glucose to ethanol.

(a) Outline the procedure you used. **2**

(9 lines)

(b) Write a balanced chemical equation for this reaction. **1**

(2 lines)

Marks

Question 10 (5 marks)

The flowchart shows the production of polyethylene.

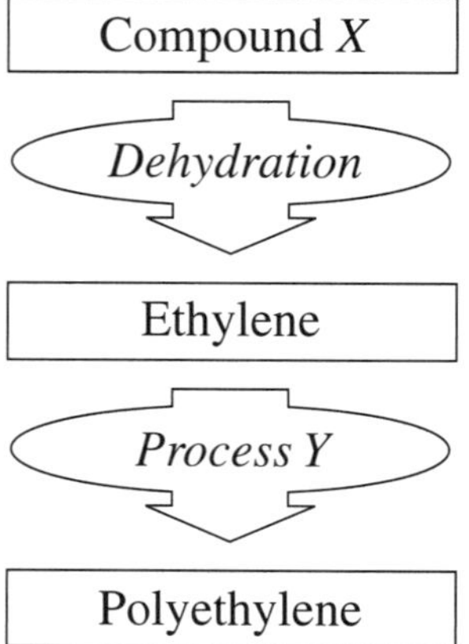

(a) Identify Compound X. **1**

(1 line)

(b) Describe *Process Y*. **3**

(9 lines)

A sample of polyethylene was produced by *Process Y*. The following graph shows the distribution of molecular weights of polymer molecules in the sample.

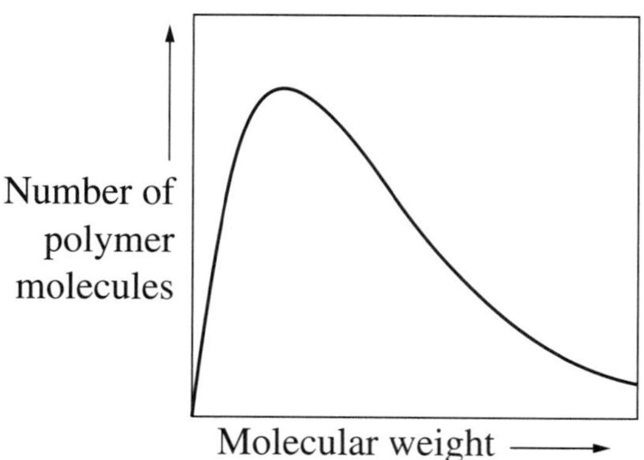

(c) Why is a range of molecular weights observed? **1**

(4 lines)

Question 11 (4 marks)

Describe how commercial radioisotopes are produced, and how transuranic elements are produced. **4**

(15 lines)

Marks

Question 12 (3 marks)

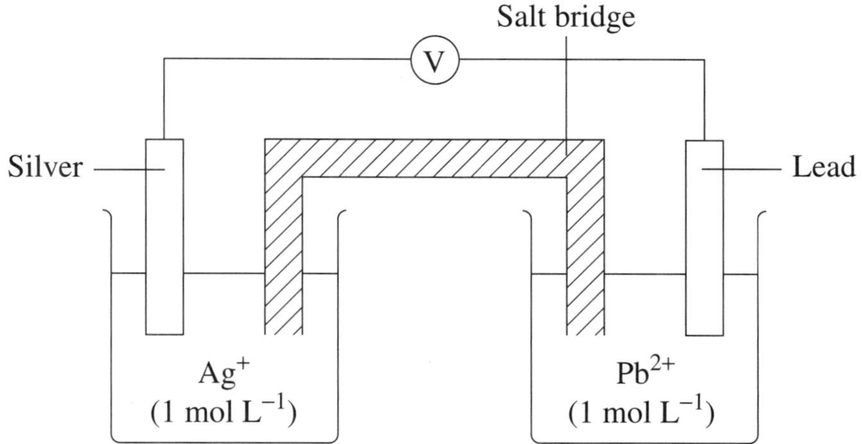

(a) Identify the cathode in this diagram. 1

(1 line)

(b) Write the net redox equation for the cell reaction, and calculate the cell potential (E°). 2

(6 lines)

Question 13 (5 marks)

Assess the suitability of biomass as a future source of energy and chemicals for industry. 5

(20 lines)

Question 14 (5 marks)

You performed a first-hand investigation to prepare an ester by reflux.

(a) Identify the products formed when propanoic acid and butanol are refluxed with acid catalyst. 1

(2 lines)

(b) Draw a fully labelled diagram of the equipment assembled for use. 2

(c) Outline the advantages of using reflux to prepare the ester. 2

(6 lines)

Marks

Question 15 (3 marks)

(a) Write a balanced chemical equation for the complete combustion of ethanol. **1**

(2 lines)

(b) A mass of 72.5 g of ethanol was burnt completely in air. Calculate the volume of carbon dioxide that was produced at 25°C and 100 kPa. **2**

(6 lines)

Core Topic
Production of Materials
Worked Answers

Multiple-choice Questions

1 D Ethene (ethylene) is the monomer used in the production of the polymer poly(ethene) (polyethylene). No molecules are lost in the process.

2 A Biomass is largely composed of plant matter. Plant matter is largely composed of cellulose and lignin.

3 B Cellulose is a polymer made up of beta-D glucose monomers. When linked together, every second monomer must be inverted.

4 A The presence of oxygen will allow decomposing bacteria to colonise the mixture and allow yeast to aerobically respire glucose, forming carbon dioxide and water instead of carbon dioxide and ethanol.

5 B Concentrated H_2SO_4 is the catalyst used to dehydrate ethanol, and dilute H_2SO_4 is used to hydrate ethylene.

6 A The anode is defined as the electrode where oxidation occurs no matter what the type of cell.

7 D Curium has an atomic number (96) beyond that of uranium (92). The others are not transuranic because they come before uranium in atomic number.

8 B

9 C M (propanol) $= 60\ \text{g mol}^{-1}$, $\therefore\ \Delta_c H = 60 \times 33.6 = 2016\ \text{kJ mol}^{-1}$.

10 D A more active metal will displace another metal's ion in solution. Z is the most active because it displaces Y^{2+} and is not displaced by X. Y is the least active because it is displaced by X and Z.

11 B The double bond opens up and carbons link up to form a long chain. Every second carbon atom will contain a $-CH_3$ branch and a $-COOCH_3$ branch.

Free-response Questions

1 Cobalt-60 is used in medicine as a treatment for cancer. This isotope decays by beta emission and also releases gamma rays. Gamma radiation kills cancer cells because it is able to penetrate quite deeply into body tissue and is high enough in energy to destroy the DNA within cells.

Cobalt-60 has a half-life of 5.3 years, which is long enough so that it doesn't need replacing often in the equipment, but is short enough that the source of the radiation emits enough energy to be effective in killing cells.

2 (a) Heat was lost by radiation from the gauze and the beaker to the surroundings.

(b) A heat shield could be placed around the apparatus to reduce heat loss to the environment.

The spirit burner could be raised so that the flame is closer to the beaker of water and the gauze not used. This should also reduce heat loss.

(c) $\Delta H \text{ (water)} = mC\Delta T$

$= 250.0 \times 4.18 \times 40$

$= 41\ 800 \text{ J}$

$= 41.8 \text{ kJ}$

$$\text{moles of ethanol} = \frac{23}{46.1} = 4.99 \times 10^{-2} \text{ mol}$$

$$\Delta H \text{ (ethanol)} = \frac{-41.8 \text{ kJ}}{4.99 \times 10^{-2} \text{ mol}}$$

$$= -838 \text{ kJ mol}^{-1}$$

Therefore, the molar heat of combustion of ethanol = 838 kJ/mol.

3 (a) The student should draw in the salt bridge to connect the two beakers (an inverted U is acceptable as long as the ends are below the surface of each solution).

(b) $Ag^+ + e^- \rightarrow Ag \quad E^o = 0.80 \text{ V}$

$Cu \rightarrow Cu^{2+} + 2e^- \quad E^o = -0.34 \text{ V}$

Cell voltage $= 0.80 - 0.34$

$= +0.46 \text{ V}$

4 Mercury button cell compared to dry cell

Both cells have a zinc anode. However the cathode in the mercury cell is C and HgO, while in the dry cell it is C and MnO_2.

Both cells use a paste-type electrolyte, KOH in the mercury cell and NH_4Cl in the dry cell.

The anode and cathode reactions are different in each cell:

Mercury: Anode reaction $Zn + 2OH^- \rightarrow ZnO + H_2O + 2e^-$
Dry cell: Anode reaction $Zn \rightarrow Zn^{2+} + 2e^-$

Mercury: Cathode reaction $HgO + H_2O + 2e^- \rightarrow Hg + 2OH^-$
Dry cell: Cathode reaction $2NH_4^+ + 2MnO_2 + 2e^- \rightarrow Mn_2O_3 + 2NH_3 + H_2O$

The dry cell has had a great impact on society being the first fully portable battery. They are also robust and easy to store. For these reasons they allowed portability of torches, radios and battery operated toys. However, the mercury button cell has also had a large influence on society. Because of its small size, it has allowed miniaturisation of calculators, cameras and watches.

One of the problems with the dry cell is that the voltage drops during use and it is therefore only useful for infrequent use, for example in torches used occasionally during a blackout. The mercury cell provides a constant voltage over a long period so that no frequent changing of cells is necessary. For these reasons it is useful in hearing aids and watches.

Both cells are relatively inexpensive, the dry cell more so, making them available to the general public.

The dry cell contains small amounts of zinc and small amounts of a slightly acidic electrolyte paste. These materials do not cause environmental problems if dry cells are placed in household garbage. In the case of the mercury cell, the batteries should be recycled so that mercury (a heavy metal) does not contaminate the environment. Mercury could leach into ground water from a tipping site.

5 (a) Cyclohexene

(b) Add 10 drops of cyclohexane to 4 mL bromine water in a test tube and shake. Repeat with cyclohexene holding all volumes and conditions constant. The one that decolourises the bromine is the cyclohexene.

(c) Cyclohexene causes the bromine water to decolourise whereas the cyclohexane does not. The bromine water hydrolyses to form HOBr, which adds across the double bond in cyclohexene.

$Br_2 + H_2O \rightarrow HOBr + HBr$

$HOBr + C_6H_{10} \rightarrow C_6H_{10\text{-}}BrOH$

6 Alkanes and their corresponding alkenes are non-polar carbon chain molecules with weak dispersion forces between the molecules. Physical properties such as melting point and solubility are due to intermolecular forces. These properties are therefore similar.

Chemical properties are due to bonding within the molecules. Alkenes have at least one reactive double bond. Alkanes have only unreactive single bonds within the molecule. Thus alkenes undergo addition reactions readily whereas alkanes are unreactive.

7 Condensation polymerisation

8 (a) Large atoms are unstable when their atomic number is greater than 83.

Smaller atoms are unstable when their neutron:proton ratio is too high or too low. For example carbon-14 is unstable because its neutron:proton ratio is greater than 1:1.

(b) A radioactive isotope can undergo decay by alpha or beta decay.

Alpha decay involves emitting a helium nucleus from the isotope nucleus.

Beta decay involves a neutron in the nucleus becoming a proton by emitting an electron.

$$^{238}_{92}U \rightarrow {}^{4}_{2}He + {}^{234}_{90}Th$$

$$^{235}_{91}Pa \rightarrow {}^{0}_{-1}e + {}^{235}_{92}U$$

Eight alpha decays and six beta decays are shown in the diagram in the question.

9 (a) A teaspoon of yeast was mixed with a sugar solution in a conical flask with cotton wool at the mouth. This was weighed and placed in an incubator at 35°C for a week. The apparatus was removed from the incubator every day to be reweighed.

(b) $C_6H_{12}O_6(aq) \rightarrow 2CO_2(g) + 2C_2H_5OH(l)$

10 (a) Ethanol

(b) Ethylene is changed from a gas to a liquid under pressure and heated in the presence of a catalyst to produce polyethylene.

The double bond within ethylene opens out and a covalent bond forms between molecules linking them together in a long chain. This is addition polymerisation.

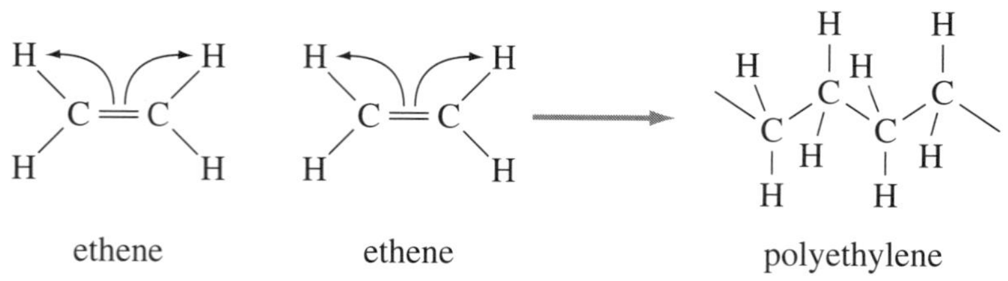

ethene ethene polyethylene

(c) Polymerisation is a random process. Few or many monomers may link together to form some long and some very long chains.

11 Commercial isotopes can be produced in both nuclear reactors and particle accelerators like cyclotrons. In a nuclear reactor, target nuclei are introduced into the reactor where they are bombarded by neutrons. These are absorbed by the atom's nucleus. The medical isotope technetium-99 is produced in a reactor.

$$^{98}_{42}\text{Mo} + {}^{1}_{0}\text{n} \rightarrow {}^{99}_{42}\text{Mo} \rightarrow {}^{99}_{43}\text{Tc} + {}^{0}_{-1}e^{-}$$

Neutron-poor isotopes are made in a particle accelerator like a cyclotron. These use electric and magnetic fields to increase the speed of charged particles so that when they collide with the target nucleus, fusion occurs. Oxygen-17 is produced in a cyclotron by bombarding nitrogen-14 with accelerated alpha particles.

$$^{14}_{7}\text{N} + {}^{4}_{2}\text{He} \rightarrow {}^{17}_{8}\text{O} + {}^{1}_{1}\text{p}$$

Transuranic elements, those elements with an atomic number greater than uranium, have been made in both reactors and cyclotrons. The target nuclei are usually already large nuclei such as uranium, plutonium or lead. Neptunium-239 is a transuranic element made in a reactor by bombarding U-238 with a neutron.

$$^{238}_{92}\text{U} + {}^{1}_{0}\text{n} \rightarrow {}^{239}_{93}\text{Np} + {}^{0}_{-1}e^{-}$$

12 (a) Silver electrode

(b)

$$\begin{array}{lll} \text{Pb} \rightarrow \text{Pb}^{2+} + 2e^{-} & E° = +0.13\text{ V} \\ 2\text{Ag}^{+} + 2e^{-} \rightarrow 2\text{Ag} & E° = +0.8\text{ V} \\ \hline 2\text{Ag}^{+}(aq) + \text{Pb}(s) \rightarrow 2\text{Ag}(s) + \text{Pb}^{2+}(aq) & E° = 0.13 + 0.8 = 0.93\text{ V} \end{array}$$

13 Biomass is a fuel source made up of cellulose and lignin from plants. The cellulose component can be converted to glucose by acid hydrolysis.

$$\text{n}H_2O + (C_6H_{10}O_5)_n \rightarrow \text{n}C_6H_{12}O_6$$

Glucose can then be fermented to ethanol using yeast.

$$C_6H_{12}O_6 \rightarrow 2C_2H_5OH + 2CO_2$$

The ethanol can be used as a fuel on its own or dehydrated to ethylene.

$$C_2H_5OH \rightarrow H_2O + C_2H_4$$

Ethylene can be used as a starting material to produce polyethylene and other plastics. The problem with using biomass is that cellulose needs to be grown. This may result in a large amount of land clearing as well as the availability of water and fertiliser. There is insufficient arable land in most countries unless cellulosic weed crops can be grown on marginal land. In addition, the acid hydrolysis of cellulose is not economically viable at present.

14 (a) Butylpropanoate and water

(b)

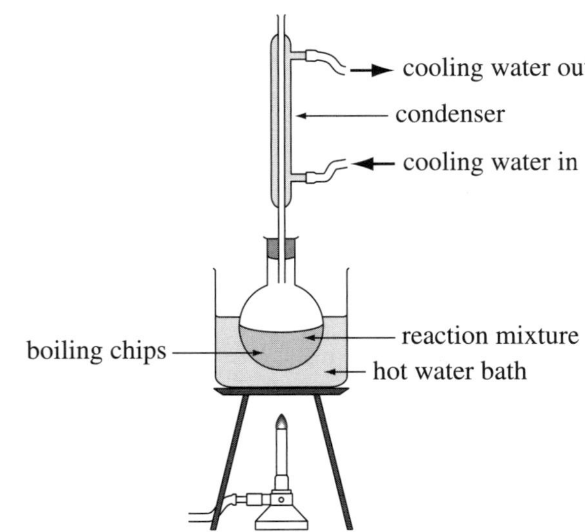

(c) The reactants are volatile and yet need to be heated to reach their activation energy. Refluxing cools the reactant gases and condenses them returning them to the reaction mixture for continued heating. The gases would otherwise escape before they reacted.

15 (a) $C_2H_6O(l) + 3O_2(g) \rightarrow 2CO_2(g) + 3H_2O(l)$

(b) $n(C_2H_6O) = \frac{72.5}{46.0} = 1.576 \text{ mol}$

$n\ CO_2$ produced $= 2 \times n(C_2H_6O) = 3.15$ mol

$V(CO_2) = 3.15 \times 24.79 = 78.1$ L

CHAPTER 2

Core Topic
The Acidic Environment

Multiple-choice Questions

Past HSC Questions

1 The pH of unpolluted rainwater is about 6.0. Which substance contributes most to this?

(A) CO_2

(B) N_2

(C) NO_2

(D) O_3

2 The graph shows the colour ranges of the acid–base indicators methyl orange, bromothymol blue and phenolphthalein.

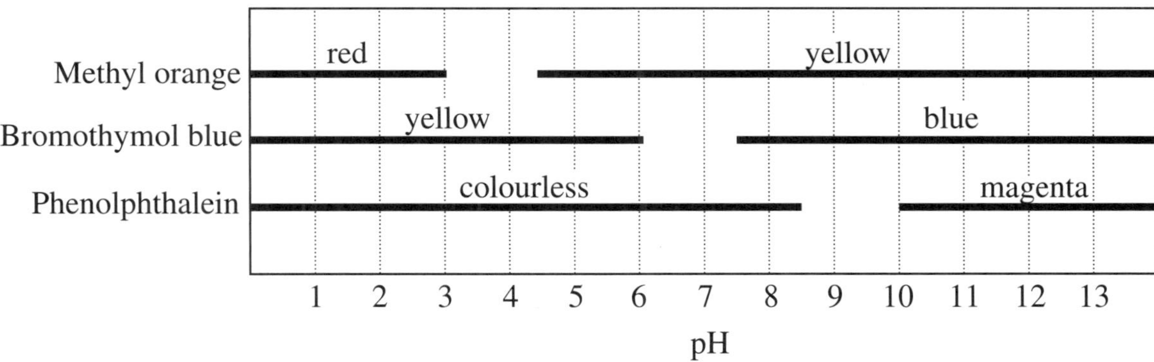

A solution is yellow in methyl orange, blue in bromothymol blue and colourless in phenolphthalein.

What is the pH range of the solution?

(A) 4.5 to 6.0

(B) 6.0 to 7.5

(C) 7.5 to 8.5

(D) 8.5 to 10.0

3 A group of students produced a red solution by boiling red cabbage leaves in water. When dilute sodium hydroxide was added to the solution, it turned purple. When dilute hydrochloric acid was added to the red solution, no colour change occurred.

Which of these substances, when added, is most likely to cause the red solution to change colour?

(A) Cleaning solution containing ammonia

(B) Concentrated hydrochloric acid

(C) Orange juice

(D) Vinegar

4 The burning of sulfur can be described by the following equation:

$$S(s) + O_2(g) \rightarrow SO_2(g)$$

What volume of sulfur dioxide gas will be released at 25°C and 101.3 kPa when 8.00 g of sulfur is burnt?

(A) 3.06 L

(B) 6.12 L

(C) 12.24 L

(D) 24.47 L

5 An understanding of Le Chatelier's principle is important in the chemical industry. Which prediction can be made using this principle?

(A) The identity of products of a chemical reaction

(B) The effect of changes in temperature on the rates of reactions

(C) The effect of catalysts on the position of equilibrium reactions

(D) The effect of changes in the concentration of chemical substances in equilibrium

6 The following equations describe some reactions in the formation of acid rain:

$$SO_2(g) + H_2O(l) \rightleftharpoons H^+(aq) + HSO_3^-(aq)$$

$$2H^+(aq) + 2HSO_3^-(aq) + O_2(g) \longrightarrow 4H^+(aq) + 2SO_4^{2-}(aq)$$

What would occur if some solid sodium sulfate (Na2SO4) were added to a sample of acid rain?

(A) The amount of $SO_2(g)$ would increase and the acidity of the solution would decrease.

(B) The amount of $SO_2(g)$ would increase and the acidity of the solution would increase.

(C) The amount of $SO_2(g)$ would be unchanged and the acidity of the solution would be unchanged.

(D) The amount of $SO_2(g)$ would be unchanged and the acidity of the solution would decrease.

7 Which is amphiprotic?

(A) H_2SO_4

(B) NH_4^+

(C) HCO_3^-

(D) SO_4^{2-}

8 What did the Brönsted–Lowry definition of acids identify that made it a significant improvement over earlier definitions?

(A) Acids contain hydrogen.

(B) Acids are proton donors.

(C) Acids contain oxygen.

(D) Acids are electron-pair acceptors.

9 In a titration, an acid of known concentration is placed in a burette and reacted with a base that has been pipetted into a conical flask.

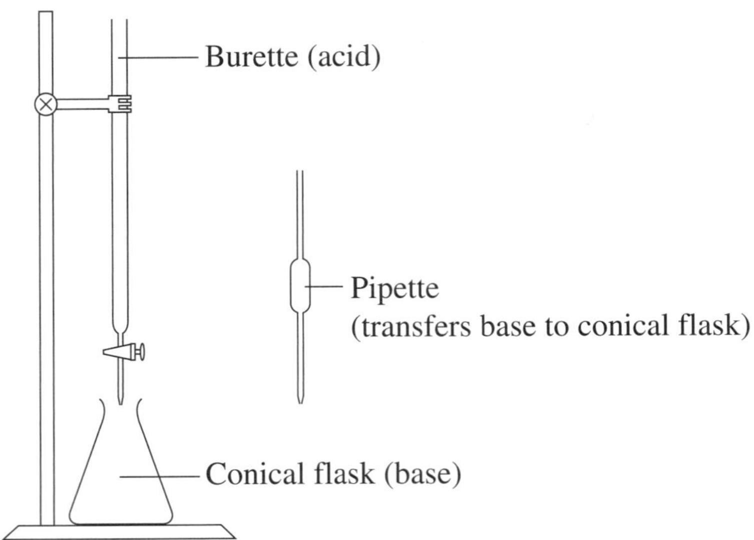

What should each piece of glassware be rinsed with immediately before the titration?

	Burette	*Pipette*	*Conical flask*
(A)	acid	base	water
(B)	water	water	water
(C)	acid	base	base
(D)	water	water	base

10 The following list of steps refers to an experimental plan for making an ester in a flask. Some of the steps in the list may NOT be required for this experiment. The steps are NOT in the correct sequence.

1. Heat the mixture under reflux.
2. Add three drops of concentrated sulfuric acid.
3. Add 1 mL of ethanol.
4. Add 1 mL of ethene.
5. Add 1 mL of ethanoic acid.
6. Distil the mixture.
7. Add three drops of phenolphthalein indicator.

Which alternative is the best sequence for making an ester?

(A) 3, 5, 7, 1

(B) 4, 3, 7, 6

(C) 5, 4, 2, 6

(D) 5, 3, 2, 1

11 Which equation represents esterification?

(A) $CH_3—O—\overset{\overset{\large O}{||}}{C}—CH_3 + H_2O \xrightarrow{H^+} CH_3—OH + CH_3—\overset{\overset{\large O}{||}}{C}—OH$

(B) $CH_3—O—\overset{\overset{\large O}{||}}{C}—H + NaOH \longrightarrow H—\overset{\overset{\large O}{||}}{C}—ONa + CH_3—OH$

(C) $CH_3—OH + CH_3—\overset{\overset{\large O}{||}}{C}—OH \xrightarrow{H^+} CH_3—CH_2—O—\overset{\overset{\large O}{||}}{C}—H + H_2O$

(D) $CH_3—OH + CH_3—\overset{\overset{\large O}{||}}{C}—OH \xrightarrow{H^+} CH_3—O—\overset{\overset{\large O}{||}}{C}—CH_3 + H_2O$

12 Which of the following is an acid–base indicator?

(A) Methanol
(B) Methyl orange
(C) Methanoic acid
(D) Methyl ethanoate

13 The diagram is a representation of the Periodic Table. The positions of six different elements are shown.

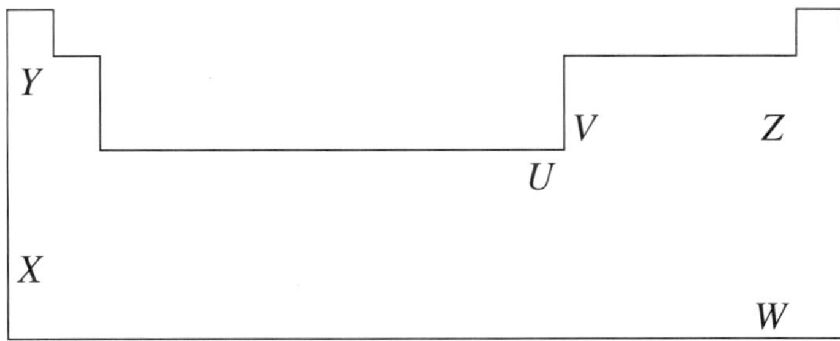

What are the reactions of oxides of these elements with acid and with base?

	Oxide reacts with acid	*Oxide reacts with base*	*Oxide reacts with acid and with base*
(A)	*Z*	*X*	*V*
(B)	*Y*	*X*	*U*
(C)	*X*	*Z*	*V*
(D)	*V*	*W*	*Y*

14 A sulfuric acid solution has a concentration of 5×10^{-4} mol L^{-1}.

What is the pH of this solution, assuming the acid is completely ionised?

(A) 3.0

(B) 3.3

(C) 3.6

(D) 4.0

15 In a titration of a strong base with a strong acid, the following procedure was used:

1. A burette was rinsed with water and then filled with the standard acid.
2. A pipette was rinsed with some base solution.
3. A conical flask was rinsed with some base solution.
4. A pipette was used to transfer a measured volume of base solution into the conical flask.
5. Indicator was added to the base sample and it was titrated to the endpoint with the acid.

Which statement is correct?

(A) The calculated base concentration will be correct.

(B) The calculated base concentration will be too low.

(C) The calculated base concentration will be too high.

(D) No definite conclusion can be reached about the base concentration.

16 Which of the following graphs shows how pH will vary when dilute HCl is added to 100 mL of dilute natural buffer solution with an initial pH of 7.0?

(A)

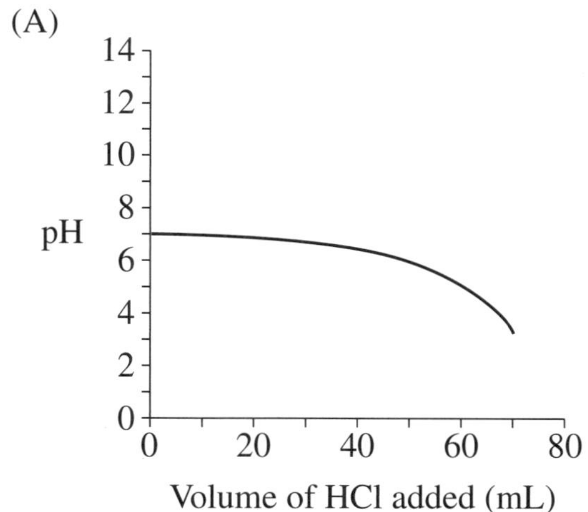

(B)

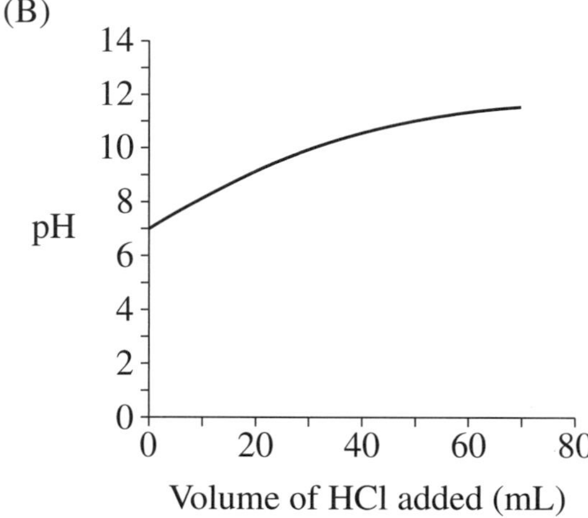

(C)

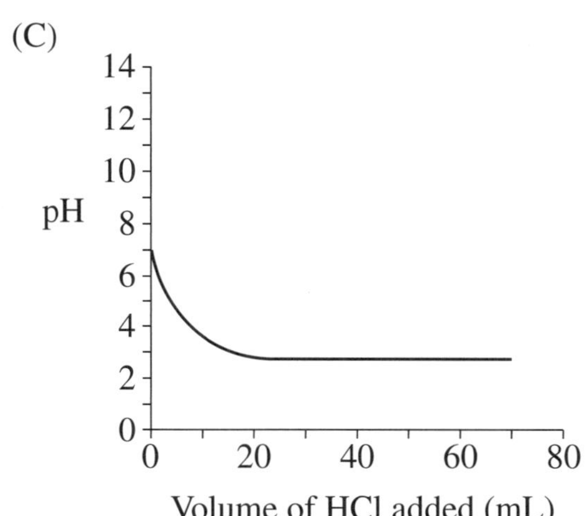

(D)

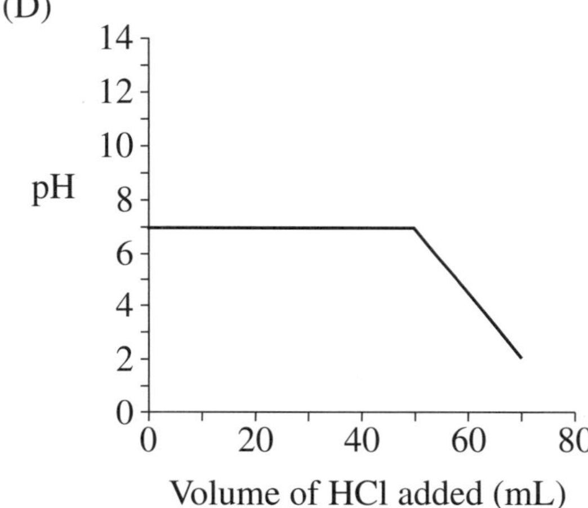

Free-response Questions

Past HSC Questions

Question 1 (4 marks) **Marks**

A 0.1 mol L^{-1} solution of hydrochloric acid has a pH of 1.0, whereas a 0.1 mol L^{-1} solution of citric acid has a pH of 1.6.

(a) State ONE way in which pH can be measured. **1**

*(1 line)**

(b) Explain why the two solutions have different pH values. **3**

(6 lines)

Question 2 (1 mark)

Barium hydroxide and sulfuric acid react according to the following equation: **1**

$$Ba(OH)_2(aq) + H_2SO_4(aq) \rightarrow BaSO_4(s) + 2H_2O(l)$$

Name this type of chemical reaction.

(1 line)

Question 3 (6 marks)

Justify the procedure you used to prepare an ester in a school laboratory. Include relevant chemical equations in your answer. **6**

(14 lines)

Question 4 (4 marks)

A household cleaning agent contains a weak base of general formula NaX. 1.00 g of this compound was dissolved in 100.0 mL of water. A 20.0 mL sample of the solution was titrated with 0.1000 mol L^{-1} hydrochloric acid and required 24.4 mL of the acid for neutralisation.

(a) What is the Brönsted–Lowry definition of a base? **1**

(2 lines)

(b) What is the molar mass of this base? **3**

(8 lines)

* Shows the number of lines available in the HSC Answer Book for this question.

Marks

Question 5 (4 marks)

(a) Identify ONE common household base. 1

(1 line)

(b) A student used indicators to determine whether three colourless solutions were acidic or basic. The indicators used are shown in the table. 3

Indicator	*Colour change*	pH *range*
Methyl orange	red to yellow	3.2–4.4
Methyl red	red to yellow	4.8–6.0
Thymol blue	yellow to blue	8.0–9.6
Alizarin	red to purple	11.0–12.4

Samples of each solution were tested with the indicators. The colours of the resulting solutions are shown in the table.

Indicator added	*Colour of solution* A	*Colour of solution* B	*Colour of solution* C
Methyl orange	yellow	yellow	yellow
Methyl red	yellow	yellow	yellow
Thymol blue	blue	blue	yellow
Alizarin	purple	red	red

The student concluded that each of the three solutions tested was basic. Assess the validity of this conclusion.

(8 lines)

Question 6 (7 marks)

Evaluate the impact of industrial sources of sulfur dioxide and nitrogen oxides on the environment, making use of appropriate chemical equations. 7

(21 lines)

Marks

Question 7 (5 marks)

Solutions of hydrochloric acid, acetic acid and sulfuric acid were prepared. Each of the solutions had the same concentration (0.01 mol L^{-1}). The pH of the acetic acid solution was 3.4.

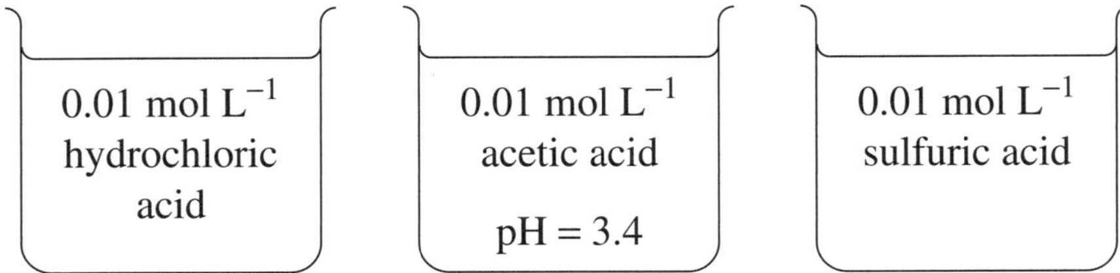

(a) Calculate the pH of the hydrochloric acid solution. **1**

(1 line)

(b) Compare the pH of the sulfuric acid solution to the pH of the hydrochloric acid solution. Justify your answer. (No calculations are necessary.) **2**

(6 lines)

(c) Explain why the acetic acid solution has a higher pH than the hydrochloric acid solution. **2**

(5 lines)

Question 8 (4 marks)

A bottle of soft drink was placed on an electronic balance and weighed. The cap was removed and placed next to the bottle on the balance. The mass of the cap, bottle and its contents was monitored. The results are shown in the graph. The experiment was conducted at 25°C and 101.3 kPa. Assume that no evaporation has occurred.

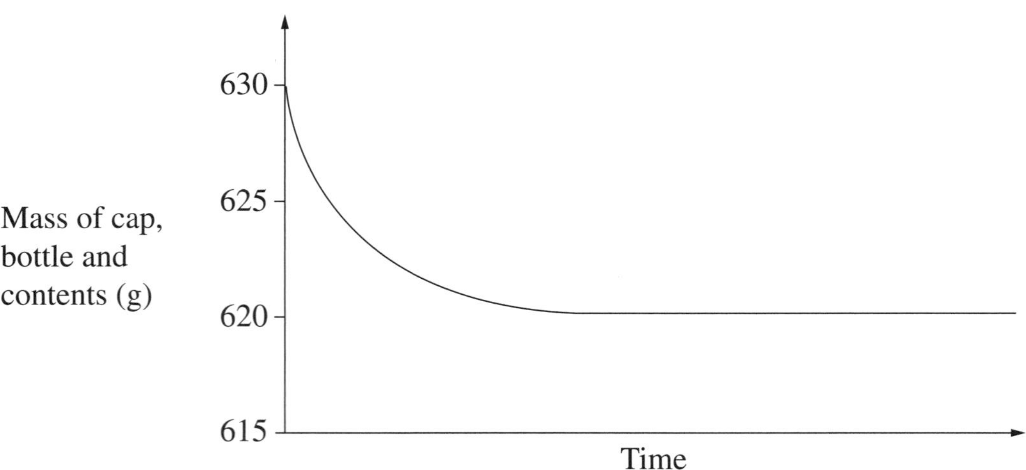

(a) Identify the gas released. **1**

(1 line)

(b) Calculate the volume of the gas released. **3**

(7 lines)

Marks

Question 9 (4 marks)

25.0 mL of 0.12 mol L^{-1} standard barium hydroxide solution was titrated with nitric acid. The results are recorded in the table.

Titration	*Volume of nitric acid used* (mL)
1	20.4
2	18.1
3	18.2
4	18.1

(a) Write a balanced chemical equation for the reaction of barium hydroxide with nitric acid. **1**

(2 lines)

(b) Calculate the concentration of the nitric acid. **3**

(8 lines)

Question 10 (4 marks)

Discuss factors that must be considered when using neutralisation reactions to safely minimise damage in chemical spills. **4**

(9 lines)

Question 11 (4 marks)

Explain the trends in boiling points shown in the graph. **4**

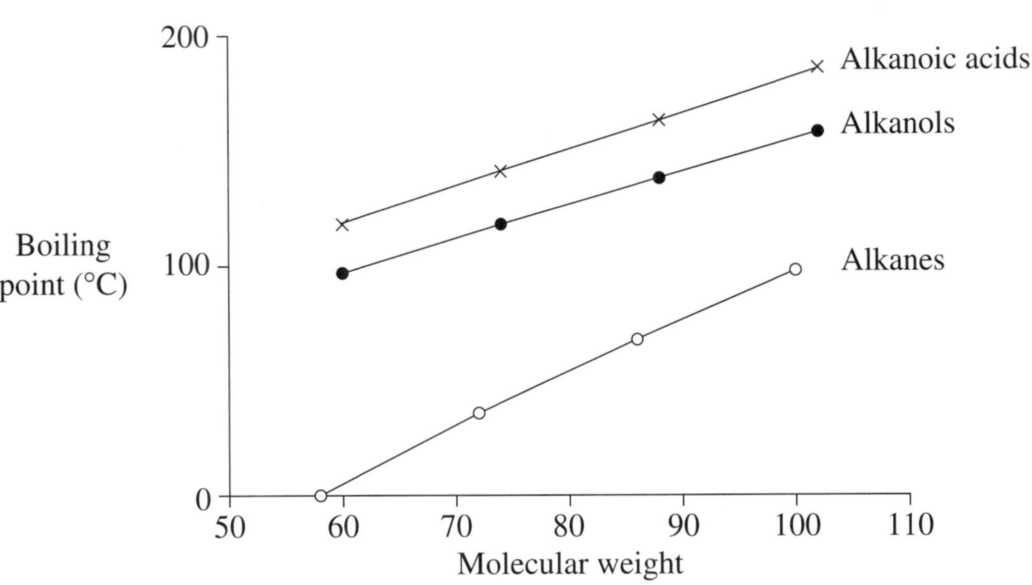

(17 lines)

Core Topic

The Acidic Environment

Worked Answers

Multiple-choice Questions

1 A A pH of 6.0 is slightly acidic. CO_2, found naturally in the atmosphere, reacts with water forming the weak acid, H_2CO_3.

2 C When a solution containing bromothymol blue is blue the pH must be higher than 7.5. When a solution containing phenolphthalein is colourless the pH must be lower than 8.5.

3 A The base NaOH turns the red indicator to purple. The acid HCl causes no colour change. Ammonia is basic. The other three options are acidic.

4 B $n\,(S) = \frac{8.00}{32.07} = 0.25\text{ mol}$

0.25 mol of S produces 0.25 mol $SO_2(g)$
$V\,(SO_2) = 0.25 \times 24.47 = 6.12\text{ L}$

5 D Le Chatelier's principle explains shifts in a system already in equilibrium due to changes in concentration or temperature.

6 C The second equation is not an equilibrium system. Therefore a change in SO_4^{2-} concentration will not cause a shift and the concentration of the hydrogen ion will not change.

7 C $HCO_3^- + H^+ \rightarrow H_2CO_3$

$HCO_3^- + OH^- \rightarrow CO_3^{2-} + H_2O$

8 B Containing hydrogen relates to Davy, containing oxygen relates to Lavoisier and being an electron pair acceptor relates to Lewis.

9 A Rinsing the burette and pipette with water would dilute the solutions so they need to be rinsed with the solutions going into them. The critical thing about the conical flask is the number of moles of solution in it, so it needs to be rinsed with water.

10 D The reactants are an alkanol and an alkanoic acid (steps 5 and 3), acid is added as a catalyst (step 2) and the mixture is heated under reflux (step 1).

11 D Esterification is a condensation reaction that requires an acid catalyst.

12 B The other alternatives are organic molecules that do not change colour in an acid or base solution.

13 C X is a basic metal oxide and will therefore react with an acid. Z is an acidic non-metal oxide and will therefore react with a base. V is amphoteric and will react with both acids and bases.

14 A H_2SO_4 is diprotic and will ionise to produce twice the concentration of H^+.

$H_2SO_4 \rightarrow 2H^+ + SO_4^{2-} \therefore [H^+] = 2 \times (5 \times 10^{-4}) = 1 \times 10^{-3}$ mol L^{-1}

$pH = -\log(1 \times 10^{-3}) = 3.0$

15 C Rinsing the flask with a base will result in too many moles of base in the flask. More moles of acid will be needed to neutralise it. Rinsing the burette with water will dilute the acid. More moles of acid will be needed to neutralise the base. Therefore the number of moles of base will appear higher than it actually is and the calculation of its concentration will appear higher than it actually is.

16 A The buffer acts to keep the pH relatively constant when an acid is added to the solution, which it does in graph A. However, over a certain volume of acid added the buffer can no longer cope and the pH begins to drop as the solution becomes acidic.

Free-response Questions

1 (a) pH meter

(b) HCl is a strong acid and therefore ionises fully. In a 0.1 mol/L solution, the concentration of H^+ ions will be 0.1 mol L^{-1}.

The pH = –log $[H^+]$, therefore pH = 1.0.

Citric acid is a weak acid and does not completely ionise; that is, it ionises less than HCl does. Therefore in a 0.1 mol/L solution the concentration of H^+ ions will be less than 0.1 mol L^{-1} and the pH more than 1.0.

2 Neutralisation or acid-base

3 

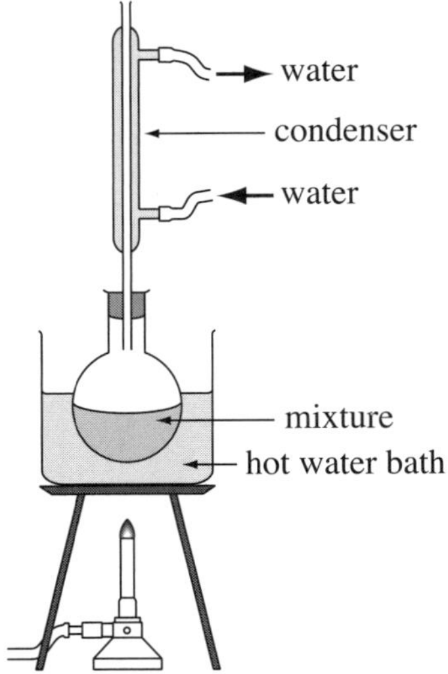

Add together 20 mL glacial acetic acid, 20 mL ethanol and a few drops of concentrated sulfuric acid in a reaction flask. Add a few boiling chips to disperse the heat and prevent the mixture from bumping.

Heat in a hot water bath under reflux for 40 minutes. The hot water bath provides a constant heating temperature of 100°C around the flask. A Bunsen flame could not be used directly because the flame is too hot and the reactants are volatile and flammable and may be ignited by the flame.

The mixture needs to be heated to increase the rate of reaction and increase the yield of the ester product.

The sulfuric acid is added to act as a catalyst.

The reason for heating under reflux is that the reactants are volatile and so they vaporise, but are returned to the mixture to continue heating when they would otherwise escape.

$$CH_3COOH + CH_3CH_2OH \rightleftharpoons H_2O + CH_3COOCH_2CH_3$$

4 (a) A B-L base is a proton acceptor.

(b) n(HCl) = 0.1000 × 0.0244
= 0.00244 mol

n(base) = 0.00244 mol, because the acid/base mole ratio is 1:1 from the equation $HCl + NaX \rightarrow H_2O + NaCl$

n(base) in 100 mL = 0.00244 × 5
= 0.0122 mol

$$n = \frac{m}{M}$$

$$\text{therefore } M = \frac{m}{n} = \frac{1.00}{0.0122}$$

$$= 82.0 \text{ g mol}^{-1}$$

5 (a) Ammonia solution (eg floor cleaner, window cleaner).

(b) The student's assessment of solutions A and B are correct because both turned thymol blue to a blue colour. This indicates a pH greater than 8.0, which is basic.

The student's assessment of solution C may be incorrect. The solution turned thymol blue yellow indicating a pH of less than 8.0. This could be basic if the pH is between 7 and 8, acidic if the pH is less than 7, or neutral if the pH is 7.

6 The industrial sources of sulfur dioxide include the burning of coal in power stations and the smelting of sulfide ores. The sources of nitrogen oxides include car engines and other high temperature combustion environments.

Both of these gases contribute to acid rain. A pH as low as 2 can occur in rainwater when these gases dissolve in atmospheric moisture.

Sulfur dioxide produced is readily oxidised in the air to form sulfur trioxide.

$$2SO_2(g) + O_2(g) \rightarrow 2SO_3(g)$$

Nitrogen reacts with oxygen in high-energy combustion conditions to form nitric oxide.

$$N_2(g) + O_2(g) \rightarrow 2NO(g)$$

The nitric oxide is readily oxidised to form nitrogen dioxide.

$$2NO(g) + O_2(g) \rightarrow 2NO_2(g)$$

When dissolved in rain these gases produce acids and subsequently acid rain.

$$SO_3(g) + H_2O(l) \rightarrow H_2SO_4(aq)$$

$$4NO_2(g) + 2H_2O(l) + O_2(g) \rightarrow 4HNO_3(aq)$$

Acid rain can affect aquatic organisms as lakes and rivers become acidic. Acid rain can damage plants in forests because the soil water can become acidic. Metal and stone buildings can be damaged because of the effect of acid on limestone, marble and metal.

$$CaCO_3(s) + 2H^+(aq) \rightarrow Ca^{2+}(aq) + CO_2(g) + H_2O(l)$$

Nitrogen oxides also contribute to photochemical smog. The brown $NO_2(g)$ gives smog its characteristic colour. Nitrogen dioxide absorbs UV radiation to form nitric oxide and oxygen atoms. The oxygen atoms combine with molecular oxygen to form ozone.

$$NO_2(g) \xrightarrow{UV} NO(g) + O(g)$$

$$O_2(g) + O(g) \rightarrow O_3(g)$$

Ozone is poisonous to humans.

The increasing smog production in big cities is reducing our quality of life, and increasing levels of acid rains are destroying forests in Europe and North America.

7 (a) pH = –log(0.01)
= 2

(b) pH of the sulfuric acid is lower than that of the hydrochloric acid because H_2SO_4 is diprotic and will ionise to produce a higher concentration of hydrogen ions compared to the HCl.

(c) Acetic acid is a weak acid and therefore only a small proportion of the molecules will ionise. This will produce a lower concentration of hydrogen ions compared to hydrochloric acid which will ionise fully because it is a strong acid. The higher concentration of hydrogen ions will give a lower pH because pH = $-\log[H^+]$.

(NB A 0.01 mol/L sulfuric acid solution does not completely ionise; it is only fully ionised in more dilute solutions.)

8 (a) Carbon dioxide

(b) $m(CO_2) = 630 - 620 = 10$ g

$$n(CO_2) = \frac{10}{44.0} = 0.227 \text{ mol}$$

$V(CO_2) = 0.227 \times 24.47 = 5.56$ L

9 (a) $Ba(OH)_2(aq) + 2HNO_3(aq) \rightarrow Ba(NO_3)_2(aq) + 2H_2O(l)$

(b) $n(Ba(OH)_2) = 0.12 \times 0.025 = 0.003$ mol

$n(HNO_3)$ needed to neutralise = 0.003×2 (from equation) = 0.006 mol

The first titre is a "rough" and should not be included in calculating an average. The average titre is thus 18.13 mL.

$$c(HNO_3) = \frac{0.006}{0.01813} = 0.33 \text{ mol L}$$

10 A weak base or acid should be used to neutralise an acid or base spill. Any excess will be weak and therefore cause little damage. A substance like sodium hydrogencarbonate is ideal as it can be used in a powdered form. A solid is easier to control and mop up later than a solution. It also has the advantage of being amphiprotic and can therefore be used to neutralise both acid and base spills.

$$HCO_3^- + H_3O^+ \rightarrow H_2CO_3 + H_2O$$

$$HCO_3^- + OH^- \rightarrow CO_3^{2-} + H_2O$$

11 The boiling point of alkanes, alkanols and alkanoic acids all increase with increasing molecular mass because the increasing mass is due to increasing carbon chain length. The longer the carbon chain the more dispersion forces exist between molecules and the more energy is needed to separate them.

As you go from alkanes to alkanols to alkanoic acids the functional group becomes more polar. Only weak dispersion forces need to be overcome to boil alkanes. However alkanols have these as well as dipole/dipole forces and hydrogen bonds due to the hydroxyl functional group. Alkanoic acids have similar dispersion forces and similar hydrogen bonds due to the hydroxyl group but they are more polar molecules owing to the double-bonded oxygen as part of the carbonyl functional group. This gives more polarity to the molecule and results in stronger dipole/dipole forces. These require the most energy to overcome.

CHAPTER 3

Core Topic

Chemical Monitoring and Management

Multiple-choice Questions

Past HSC Questions

1 Why is chlorine used to treat local water supplies?

(A) To make water suitable for swimming
(B) To kill micro-organisms living in the water
(C) To promote sedimentation of finely suspended solids
(D) To precipitate heavy metal ions such as lead and mercury

2 The atomic absorption spectrophotometer was developed by Sir Alan Walsh and his team at CSIRO in the 1950s. Its development was one of the most significant in Australian chemical technology. What did it provide?

(A) A rapid method to monitor chemical pollutants in water supplies
(B) The first method for determining the concentrations of metal ions in water supplies
(C) A method for determining the concentrations of hydrocarbons at very low concentrations
(D) A method for determining the concentrations of metal ions at very low concentrations

3 Four students analysed a sample of fertiliser to determine its percentage of sulfate.

Each student:

- weighed an amount of fertiliser;
- dissolved this amount in 100 mL of water;
- added aqueous barium nitrate;
- filtered, dried and weighed the barium sulfate precipitate.

Their results and calculations are shown in the table.

Student	*Mass of fertiliser used* (g)	*Mass of* $BaSO_4$ *weighed* (g)	*Percentage of sulfate in fertiliser* (%)
A	11.6	19.5	69.2
B	10.4	16.9	66.9
C	10.268	22.612	90.6
D	11.1	18.2	67.5

The percentage of sulfate calculated by Student *C* was significantly higher than that of the other students. Which is the most likely reason for this?

(A) Student *C* did not dry the sample for long enough.

(B) Student *C* added more $Ba(NO_3)_2$ solution than the other students.

(C) Student *C* used a balance capable of measuring weight to more decimal places.

(D) Student *C* waited longer than the other students for the $Ba(NO_3)_2$ to react completely with the sulfate.

4 Which diagram represents the most effective design for a microscopic membrane filter to purify contaminated water?

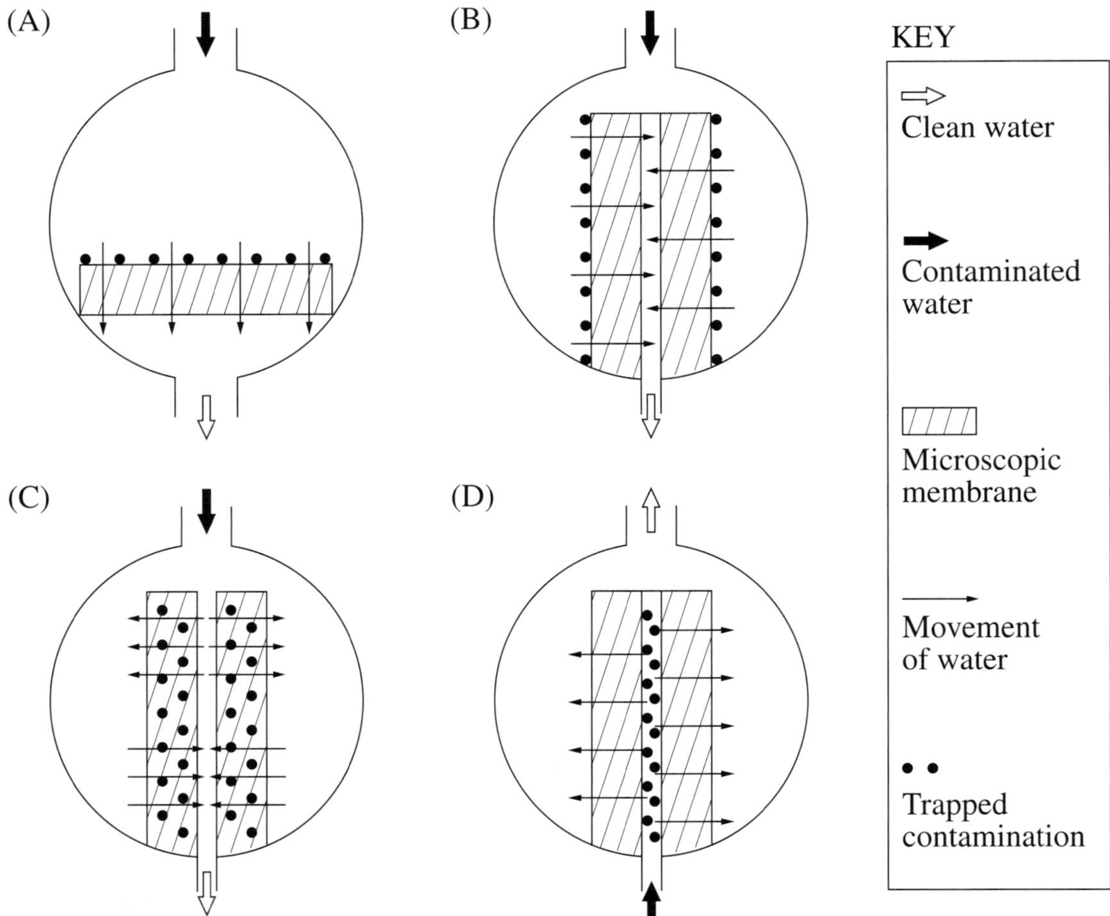

5 A car engine burns fuel with insufficient air. Which substance would be emitted in the exhaust in higher levels than from an engine with a correct fuel to air ratio?

(A) Carbon dioxide

(B) Carbon monoxide

(C) Oxygen

(D) Water

6 Oxides of nitrogen are produced by the combustion of coal. The percentage of these compounds remaining in emissions can be reduced using the methods named in the graph.

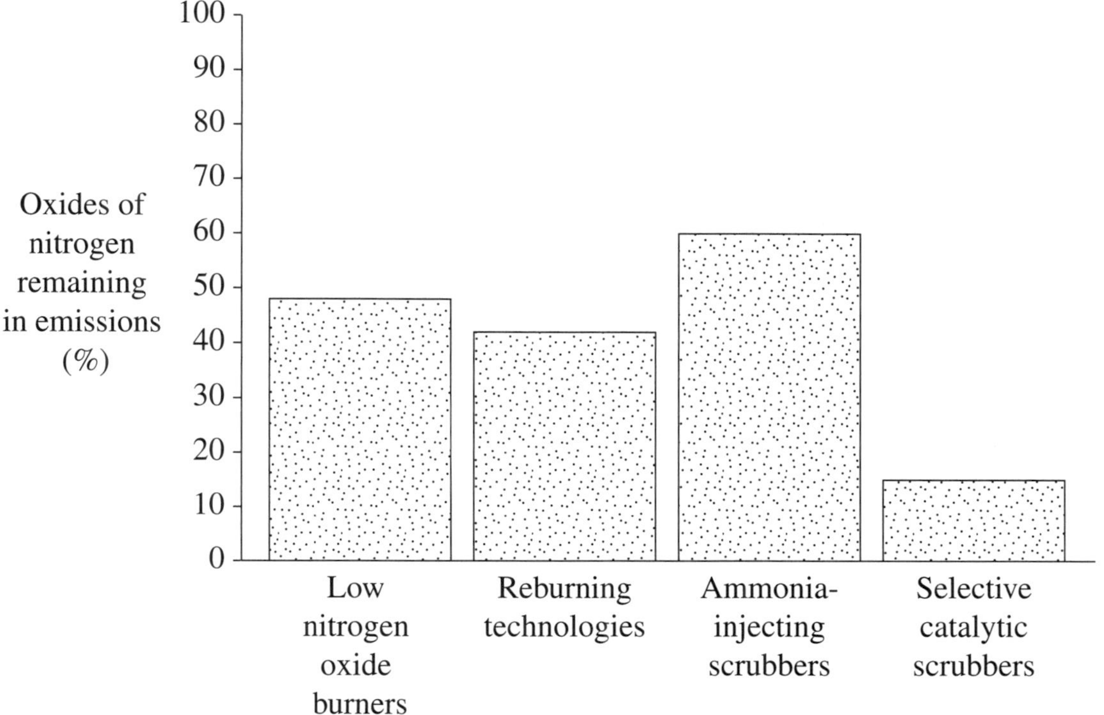

Which of the following is the most effective method for removing oxides of nitrogen produced by the combustion of coal?

(A) Low nitrogen oxide burners

(B) Reburning technologies

(C) Ammonia-injecting scrubbers

(D) Selective catalytic scrubbers

7 The Haber Process for producing ammonia was developed early in the twentieth century. What was the major advantage of its development?

(A) A government sold the process to other governments.

(B) The inventor sold the process for a great deal of money.

(C) It provided a source of nitrogen for farming and industry.

(D) It provided jobs for many who were unemployed.

8 Ammonia is produced from hydrogen and nitrogen, according to the equation:

$$N_2(g) + 3H_2(g) \rightleftharpoons 2NH_3(g)$$

The graph shows the yield of ammonia produced at 200°C and 100 kPa.

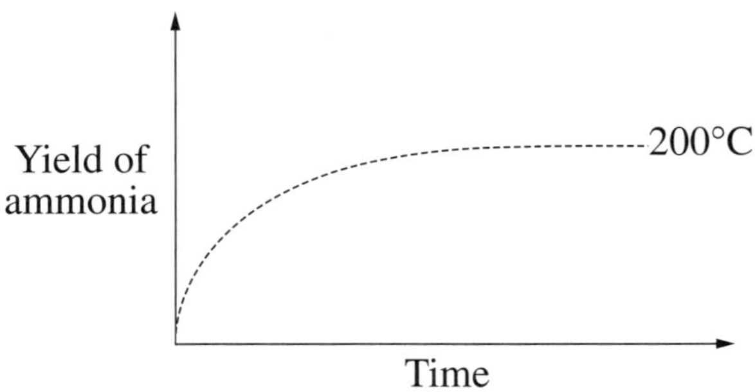

Which graph shows a correct comparison of the yield of ammonia produced at a temperature of 400°C and 100 kPa with the yield produced at 200°C and 100 kPa?

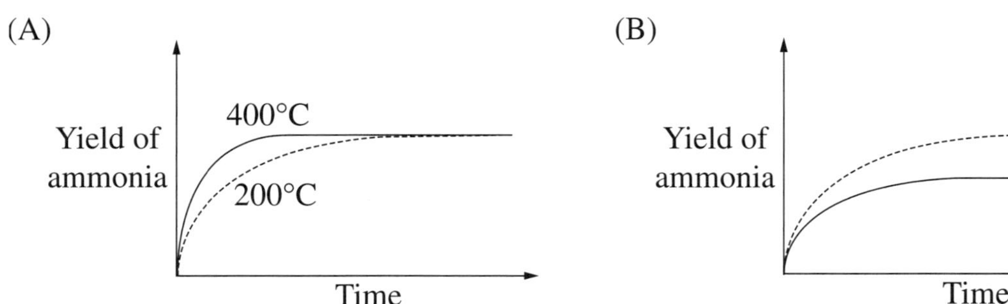

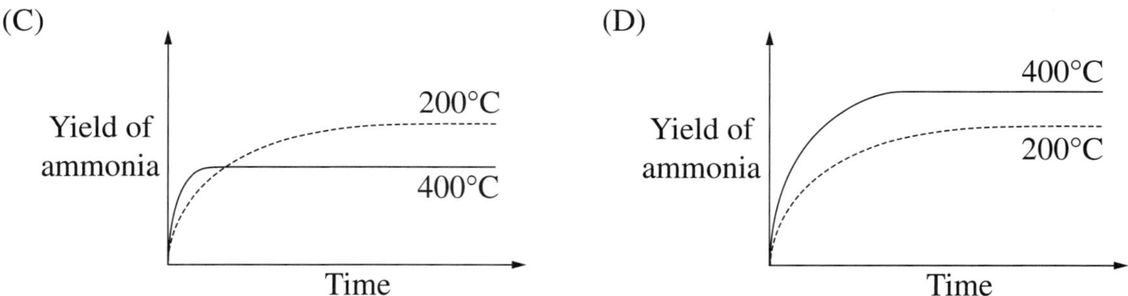

9 The table gives the results of chemical tests for some cations and anions.

(ppt = precipitate)

Ion	*Add cold* 0.1 M HCl	*Add* 0.1 M KSCN	*Add* 0.1 M Na_2CO_3	*Add* 0.1 M $AgNO_3$
Ca^{2+}	no change	no change	white ppt	no change
Fe^{3+}	no change	red colour	brown ppt	no change
Ba^{2+}	no change	no change	white ppt	no change
Pb^{2+}	white ppt	no change	white ppt	no change
Cl^-	no change	no change	no change	white ppt

When a group of students performed the above tests on an unknown solution they obtained the following results:

Add cold 0.1 M HCl	*Add* 0.1 M KSCN	*Add* 0.1 M Na_2CO_3	*Add* 0.1 M $AgNO_3$
no change	no change	white ppt	white ppt

Which conclusion is consistent with these results?

(A) The sample contained both $CaCl_2$ and $BaCl_2$.

(B) The sample contained both $CaCl_2$ and $PbCl_2$.

(C) The sample contained both $FeCl_3$ and $PbCl_2$.

(D) The sample contained both $FeCl_3$ and $BaCl_2$.

10 In which layer of the atmosphere does ozone act as a UV radiation shield?

(A) Mesosphere

(B) Stratosphere

(C) Thermosphere

(D) Troposphere

11 Which of the following could be used to determine the total dissolved solids in a sample of muddy river water?

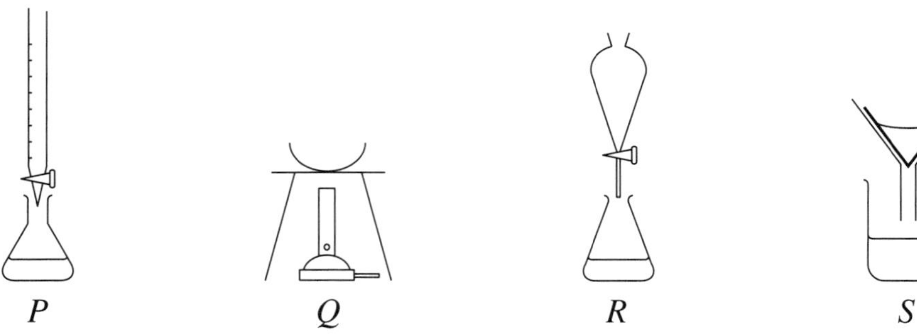

(A) *P* and *Q*

(B) *R* and *S*

(C) *P* and *R*

(D) *Q* and *S*

12 What is the name of the compound shown?

```
     H   F   Cl
     |   |   |
H—C—C—C—H
     |   |   |
     H   H   F
```

(A) 1-chloro-1,2-difluoropropane

(B) 3-chloro-2,3-difluoropropane

(C) 1,2-difluoro-1-chloropropane

(D) 1-chloro-1,2-difluoropentane

13 The graph shows October ozone concentrations above Halley Bay in Antarctica between 1956 and 1998.

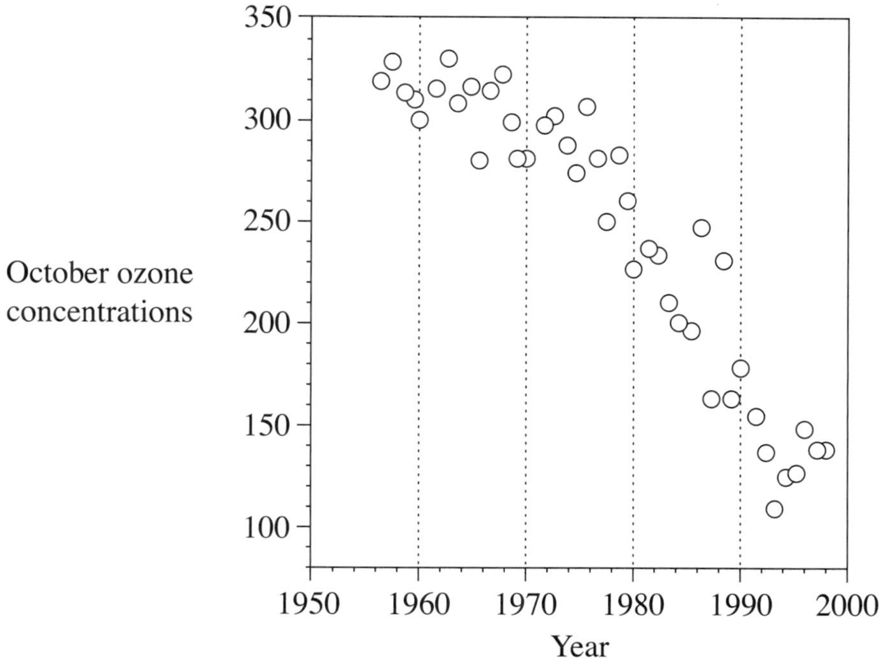

Reproduced with the permission of Oxford University Press, London

Based on these data alone, which of the following is a valid statement about the concentration of ozone above Halley Bay?

(A) It was greater in 1998 than in 1993.

(B) It will be greater in 2004 than in 1998.

(C) The variation in ozone concentration between 1960 and 1970 was due to changes in atmospheric CFC concentrations.

(D) The variation in ozone concentration from one year to the next is due only to changes in atmospheric CFC concentrations.

14 The graph shows the mass and amount of carbon, fluorine and chlorine atoms in one mole of a compound.

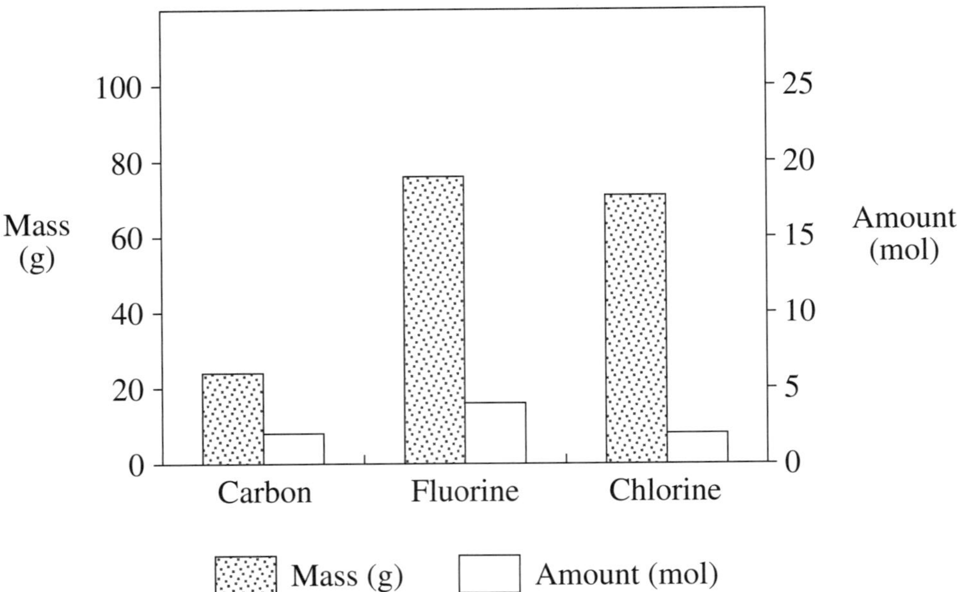

What is the molecular formula for this compound?

(A) CF_2Cl

(B) CF_2Cl_2

(C) $C_2F_3Cl_3$

(D) $C_2F_4Cl_2$

Free-response Questions

Past HSC Questions

Question 1 (6 marks) **Marks**

In the early twentieth century, Fritz Haber developed a method for producing ammonia, as shown by the equation:

$$N_2(g) + 3H_2(g) \rightleftharpoons 2NH_3(g)$$

(a) Ammonia is used as a cleaning agent. State ONE other use of ammonia. **1**

*(1 line)**

(b) Explain the effect of liquefying the ammonia on the yield of the reaction. **2**

(4 lines)

(c) Explain why it is essential to monitor the temperature and pressure inside the reaction vessel. **3**

(8 lines)

Question 2 (6 marks)

Explain the need for monitoring the products of a chemical reaction such as combustion. **6**

(12 lines)

* Shows the number of lines available in the HSC Answer Book for this question.

Marks

Question 3 (4 marks)

A university student decided to measure the concentration of lead (Pb) in the soil around his home. He prepared five standard lead solutions of known concentration. The absorbance of these solutions was measured. These results are shown in the table.

Concentration of lead standard (ppm)	*Absorbance*
0	0.00
1	0.15
2	0.31
3	0.44
4	0.59
5	0.75

(a) Draw a line graph of these data. **1**

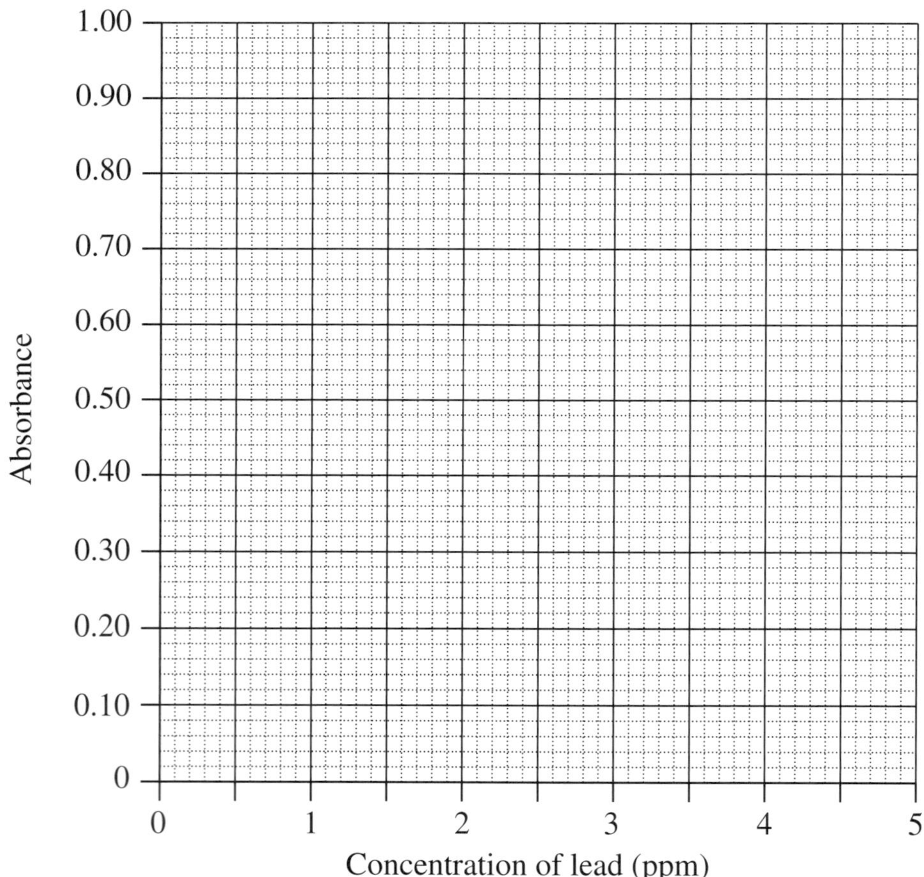

Question 3 continues

Marks

Question 3 (continued)

(b) The student prepared solutions from four different soil samples around his home. These solutions were also analysed using the same method. The results are shown in the table. **1**

Solutions made from soil samples	
Area sampled	*Absorbance*
Front garden bed	0.19
Back garden bed	0.09
Mail box	0.22
Back fence	0.11

Determine the highest concentration of lead in the soil around the home.

(1 line)

(c) State an hypothesis to account for the variation in lead concentration around the student's home. **2**

(4 lines)

End of Question 3

Marks

Question 4 (4 marks)

Oxygen exists in the atmosphere as the allotropes oxygen and ozone. The graph shows a typical change in ozone concentration with changing altitude. **4**

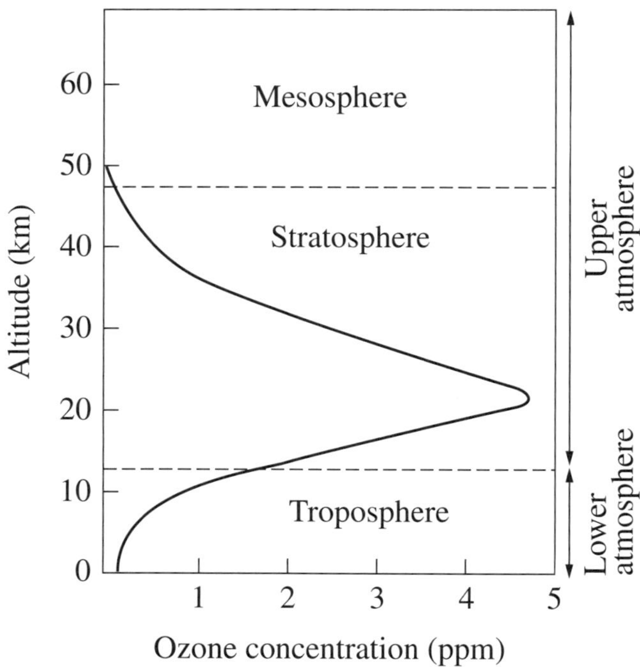

Compare the environmental effects of the presence of ozone in the upper and lower atmosphere.

(7 lines)

Question 5 (4 marks)

Assess the impact of atomic absorption spectroscopy (AAS) on the scientific understanding of the effects of trace elements. **4**

(12 lines)

Question 6 (6 marks)

(a) What is the systematic name of the CFC in the diagram? **1**

```
    Cl  Cl
    |   |
F — C — C — F
    |   |
    F   F
```

(1 line)

(b) Identify the bonding within ozone, using a Lewis electron-dot diagram. **2**

(c) Discuss how CFCs damage the ozone layer, using relevant equations. **3**

(6 lines)

Marks

Question 7 (3 marks)

Water can be described as either 'hard' or 'soft'.

A sample of hard water contains 6×10^{-4} mol L^{-1} of magnesium carbonate. Calculate the mass, in mg, of magnesium carbonate in 150 mL of this sample. **3**

(6 lines)

Question 8 (5 marks)

Describe the physical and chemical processes needed to purify and sanitise a town water supply. **5**

(10 lines)

Question 9 (4 marks)

Describe the process of eutrophication, and assess the suitability of water quality tests used to monitor it. **4**

(14 lines)

Question 10 (5 marks)

A student carried out an investigation to analyse the sulfate content of lawn fertiliser. The student weighed out 1.0 g of fertiliser and dissolved it in water. 50 mL of 0.25 mol L^{-1} barium chloride solution was then added. A white precipitate of barium sulfate formed, which weighed 1.8 g.

(a) Calculate the percentage by mass of sulfate in the fertiliser. **2**

(6 lines)

(b) Evaluate the reliability of the experimental procedure used. **3**

(11 lines)

Marks

Question 11 (4 marks)

The results of analysis of a set of standard cadmium solutions are presented in the table.

Concentration of cadmium standard solution (ppm)	*Absorbance*
0	0.00
3	0.22
6	0.38
9	0.62
12	0.83

(a) Draw an appropriate graph of the data. **2**

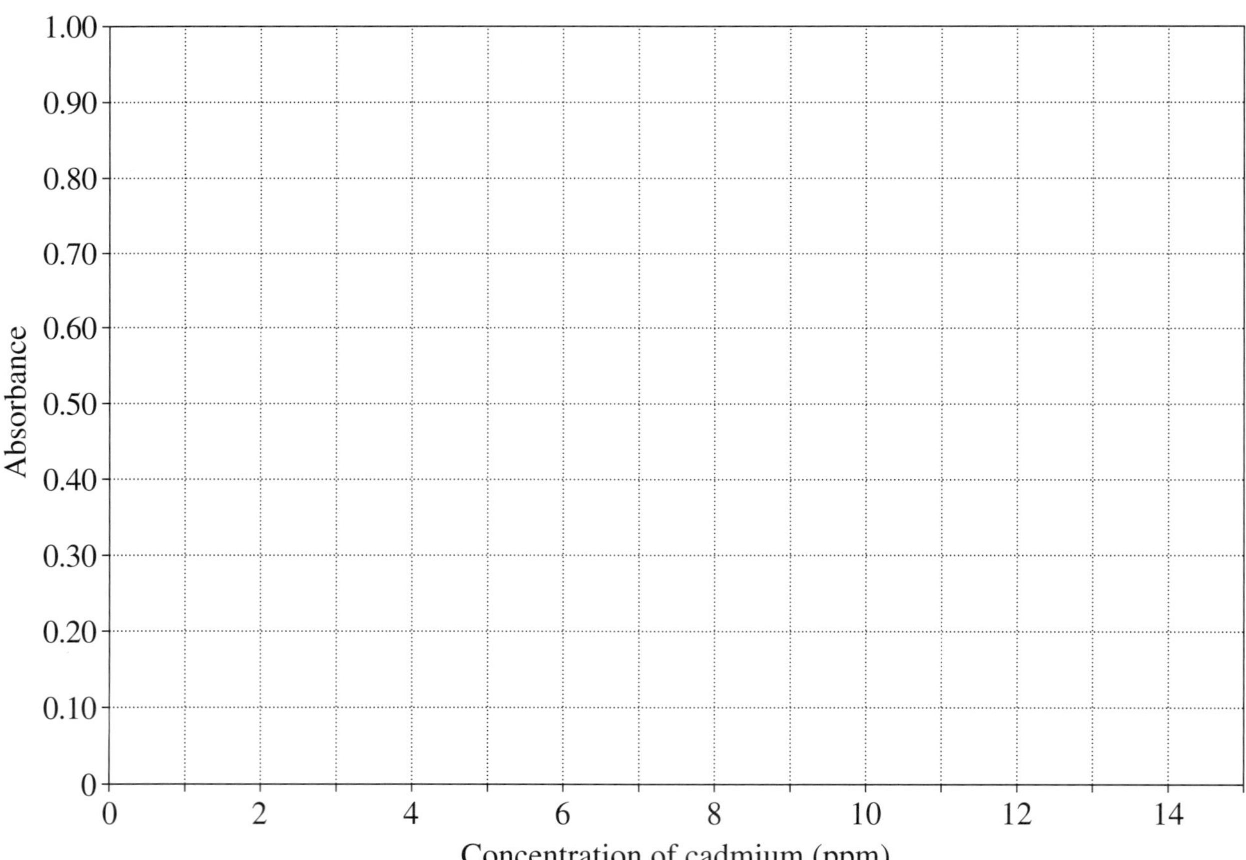

Question 11 continues

Marks

Question 11 (continued)

The map shows a catchment area. There is an industrial plant, a sewage treatment plant and a small town, all of which discharge water into the river. Water samples were collected at four sites.

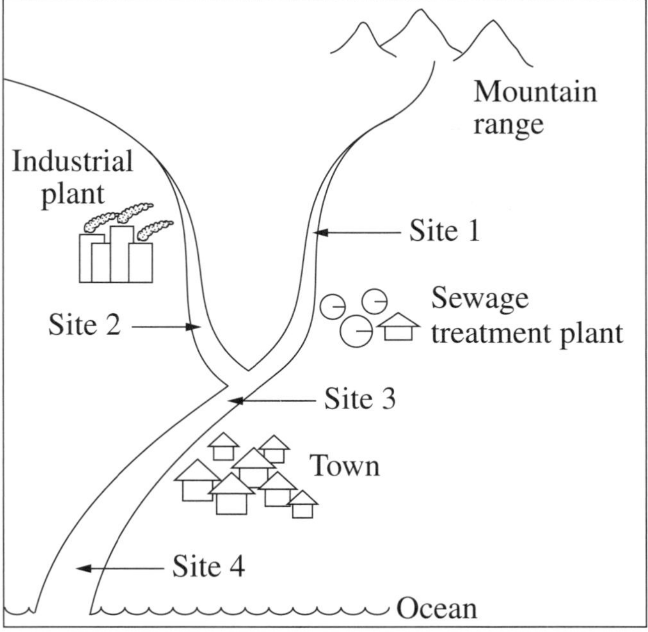

The results of analysis of cadmium levels from these four sites are given in the table.

Sample site	*Absorbance*
Site 1	0.08
Site 2	0.15
Site 3	0.55
Site 4	0.40

(b) Justify your conclusion about the most likely source of cadmium pollution. **2**

(8 lines)

Question 12 (7 marks)

Evaluate the importance of monitoring and managing the conditions used in the Haber process. **7**

(26 lines)

Core Topic

Chemical Monitoring and Management

Worked Answers

Multiple-choice Questions

1 B

2 D AAS is only used to determine low metal ion concentrations.

3 A To get an accurate mass of precipitate, it must be dried thoroughly. Any moisture present will add to the mass giving an unreliable result.

4 B The contaminated water must pass through the filter. The filter with the highest surface area would be most effective. Contaminants do not pass through the pores of the filter.

5 B The products of incomplete combustion of a hydrocarbon are a mixture of carbon dioxide, carbon monoxide, carbon and water. Carbon dioxide and water are formed in complete combustion. Water levels do not change.

6 D

7 C Haber developed a process that fixed nitrogen in a form that could be used as a fertiliser for plants and a reactant and catalyst in industry.

8 C The production of ammonia is an exothermic reaction. When the reaction is run at higher temperatures, the equilibrium will shift in reverse reducing the yield of ammonia. However, at the higher temperature the reaction will be faster and will reach equilibrium earlier.

9 A When HCl was added and no precipitate formed, Pb^{2+} could not have been present. When the addition of KSCN did not result in a precipitate, Fe^{3+} could not have been present. Only one alternative answer contains neither of these ions.

10 B The ozone layer lies about 25 km above the Earth's surface. It lies within the stratosphere which occupies the part of the atmosphere from about 15 to 50 km above the Earth's surface.

11 D Insoluble substances would be filtered by the equipment in S leaving the dissolved solids. The equipment in Q is used to evaporate the river water leaving the dissolved solids in the evaporating basin.

12 A The longest straight chain is numbered to give the lowest locant set for the functional groups (ie 1,1,2 and not 2,3,3) and the molecule is named with the functional groups in alphabetical order.

13 A Answer (B) is a prediction and is therefore not based on the *data alone*. Answers (C) and (D) are possible explanations for the data and are therefore not based on the data alone. Answer (A) could be a valid statement based on the data with close examination of the data.

14 D The molecular formula gives the number of atoms of each element.

The graph shows:

$$n(C) = \frac{24 \text{ g}}{12 \text{ g}} = 2; n(F) = \frac{76 \text{ g}}{19} = 4; n(Cl) = \frac{71 \text{ g}}{35.45 \text{ g}} = 2$$

Free-response Questions

1 (a) Ammonia is used in the production of fertilisers.

(b) When ammonia gas is liquefied, the concentration of the gas is reduced. The equilibrium will shift forward to partially counteract the change, increasing the yield of ammonia.

(c) Temperature needs to be monitored because if temperature rises too high the reaction will shift in reverse because the forward reaction is exothermic. If temperature falls too low the molecules may not have enough energy to overcome the activation energy barrier. Therefore the temperature needs to be a compromise such that there is a sufficient rate and high enough yield.

Pressure needs to be monitored to keep it high. A high pressure causes the equilibrium to shift to the right (products) because there are more moles of gas reactant than product. The forward reaction counteracts an increase in pressure.

2 Combustion is the reaction of a fuel with oxygen to release energy. When a plentiful supply of oxygen is available, complete combustion occurs. Carbon dioxide and water are the only products.

$$C_8H_{18} + \frac{25}{2}O_2 \rightarrow 8CO_2 + 9H_2O$$

The products need to be monitored because if oxygen becomes limiting, alternative substances are produced that are more harmful to the environment and human population than carbon dioxide.

$$C_8H_{18} + \frac{17}{2}O_2 \rightarrow 8CO + 9H_2O$$

or

$$C_8H_{18} + \frac{9}{2}O_2 \rightarrow 8C + 9H_2O$$

Carbon monoxide (CO) is a poisonous gas and can therefore affect human health. Carbon (soot) is carcinogenic to humans and can be irritating to the lungs. Both of these alternative products can also signal a decrease in fuel efficiency which will result in a decreased energy yield from the fuel.

3 (a)

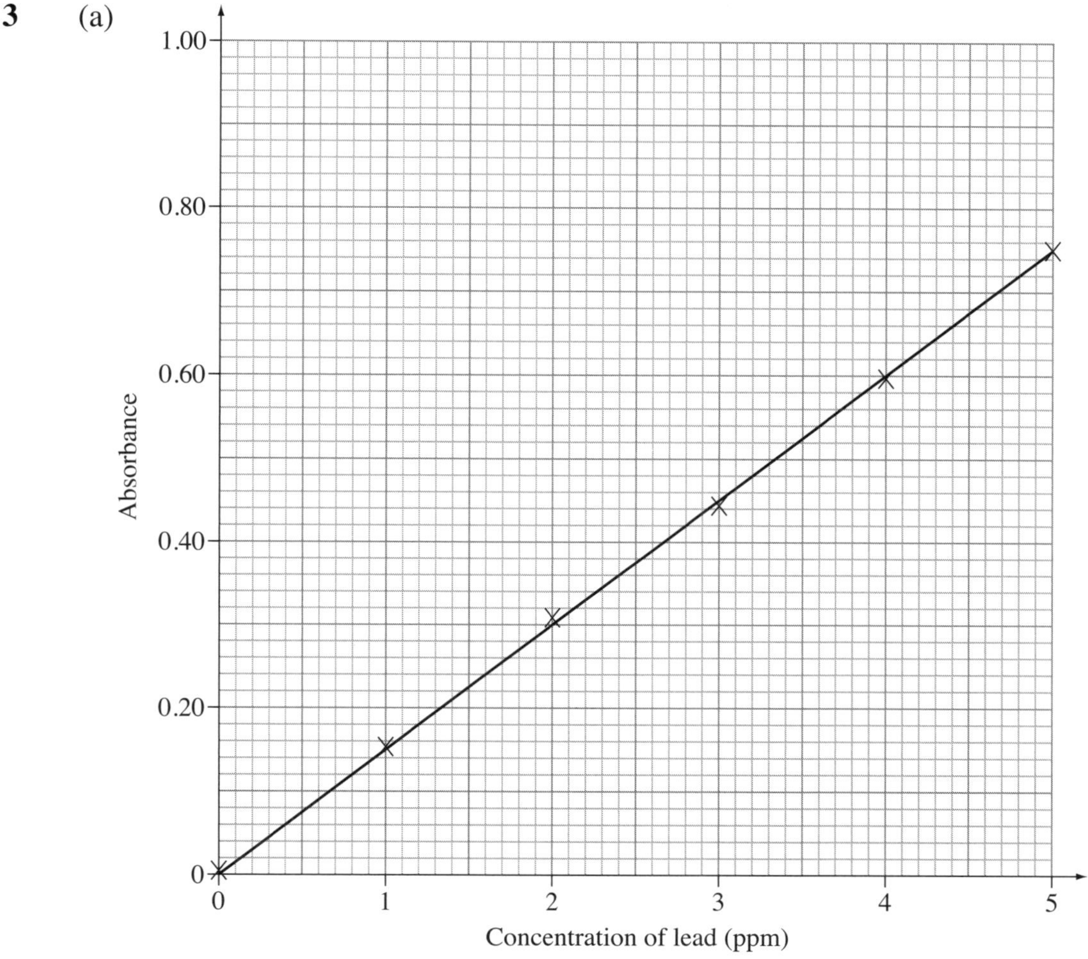

(b) 1.5 ppm (when A = 0.22)

(c) That the higher levels of lead in the front garden and around the mail box compared to the back garden and back fence are due to proximity to the main road. Cars (prior to the use of unleaded petrol) emitted lead aerosols from their exhaust pipes.

4 Ozone is found in high concentrations in the upper atmosphere (4–5 ppm in the stratosphere). Here it is important to the environment as it absorbs UV-B radiation from the sun. This form of radiation would be damaging to the environment and human health if it were to pass through the atmosphere to earth. It would cause increased melanomas and cataracts in humans, reduce phytoplankton populations and damage crops.

Ozone is found in low concentrations (<1 ppm) in the lower atmosphere. Here it is poisonous to humans but because it is in such low concentrations, the effect is minimal.

5 AAS spectroscopy is used to measure the concentration of trace metal ions to ppm or ppb. Trace elements are elements needed in very small amounts by living things. Zinc, copper and iron are all trace elements found in soil that when limited can affect agricultural productivity. Therefore the use of AAS has allowed the monitoring of levels of trace elements in soil to levels not possible before using the older procedures of gravimetric and volumetric analysis. When levels are found to be low, additional chemicals can be added to the soil to increase the amounts of the trace element.

6 (a) 1,2-dichloro-1,1,2,2-tetrafluoroethane

(b)

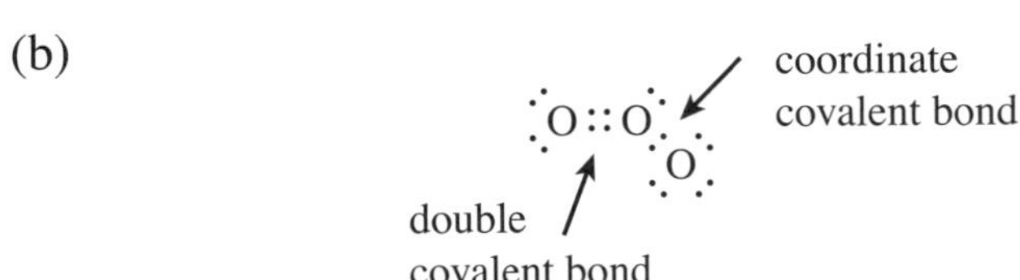

(c) When CFCs reach the stratosphere, UV radiation splits a chlorine radical from the CFC molecule.

$$CF3Cl\ UV \rightarrow CF3 + Cl$$

This Cl radical is very reactive and reacts with an ozone molecule to form an oxygen molecule and an oxygen chloride molecule.

$$O_3 + Cl \rightarrow O_2 + OCl$$

The OCl molecule then reacts with an oxygen radical to form an oxygen molecule releasing the chlorine radical to react with many more ozone molecules.

$$OCl + O \rightarrow O_2 + Cl$$

Thousands of ozone molecules can be decomposed in this manner by one CFC molecule reducing the thickness of the ozone layer.

7 $n(MgCO_3) = 6 \times 10^{-4} \times 0.150 = 9 \times 10^{-5}$ mol

$m(MgCO_3) = 9 \times 10^{-5} \times 84.32 = 7.59 \times 10^{-3}$ g

7.59×10^{-3}g $\times$ 1000 = 7.59 mg

8 There are various physical and chemical steps in the purification of water. These are:

Physical processes:

- screening out large debris from the raw water
- coagulation of colloidal (clay) particles to increase particle size to ensure rapid sedimentation
- sedimentation to remove larger suspended particles
- filtration to remove finer suspended particles.

Chemical processes:

- aeration to oxidise soluble iron and manganese salts to insoluble compounds that will precipitate
- oxidation (eg by potassium permanganate) of iron and manganese salts to produce insoluble compounds

  eg $3Mn^{2+} + 2MnO_4^- + 2H_2O \rightarrow 5MnO_2(s) + 4H^+$
- lime softening to precipitate out high levels of calcium and magnesium ions using carbonate ions

  $Ca^{2+} + CO_3^{2-} \rightarrow CaCO_3(s)$
- addition of chlorine and ammonia to kill microbes (biochemical).

9 Eutrophication is the abundant growth of algae and water plants due to the presence of high levels of nitrogenous and phosphate nutrients in the waterway. The presence of algal blooms has several effects. Excessive surface algal growth blocks the sunlight required for water plants that are deeper in the water. Algal blooms also interfere with diffusion of oxygen from the air into the water. Eventually the population of algae dies due to lack of nutrients and the dissolved oxygen levels in the water become severely depleted as aerobic decomposers break down their remains. Once the DO levels drop to near zero anaerobic decomposers emerge from the mud and become active as they decompose the waste.

To avoid eutrophication, the following maximum levels are recommended in lakes and (raw water) storage reservoirs: phosphorus = 10 ppb; nitrogen = 350 ppb. In addition the N:P ratio is often used to determine whether a body of water is eutrophic. Values of N:P greater than 10:1 protect waterways from eutrophication.

10 (a) $n(BaSO_4) = \dfrac{1.8}{233.37} = 7.7 \times 10^{-3}$ mol

$n(SO_4^{2-}) = n(BaSO_4) = 7.7 \times 10^{-3}$

$\%\text{ mass } SO_4^{2-} = \dfrac{0.74}{1.0} \times 100 = 74\%$

(b) The fertiliser may not have been completely soluble in water. Therefore, filter the mixture. Acid should have been used to help dissolve it. This will remove carbonate ions as well as prevent phosphate ions from precipitating with barium ions.

Using barium chloride to precipitate sulfate is reliable because barium chloride is soluble but barium sulfate is insoluble.

The most unreliable aspect of this procedure probably involved the technique used to collect the precipitate. Barium sulfate is very fine and much of it would have passed through filter paper if it was used. This would have resulted in a lower than true value for the mass of barium sulfate. Boiling the mixture to coagulate the precipitate is recommended prior to filtration. If a sintered glass funnel had been used to filter the precipitate a better result would have been obtained, but some precipitate may have passed through this as well.

The result would have been more reliable if repeat trials had been conducted and averaged. There is also no proof that the 50 mL of barium chloride was sufficient to precipitate all the sulfate ions.

11 (a)

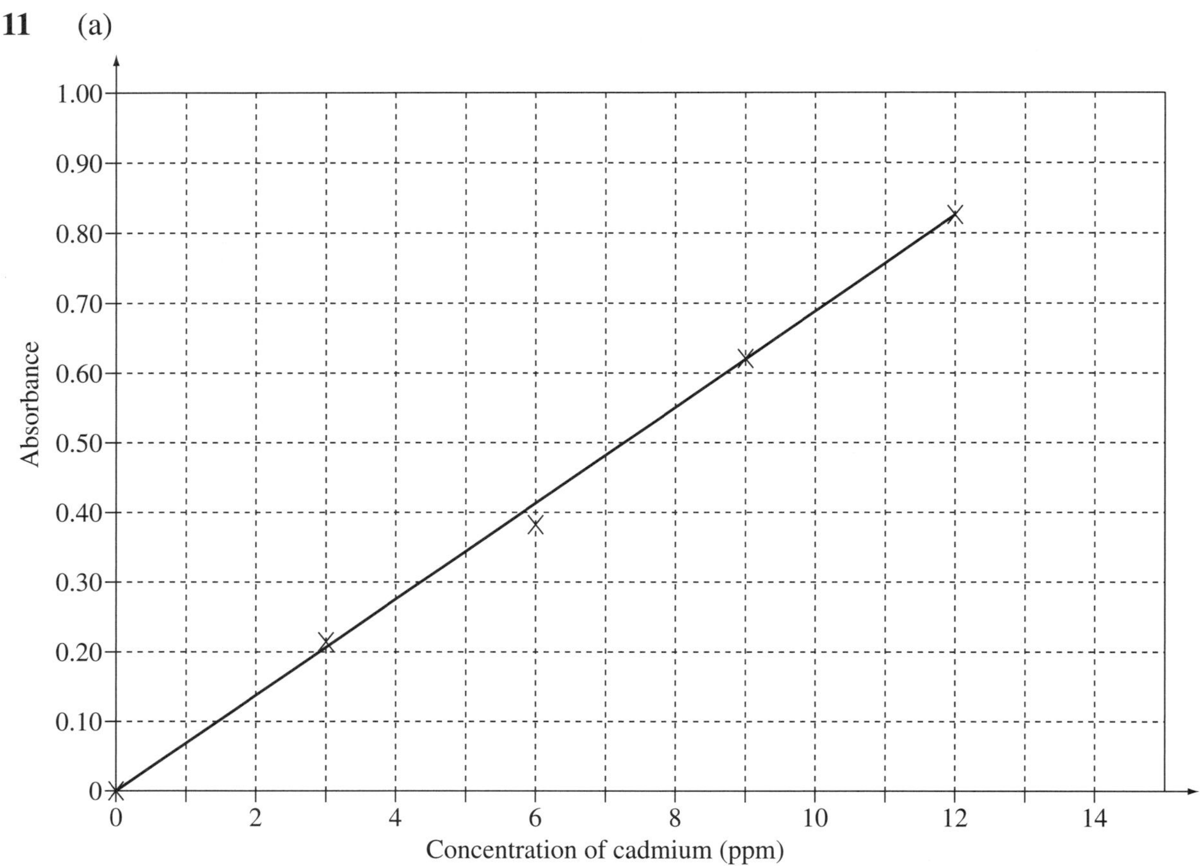

(b) The most likely source of cadmium pollution is the sewage treatment plant.

Water is flowing toward the ocean. Site 3 recorded the highest levels (0.55) so that cadmium could have originated from either the industrial plant or the sewage treatment plant. However, if it was the industrial plant then site 2 would have recorded the highest levels because it is closer to the industrial plant and the cadmium would not have been diluted. In fact the recording at site 2 was quite low (0.15). It could not have originated further upstream from the sewage treatment plant because the recording at site 1 was the lowest (0.08).

12 The Haber process is the process of producing ammonia from gaseous hydrogen and nitrogen.

$$3H_2(g) + N_2(g) \rightleftharpoons 2NH_3(g)$$

Because the system can come to equilibrium and is exothermic, conditions need to be monitored and kept at levels that favour the formation of ammonia.

Temperature needs monitoring to maintain it at about 500°C. If it falls below this the reaction rate becomes too low and if it rises above this the equilibrium shifts in reverse and the yield of ammonia drops. A porous iron catalyst is used to raise the reaction rate at this moderate temperature.

A high pressure of 15–25 MPa is used to optimise the yield of ammonia because its production is accompanied by a decrease in the number of gaseous molecules. Pressure needs monitoring because if it drops, the yield of ammonia will also drop but if rises, the reaction vessel may not withstand the pressure.

The ratio of reactant gases must be monitored to keep it at 3:1 H_2:N_2 because this is the ratio at which these gases react. If this ratio is not adhered to, a limiting reactant will slow down the reaction and be costly.

For these reasons, it is extremely important to monitor the conditions used in the Haber process. If any of these conditions change the yield of ammonia will not be optimal.

CHAPTER 4

Option Topic
Industrial Chemistry

Past HSC Questions

Question 1 (25 marks) **Marks**

Electrolysis is an important industrial process.

(a) (i) Define *electrolysis*. 1

(ii) Compare the reaction products from the electrolysis of molten sodium chloride and concentrated aqueous sodium chloride. 2

(b) Carbonyl chloride, $COCl_2$, is a colourless, poisonous gas that is also known as phosgene. It is needed for the production of insecticides, polyurethane plastics and polycarbonate. It is produced from the exothermic equilibrium reaction of carbon monoxide gas and chlorine gas. When the reaction vessel is cooled below 8°C the phosgene is a liquid.

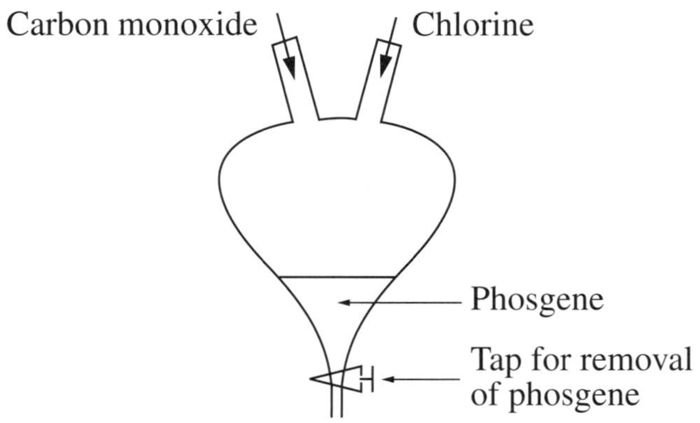

Carbon monoxide

Chlorine

Phosgene

Tap for removal of phosgene

Reaction vessel for the formation of phosgene at 4°C

(i) Write a balanced equation for the formation of phosgene. 2

(ii) Explain how industry could maximise the production of phosgene. 2

(c) Explain why sulfuric acid is an important industrial chemical. Include balanced chemical equations in your answer. 5

(d) (i) Name the chemical process used to make soap. 1

(ii) Outline the procedure for making soap in the school laboratory. 2

(iii) Describe a safety risk associated with the procedure outlined in part (ii), and suggest a safe work practice to minimise the risk. 3

(e) Evaluate how environmental issues are addressed in the Solvay process. 7

Question 2 (25 marks) **Marks**

(a) (i) Define *saponification*. **1**

(ii) Account for the cleaning action of soap. **3**

(b) One of the reactions used to form sulfuric acid is the reaction of oxygen with sulfur dioxide under equilibrium conditions to form sulfur trioxide. **4**

Before the reaction, the concentration of sulfur dioxide was 0.06 mol L^{-1} and the concentration of oxygen was 0.05 mol L^{-1}. After equilibrium was reached, the concentration of sulfur trioxide was 0.04 mol L^{-1}.

Calculate the equilibrium constant, K, for the reaction. Show relevant working.

(c) (i) Use a chemical equation to describe what happens when sulfuric acid is added to water in a laboratory. **2**

(ii) Describe the use of sulfuric acid as an oxidising agent, as a dehydrating agent and as a means of precipitating sulfates. Use chemical equations to illustrate your answer. **3**

(d) During your practical work, you performed a first-hand investigation involving an equilibrium reaction.

(i) Outline the procedure you used. **2**

(ii) Explain how you analysed the equilibrium reaction qualitatively. **4**

(e) Evaluate changes in industrial production methods for sodium hydroxide. **6**

Question 3 (25 marks) **Marks**

(a) (i) Identify ONE use of sulfuric acid in industry. **1**

(ii) One of the starting materials used for preparing sulfuric acid is sulfur. Describe the process used to extract sulfur from mineral deposits. **3**

(b) During your practical work you performed a first-hand investigation to identify the products of electrolysis of sodium chloride.

(i) Describe ONE precaution you took to minimise hazards, or to dispose of reactants and products safely. **1**

(ii) Outline the procedure you used to identify the products of electrolysis of sodium chloride. **3**

(c) Analyse how an understanding of the structure and cleaning action of soaps led to the development of synthetic detergents. **5**

(d) The Ostwald process is used for making nitric acid from ammonia, and involves several equilibrium steps.

(i) Identify the only factor that changes the value of an equilibrium constant. **1**

(ii) One step in the process produces nitrogen dioxide according to the equation: **2**

$$2NO(g) + O_2(g) \rightleftharpoons 2NO_2(g).$$

This reaction is exothermic. Describe TWO methods that could be used to increase the yield of nitrogen dioxide.

(iii) A 1 L reaction vessel initially contained 0.25 mol NO and 0.12 mol O_2. After equilibrium was established there was only 0.05 mol NO. **3**

Calculate the equilibrium constant for the reaction. Show all relevant working.

(e) Assess how environmental issues have been addressed in an industrial method of production of an acid, and an industrial method of production of a base. **6**

Option Topic
Industrial Chemistry
Worked Answers

Question 1

(a) (i) Electrolysis is the process in which electricity is used to produce a chemical reaction that does not occur spontaneously.

(ii) In the electrolysis of molten sodium chloride, sodium metal and chlorine gas are the only products.

When concentrated aqueous sodium chloride is electrolysed, due to the presence of water, sodium hydroxide forms instead of sodium metal and hydrogen gas is also produced as well as chlorine gas.

(b) (i) $CO(g) + Cl_2(g) \rightleftharpoons COCl_2(g)$

(ii) Cooling the reaction vessel would lead to more phosgene forming as the forward reaction is exothermic. Therefore, according to Le Chatelier's principle, lowering the temperature would be counteracted by the equilibrium shifting to the right.

Also, if the temperature is kept below 8°C the phosgene condenses to a liquid that can be drawn off, further pushing the equilibrium to the right.

(c) Sulfuric acid is important owing to its many applications and uses.

Concentrated sulfuric acid has a strong affinity for water making it a good dehydrating agent. This is useful in the production of ethylene from ethanol.

$$CH_3CH_2OH \rightarrow CH_2CH_2 + H_2O$$

Sulfuric acid is used in the production of esters in which it acts as a catalyst to increase the rate of the reaction that is otherwise slow.

$$CH_3CH_2OH + CH_3COOH \rightleftharpoons CH_3CH_2COOCH_3 + H_2O$$

The presence of ions such as Ba^{2+} and Pb^{2+} in water can be tested by precipitating them with sulfuric acid.

$$Ba^{2+}(aq) + SO_4^{2-}(aq) \rightarrow BaSO_4(s)$$

Because it is a strong acid which readily ionises, sulfuric acid can be used as an electrolyte in lead acid batteries. It also has useful properties as an oxidising agent in reactions with various metals, e.g. in oxidising copper to copper ions.

$$Cu(s) + 2HSO_4^-(aq) + 2H^+(aq) \rightarrow SO_2(g) + Cu^{2+}(aq) + SO_4^{2-}(aq) + 2H_2O(l)$$

(An alternative answer would outline the production of ammonium sulfate fertiliser.)

(d) (i) Saponification

(ii) 5 mL canola oil is added to 20 mL 4.0 M NaOH in a 500-mL beaker. The mixture is heated gently for 30 minutes on a hot plate. Water is added from time to time so that the volume of the mixture remains the same. 10 g NaCl is added to precipitate the soap formed. The mixture is further heated for 3 minutes. Soap is recovered from the mixture by filtration.

(iii) Safety risk: NaOH is corrosive and can burn the skin. To minimise the risk gloves and safety glasses must be worn when handling NaOH. Heat the mixture in a fume cabinet to avoid breathing fumes that may contain NaOH.

(e) The Solvay process involves reacting ammonia with concentrated brine in a Solvay tower. The process is exothermic, therefore the system is cooled by cold water pipes to achieve a greater yield. This ammoniacal brine is reacted with carbon dioxide and water to form hydrogencarbonate and ammonium ions. The solid compound sodium hydrogencarbonate crystallises within the tower forming sodium hydrogencarbonate which is filtered and washed to remove ammonium and chloride ions. The sodium hydrogencarbonate is heated strongly in a furnace to decompose it to sodium carbonate.

Environmental considerations
Ammonia and carbon dioxide are recycled and need monitoring. Ammonia can be toxic in closed areas. Carbon dioxide may contribute to global warming.

Heat: If cooling water is released into the ocean, there is little problem as heat is dissipated easily. However when released into lakes, the warm water can lead to eutrophication by reducing the dissolved oxygen in the water and must first be cooled in a cooling tower.

Calcium chloride: This can be dumped in the ocean because the calcium and chloride ions will not affect the balance of ions there. Calcium ions may precipitate out carbonate in the ocean but this is not a large problem.

$CaCl_2$ cannot be dumped in inland lakes because Ca^{2+} will contribute to hardening of the water and Cl^- will kill organisms. It can be dumped in deep pits as $CaCl_2$ solid but this may leach out into the water table, creating future salinity problems.

Evaluation: Although it has been shown that serious considerations need to be made when locating a Solvay processing plant, it can be done and still protect the environment. The best solution is a location near coastal waters.

The Solvay process is being phased out as improved methods of developing sodium carbonate are found.

Question 2

(a) (i) Saponification is the hydrolysis of a fat or oil, under alkaline conditions, to produce a soap and glycerol.

(ii) The way in which soaps clean surfaces can be explained in terms of the solubility of polar and non-polar substances. When soaps dissolve in water, they dissociate into a negatively charged ion, derived from the fatty acid or oil, and a sodium or potassium ion, derived from the alkali used to make it. The negatively charged ion contains a long non-polar hydrocarbon chain, called a hydrophobic end, and a polar carboxylic (COO–)group, which is hydrophilic, and can form ion-dipole attractions with water. This negatively charged fatty acid ion is called a surfactant. The hydrocarbon chain strongly attracts non-polar grease molecules, due to dispersion forces, while the hydrophilic end readily dissolves in water. When water is agitated, the oil and grease are removed from the surface being cleaned because they are attached by dispersion forces to the hydrocarbon chain of the surfactant ion. The anionic heads of the soap are present on the surface of the stabilised grease droplets and so can interact with the polar water molecules. A stabilised emulsion of grease and water results and there is no tendency for grease to redeposit on the clothes.

(b)

$$2SO_2 + O_2 \rightleftharpoons 2SO_3$$

	$2SO_2$	O_2	$2SO_3$
initial:	0.06	0.05	0.04 mol L^{-1}
change:	0.04	0.02	–
final:	0.02	0.03	0.04

$$K = \frac{[SO_3]^2}{[SO_2]^2[O_2]}$$

$$= \frac{(0.04)^2}{(0.02)^2(0.03)}$$

$$= 133$$

(c) (i) When sulfuric acid reacts with water, the reaction is extremely exothermic. This is because even though the acid ionises strongly, it ionises incompletely, and this produces a high concentration of hydronium ions.

$$H_2SO_4 + H_2O \rightarrow H_3O^+ + HSO_4^-$$

$$HSO_4^- + H_2O \rightleftharpoons H_3O^+ + SO_4^{2-}$$

(ii) Hot, concentrated sulfuric acid can act as an oxidising acid with many metals to produce sulfur dioxide.

$$Cu(s) + 2H_2SO_4(aq) \rightarrow Cu^{2+}(aq) + SO_4^{2-}(aq) + SO_2(g) + 2H_2O(l)$$

Sulfuric acid is a strong dehydrating agent, removing water from sucrose to leave carbon.

$$C_{12}H_{22}O_{11}(s) \rightarrow 12C(s) + 11H_2O(l)$$

Sulfuric acid can also be used to precipitate heavy metal cations, such as barium, as their sulfates.

$$Ba^{2+} + SO_4^{2-} \rightarrow BaSO_4(s)$$

(d) (i) A solution of cobalt chloride is divided up equally into two test tubes. To one test tube, dilute hydrochloric acid is added until a colour change occurs. The second test tube is alternately heated in a Bunsen flame and cooled under running water, until a colour change is seen in each procedure.

(ii) The procedure shown in part (i) can be represented by the following equation:

$$\underset{\text{Pink}}{Co(H_2O)_6^{2+}} + 4Cl^- \rightleftharpoons \underset{\text{Blue}}{CoCl_4^{2-}} + 6H_2O$$

It can be seen that the addition of Cl^- ions, from the HCl, forces the reaction to the right, according to Le Chatelier's Principle, which states that when a system is at equilibrium, the system reacts to any change imposed by partially counteracting the change. Thus, the liquid in the test tube appears bluer. Similarly, when heated, the liquid in the test tube appears more blue, because the forward reaction is endothermic, and the system has reacted to the change imposed by shifting to the right. When cooled, the reverse reaction occurs, producing more of the pink $Co(H_2O)_6^{2+}$.

(e) Three main methods have been used to produce sodium hydroxide by electrolysis. These are the diaphragm process, the mercury process and the membrane process. NaOH was originally produced by the diaphragm process, but this method employed a diaphragm made of asbestos which was later found to be highly dangerous. This process also produced NaOH which was contaminated with chloride ions and the strong oxidising agent, sodium hypochlorite. The diaphragm is made of a polymer-metal oxide composite in modern electrolysis plants.

The next process employed was the mercury cell, which involved the use of a flowing liquid mercury cathode. The advantage of this method was that sodium ions were reduced to sodium metal which dissolved in the mercury to produce an amalgam. The mercury could be recycled and reused, and the chlorine gas produced at the anode could be collected. This process had the advantage that the sodium hydroxide produced was of high quality, but in the process some poisonous mercury invariably escaped into nearby waterways, and for environmental reasons, this process was phased out and replaced by the membrane process.

The membrane process employs a synthetic ion selective polymer membrane to separate the anode and cathode compartments. This type of cell produces very pure NaOH, without the health risks associated with both asbestos and mercury, and with no release of dangerous chemicals to the environment. For this reason, the membrane cell is currently the preferred option for the industrial production of sodium hydroxide.

Question 3

(a) (i) Your answer should give one of the following:

- manufacture of fertilisers
- manufacture of soaps and detergents
- metal processing
- textile processing

(ii) Sulfur is extracted from mineral deposits using the Frasch process. Superheated water is forced down a pipe into the deposit (1 mark) where it melts the sulfur (1 mark). Compressed air then forces the molten sulfur up a pipe to the surface (1 mark) where it cools and solidifies.

(b) (i) Your answer should give one of the following:

Precautions

- Performed electrolysis in the fume hood – to minimize inhalation of $Cl_2(g)$.
- Used small amounts – to minimize dangers and hazards.
- Wore eye goggles – to prevent splashes in the eyes of $Cl_2(aq)$, NaOH solution, conc. salt solution.
- Wore latex gloves to avoid contact of alkalis and acids with skin.

Safe disposal of reactants and products

- Used only small quantities to minimize waste.
- Tested the pH of resultant mixture. Adjusted pH to a range 7–10. Disposal down the sink with copious flushing of water.

(ii) The three products were identified in the following ways:

Chlorine gas Cl_2(g) and chlorine water (1 mark)

Identified by its yellow-green colour and distinctive 'swimming pool' smell. Moist litmus placed near the electrode went red then was bleached white. Another way is to shake some chlorine water with sodium iodide (NaI) solution and a few millilitres of trichoroethane as a solvent. The $I^-(aq)$ is oxidized to I_2 which dissolves in the organic solvent to give a pink-purple solution.

Sodium hydroxide solution (1 mark)

Produced at the cathode. Add a few drops of phenolphthalein to the brine solution. Pink colouration around the cathode indicates the hydroxide ion.

Hydrogen gas (1 mark)

Clear bubbles which form at the cathode. Collect a test tube of the gas by downward displacement of water. When a lighted taper is placed into the gas, it explodes with a 'pop'.

(c) *Structure of soap* (1 mark)

Soap is the long chain salt of a fatty acid.

Its structure has a long hydrocarbon chain tail with a negatively charged ionic acid end. A positive metal ion, usually sodium, completes the molecule.

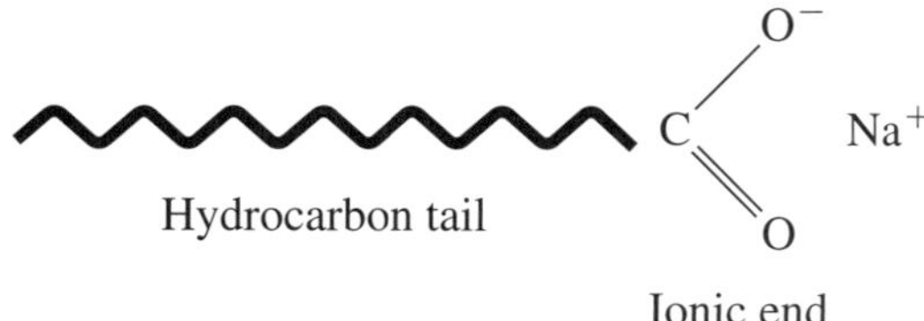

Cleansing action of soap (2 marks)

Dirt is usually mixed with grease. The non-ionic hydrocarbon end of the molecule dissolves in the grease and the hydrophilic (water-loving) ionic end dissolves in the water. Thus the soap molecules help to emulsify the grease, making it into small droplets which can then mix with water and be washed away.

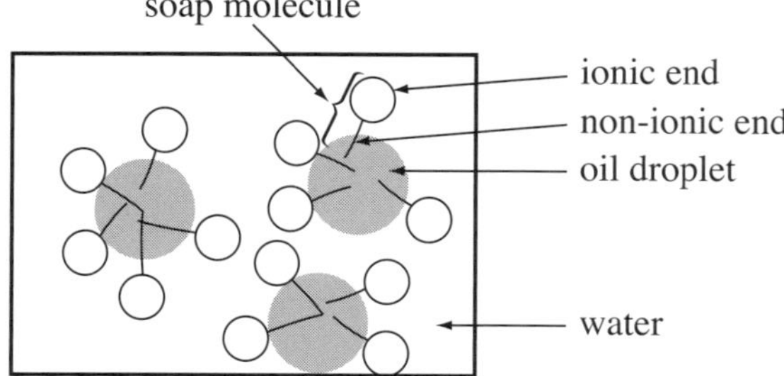

Development of synthetic detergents (2 marks)

Synthetic detergents were designed to mimic the structure and thus the action of soaps. The detergent molecule has a long hydrocarbon tail like soap and a negative ion at the end. The negative ion is usually a sulfate, not an acid group however.

The first detergents had a branched hydrocarbon tail but were not biodegradable in the environment. The development of a straight-chain tail like the structure of regular soaps led to the second generation of detergents being biodegradable.

(d) (i) Temperature change (1 mark) is the only factor that changes the value of an equilibrium constant.

(ii) Two methods that could be used to increase the yield of nitrogen dioxide are

- Increase the pressure (1 mark) of the gases
- Lower the temperature (1 mark)

(iii)

	reaction	$2\,NO(g) + O_2(g) \rightleftharpoons 2NO_2(g)$		
	moles initially	0.25	0.12	0
	moles at equilibrium	$0.25 - 2x$	$0.12 - x$	$2x$

Since $0.25 - 2x = 0.05$, $x = 0.1$ and moles L^{-1} at equilibrium:

0.05 0.02 0.2 moles L^{-1}

$$K = \frac{[NO_2]^2}{[NO]^2[O_2]}$$

$$K = \frac{[0.2]^2}{[0.05]^2[0.02]} = \frac{0.04}{5 \times 10^5} = 800\ (\text{mol } L^{-1})^{-1} \quad (1 \text{ mark})$$

(e) The general answer to this question requires the student to:

- select an environmental problem in a named acid production (1 mark)
- state how this problem has been solved or addressed (1 mark)
- assess how effective the remedy has been (1 mark)

These should then be repeated for the base.

Sample answers:

Acid production

The contact process is used to oxidise sulfur dioxide (SO_2) to sulfur trioxide (SO_3) in the production of sulfuric acid. An environmental problem is the release of small amounts of SO_2 into the atmosphere. This is a problem as it is poisonous to plants and animals, increases corrosion and adds to acid rain and smog.

A method to address this is to pass the exiting gases from the contact process through a 'scrubber' to mix with a solution of Caro's acid (H_2SO_5). This powerful oxidizer converts SO_2 to sulfuric acid (H_2SO_4). This is an excellent way to address the problem as over 90% of the already small residual SO_2 is removed in this way to form usable product.

OR

The source sulfur material in the manufacture of sulfuric acid was once just elemental sulfur. However, use of waste gases from the roasting of sulfide ores is now being used for almost half the source material.

For example, roasting of Zn ore: $ZnS + \frac{3}{2} O_2 \rightarrow ZnO + SO_2(g)$

The roasting of ores produces large amounts of dangerous sulfur dioxide gas. This is now effectively contained and used as a cheap starting material for sulfuric acid production and has reduced gaseous pollution in the metal ore roasting industry.

OR

Sulfur exists as a pollutant in natural gas and petroleum. As these fuels burn they produce $SO_2(g)$ which is a pollutant in industrial areas and cities. From the 1970s onward, it has been possible to extract sulfur from the natural gas and petroleum to use as a cheap starting material for sulfuric acid production. This has led to a greatly reduced amount of acid rain and smog in industrial areas.

AND

Base production

In the mercury cell for the electrolytic production of sodium hydroxide (NaOH) from salt brine, liquid mercury is used as the cathode. Sodium metal forms at this cathode and dissolves in the mercury. The mercury is transferred to the denuding chamber to react with water and produce NaOH. However, small amounts of mercury are lost in wastewater. This can cause severe environmental damage because it accumulates in the aquatic food chain. This problem has been addressed by the development and use of the diaphragm cell. This uses a steel cathode and separation of the products is achieved using a diaphragm. The mercury pollutant has been completely eliminated with the use of the newer diaphragm cells.

OR

In the production of sodium hydroxide (NaOH) using a diaphragm cell, asbestos has been used most widely for the diaphragm separating the two half-cells. When the diaphragm needs replacing, workers would be exposed to the asbestos fibres that cause asbestosis. Waste asbestos in used diaphragms was also difficult to dispose of in environmentally acceptable ways. In modern plants the asbestos diaphragm has been replaced by a safe polymer-metal oxide composite material. The final product in contaminated by about 2% salt. In the membrane cell the diaphragm is replaced by a semi-permeable membrane that allows the passage of Na^+ ions but not OH^- ions. The sodium hydroxide produced in the membrane cell is highly pure and the energy expended in its production is less than the diaphragm process. The chlor-alkali industry does not contaminate the environment with waste but the huge amounts of electricity required means that significant amounts of carbon dioxide are released into the atmosphere from the coal burning in power stations.

CHAPTER 5

Option Topic

Shipwrecks, Corrosion and Conservation

Past HSC Questions

Question 1 (25 marks)			**Marks**
(a)	(i)	Identify the main metal used to construct ships.	**1**
	(ii)	Although aluminium is a very reactive metal, with a very low reduction potential, it is used in many structures exposed to oxidising conditions. Explain why aluminium can be used in this way.	**2**
(b)	(i)	Give an example of a metal commonly used as a sacrificial anode.	**1**
	(ii)	Explain why sacrificial anodes are added to metal-hulled ships.	**3**
(c)		Describe the effect of adding other elements to iron on the properties and uses of steels.	**5**
(d)	(i)	Define *corrosion*.	**1**
	(ii)	Outline a procedure that could be used to compare the corrosion rates of different metals or alloys in the school laboratory.	**2**
	(iii)	Describe ways in which accuracy and reliability could be improved in the procedure described in part (ii).	**3**
(e)		For ONE specific metal, evaluate the steps that can be used to clean, stabilise and preserve artefacts recovered from shipwrecks.	**7**

Question 2 (25 marks) **Marks**

(a) (i) Name the type of electrochemical cell that produces a spontaneous reaction. **1**

(ii) Calculate the voltage required to operate the cell shown in the diagram as an electrolytic cell, showing relevant half-equations in your working. **3**

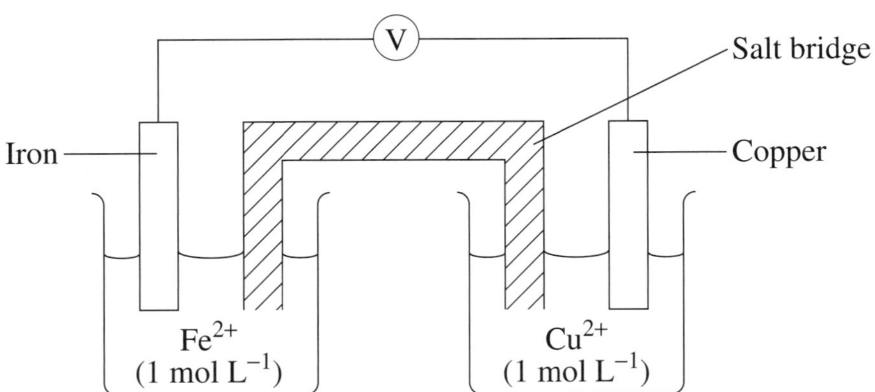

(b) Describe how the work of early scientists increased our understanding of electron transfer reactions. **4**

(c) (i) Name ONE method for removing salt from an artefact recovered from a wreck. **1**

(ii) Explain, using an example, chemical procedures used to clean and preserve artefacts from wrecks. **4**

(d) During your practical work you performed a first-hand investigation to compare and describe the rate of corrosion of materials in different acidic and neutral solutions.

(i) Outline the procedure used. **2**

(ii) It is hypothesised that acidic environments accelerate the corrosion of shipwrecks. **4**

Explain how data obtained from the procedure in part (d) (i) does or does not support this hypothesis.

(e) Analyse the effect of ocean depth on corrosion of metallic objects. **6**

Question 3 (25 marks) **Marks**

(a) (i) Identify ONE passivating metal. **1**

(ii) Account for the differences in corrosion of active and passivating metals. Include a relevant balanced chemical equation in your answer. **3**

(b) During your practical work you performed a first-hand investigation to identify the factors that affect the rate of an electrolysis reaction.

(i) Describe ONE precaution you took to minimise hazards, or to dispose of reactants and products safely. **1**

(ii) Outline the procedure you used to show how ONE factor affects the rate of an electrolysis reaction. **3**

(c) The Titanic struck an iceberg in 1912 and sank to a depth of more than three kilometres. **5**

Analyse how theories about corrosion at great ocean depth have changed since the recent discovery of extensive corrosion on wrecks such as the Titanic.

(d) Fishermen face a problem of limiting corrosion of their steel fish hooks. A fisherman has the choice of storing his steel fish hooks in a plastic, copper or aluminium container.

(i) What is the main metal in steel? **1**

(ii) Compare the effectiveness of these containers in limiting corrosion of steel fish hooks. **5**

(e) Assess how an understanding of electrolysis has contributed to the development of methods for cleaning and restoring marine artefacts. **6**

Option Topic

Shipwrecks, Corrosion and Conservation

Worked Answers

Question 1

(a) (i) Iron

(ii) Aluminium is a passivating metal. It readily reacts with oxygen to form an impermeable oxide coating. This coating prevents oxygen from coming into contact with the underlying aluminium metal.

(b) (i) Zinc

(ii) The sacrificial anode has a greater tendency to oxidise than the iron ship to which it is attached.

$$Zn(s) \rightarrow Zn^{2+}(aq) + 2e^- \quad E° = +0.76 \text{ V}$$
$$Fe(s) \rightarrow Fe^{2+}(aq) + 2e^- \quad E° = +0.41 \text{ V}$$

Because the metals are touching, the zinc will become an anode and corrode, protecting the iron from corroding. The iron becomes cathodic.

(c) Chromium when added to iron increases its resistance to corrosion because the chromium atoms react to form a passivating layer that prevents oxygen from coming into contact with iron. Chromium also increases the metal's resistance to abrasion and heat. The alloy stainless steel is therefore used to make kitchen sinks and cutlery.

Steel is an alloy of iron with 2% or less carbon. The properties and uses of steel vary according to the percentage carbon in the steel. Mild steel contains less than 0.2% carbon. It is soft and malleable but has a high tensile strength and is therefore used for car bodies, shipbuilding and roofing. Structural steel contains 0.3% to 0.6% carbon. It is harder with higher tensile strength but maintains some malleability. It is therefore used for beams and girders as well as railway lines.

(d) (i) Corrosion is the oxidation of a metal due to the attack by environmental agents such as oxygen, water or acids.

(ii) Weigh three different metals such as iron, zinc and copper and place in separate test tubes. Half-cover each with salt water and leave for three days. After this time, lightly remove traces of corrosion and reweigh. Repeat the experiment in aqueous solutions of different salts.

(iii) Accuracy: all variables need to be very carefully controlled (eg same surface area; solutions of the same concentration; constant temperature) and the electronic balance will need to be very sensitive (eg 3 decimal places). It will be important to remove all corrosion with steel wool but not to abrade the metal beneath.

Reliability: the experiment needs to be repeated a minimum of five times with all variables controlled.

(e) Metal: Iron

Stabilising: Iron artefacts are often saturated with dissolved chlorides and sulfates. They must be immersed in seawater to prevent evaporation and crystallisation of these salts or the crystals may expand and destroy the structure of the artefact.

The artefact is then stored in a weak sodium hydroxide solution. This will passivate the metal against further corrosion by forming a protective oxide coating.

Cleaning: The artefact must now be X-rayed to identify the extent of calcium carbonate encrustation. The majority can then be chiselled off and more carefully removed with a dental drill. The artefact is then soaked in a weak acid bath for several hours to dissolve any remnant calcium carbonate.

To remove chlorides electrolysis is used. The artefact is the cathode and is surrounded by a stainless steel mesh anode. The electrolyte consists of a sodium hydroxide solution. The chloride ions will be extracted from the artefact. At the cathode some iron ions will be reduced to metallic iron.

Anode: $2Cl^- \rightarrow Cl_2(g) + 2e^-$
Cathode: $Fe^{2+} + 2e^- \rightarrow Fe(s)$
$2H_2O + 2e^- \rightarrow 2OH^- + H_2(g)$

Preservation: The artefact is then rinsed in de-ionised water to remove residual chlorides and is dehydrated in acetone. It is coated in clear polyurethane polymer to preserve it. This forms a barrier so that oxidants such as water and oxygen do not come in contact with the iron.

Question 2

(a) (i) Galvanic cell

(ii) $Fe \rightarrow Fe^{2+} + 2e^-$ $E° = 0.44$ V

$Cu^{2+} + 2e^- \rightarrow Cu$ $E° = 0.34$ V

Total $E° = 0.44 + 0.34 = 0.78$ V

Connect the negative terminal of the power supply to the iron and the positive terminal to the copper. The applied voltage must be greater than 0.78 V.

(b) Galvani electrically stimulated the muscle fibres in frogs' legs. These experiments showed that muscular contractions could be produced by static electricity. He also showed that continual muscle contractions occurred when frogs' spinal cords were connected by copper hooks to an iron railing. He concluded that animal tissue contained a 'vital force' he called 'animal electricity'.

Davy worked in electrochemistry using the voltaic pile and electric batteries. He constructed the largest battery ever built at that time and passed strong electric currents through molten salts suspected of containing undiscovered elements. He isolated potassium and sodium and the alkaline earth elements barium, calcium, strontium and magnesium.

Faraday worked in electromagnetism and discovered the principle of the electric motor and the phenomenon of electromagnetic induction. He was the first to liquefy chlorine and he isolated benzene. He established the laws of electrolysis that govern the quantitative relationships in electrolytic processes. Faraday showed how to calculate the amount of product that would form at an electrode given the voltage and the time the current was flowing.

(c) (i) leaching in fresh water OR electrolysis

(ii) To remove chlorides from artefacts electrolysis is used. The artefact is placed in a plastic vat and is submersed in dilute NaOH electrolyte. The artefact is made to be the cathode and it is wrapped in a steel mesh anode. The chloride ions are slowly extracted from the corroded metal as they are attracted to the oppositely charged electrode and are oxidised at the anode.

$$FeCl_2 \rightarrow Fe^{2+} + 2Cl^-$$

Hydroxide ions and water are also oxidised at the anode. Water is reduced to hydrogen at the cathode. The bubbles of hydrogen assist the cleaning process.

Anode: $2Cl^- \rightarrow Cl_2 + 2e^-$

Cathode: $2H_2O + 2e^- \rightarrow 2OH^- + H_2$

Once the chlorides are removed the artefact can be dehydrated in alcohol and then coated with clear polyurethane polymer or wax to preserve it. These form a physical barrier so that oxidising agents such as oxygen and water cannot come into contact with the iron.

(d) (i) A piece of steel was placed in a test tube half submerged in 1 mol/L hydrochloric acid solution. An identical piece was placed in water under the same conditions of temperature. Two pieces of zinc were set up under similar conditions. All metals were first weighed, left for two days in the solutions, then removed, cleaned and reweighed. The experiment was repeated with solutions of decreasing HCl molarity (eg 0.1 mol/L, 0.01 mol/L etc.).

(ii) The mass lost from the zinc and iron was found to be greater than the mass lost from the metals in water. The greater the hydrogen ion concentration, the greater the mass of metal lost. The hypothesis is supported because when metals corrode they form soluble metal ions which dissolve in the solution and do not add to the mass, unlike insoluble metal ions.

$$Zn(s) \rightarrow Zn^{2+}(aq) + 2e^-$$

$$Fe(s) \rightarrow Fe^{2+}(aq) + 2e^-$$

(e) As the ocean becomes deeper the following factors change:

The concentration of dissolved oxygen decreases with depth

The temperature decreases with depth

The amount of sulfate-reducing bacteria increases with depth

Decreased oxygen concentrations would be expected to reduce corrosion of metals because, for the metal to be oxidised, an oxidising agent (usually oxygen) needs to be present, for example:

$$Fe \rightarrow Fe^{2+} + 2e^-$$

$$O_2 + 2H_2O + 4e^- \rightarrow 4OH^-$$

Lower temperatures would be expected to decrease corrosion of metals because there is a direct relationship between temperature and reaction rate. The lower the temperature the slower the reaction rate.

At depth an increase in the amount of sulfate-reducing bacteria in the anaerobic environment around a deep wreck would be expected to increase corrosion because the lack of oxygen at depth does not affect these bacteria. They reduce sulfate from sea water rather than oxygen. They obtain the electrons for the reduction by oxidising metals on shipwrecks.

$$Fe \rightarrow Fe^{2+} + 2e^-$$

$$SO_4^{2-} + 10H^+ + 8e^- \rightarrow H_2S + 4H_2O$$

Question 3

(a) (i) Aluminium

(ii) An active metal such as magnesium will be readily oxidised by oxygen gas in the air as the oxygen comes into contact with the metal surface.

$$2Mg(s) + O_2(g) \rightarrow 2MgO(s)$$

The MgO is porous and crumbly and oxygen can penetrate to continue the corrosion of the metal. The metal will lose its structure and eventually completely corrode away.

A passivating metal such as aluminium reacts with oxygen to form a thin impermeable layer which protects it from corrosion. Aluminium will therefore maintain its structure and not corrode over time. If the thin passivating layer is removed by abrasion it rapidly reforms.

(b) (i) Iodine is produced in this investigation. It is toxic by all routes of exposure so that eye and skin protection are recommended. Goggles will prevent a splash into the eyes and latex gloves will prevent skin contact.

(ii) To test the effect of concentration on the rate of the electrolysis of potassium iodide solution. Two graphite electrodes were placed into 100 mL 0.5 mol L^{-1} KI solution in a beaker. These were connected to a power pack with 2 volts applied for 5 minutes. The rate at which bubbles formed at the cathode was observed and scored on a 1 to 5 scale, 5 being vigorous and 1 being nil. The rate of bubbles forming related directly to reaction rate.

The procedure was repeated using 1 mol L^{-1} and then 2 mol L^{-1} KI solutions. All other conditions such as temperature, volumes and voltage were held constant.

(c) Previous to the discovery of the Titanic and other wrecks at great depth it was thought that the major contributors to corrosion were availability of oxygen, temperature and the pH of the environment. The higher the concentration of oxygen the greater the rate of corrosion because oxygen is an oxidising agent and with greater concentration there are more successful collisions between oxygen and iron that result in oxidation of the iron. The higher the temperature the greater the rate of corrosion because as temperature increases the rate of collisions also increases. As pH decreases the concentration of H^+ increases. H^+ acts as a catalyst for corrosion increasing its rate.

At great depth oxygen levels were thought to be low because oxygen enters at the surface and needs to diffuse many kilometres, by which time most has been absorbed by living organisms. Temperature was thought to be low as it decreases from the surface waters down. pH was thought to be close to neutral because there were no factors known to raise or lower pH at depth. It was therefore theorised that the corrosion rate at depth would be very low and that wrecks such as the Titanic would be in very good condition.

In fact the Titanic was found in very bad condition with extensive corrosion. It was evidence that other factors are responsible. Analysis of the rusticles on the Titanic found them to be saturated with anaerobic bacteria. Current theory suggests that these are largely responsible for the extensive corrosion seen. These bacteria reduce dissolved sulfate ions in the surrounding waters.

$$SO_4^{2-}(aq) + 10H^+(aq) + 8e^- \rightarrow H_2S(aq) + 4H_2O(l)$$

To provide electrons for this process they facilitate the oxidation of iron.

$$Fe(s) \rightarrow Fe^{2+}(aq) + 2e^-$$

A wreck in deep anaerobic conditions provides the perfect environment for these bacteria: a lack of competition from aerobic bacteria and a good source of iron in the wreck. In addition these bacteria produce the acid hydrogen sulfide. This acts as a catalyst for the corrosion process.

Galvanic action caused by the presence of less reactive metals touching the iron ship could also accelerate corrosion.

(d) (i) Iron

(ii) Both oxygen and water are required for iron to corrode.

$$O_2(g) + 2H_2O(l) + 4e^- \rightarrow 4OH^-$$

As it is water resistant, a plastic box will be effective in preventing corrosion as long as there is no water in the box already. This is unlikely as fish hooks would be wet when they are placed into the box after fishing.

Therefore the air in a plastic box would be humid with a plentiful supply of oxygen and water and would not limit corrosion of steel fish hooks.

Because copper is a less reactive metal than iron the copper box would act as a cathode and the iron hook would act as an anode accelerating corrosion of the hook.

$$Cu(s) \rightarrow Cu^{2+}(aq) + 2e^- \qquad E° = -0.34\ V$$

$$Fe(s) \rightarrow Fe^{2+}(aq) + 2e^- \qquad E° = +0.44\ V$$

Aluminium is a more reactive metal than iron.

$$Al(s) \rightarrow Al^{3+}(aq) + 3e^- \qquad E° = +1.68\ V$$

Therefore the aluminium box will act as the anode and oxidise and the iron hook will act as the cathode and be protected from corrosion.

(e) Electrolysis is the process of using an electric current to cause a chemical change. An external source of electrical energy is applied across two electrodes that are immersed in an electrolyte solution. Reduction takes place at the cathode and oxidation at the anode. This knowledge can be used to both clean and restore marine artefacts.

Metal artefacts need to be cleaned by removing sulfate and chloride salts from within the structure, otherwise they may crystallise and the growing crystals will weaken the metal structure. It is very difficult to remove these ions, particularly chloride, from iron by desalination. Much of the chloride is present as insoluble hydroxychlorides trapped within the iron oxide deposits. A more efficient way of removing chlorides is by electrolysis. The artefact is placed as the cathode in a 0.5 mol L^{-1} sodium hydroxide electrolyte. A stainless steel anode is used as this is unlikely to react. When a low voltage is applied, electrons are forced into the cathode. The negatively charged sulfate and chloride ions are repelled from the negatively charged cathode and move toward the anode where they are oxidised. The concentration of these ions is monitored and the solution replaced when the concentration becomes too high.

At the cathode water is reduced.

$$2H_2O(l) + 2e^- \rightarrow 2OH^- + H_2(g)$$

At the anode chloride, hydroxide and water are oxidised.

$$2Cl^-(aq) \rightarrow Cl_2(g) + 2e^-$$

$$4OH^-(aq) \rightarrow 2H_2O(l) + O_2(g) + 4e^-$$

$$2H_2O(l) \rightarrow 4H^+(aq) + O_2(g) + 4e^-$$

Electrolysis can be used to partly restore artefacts by returning metal ions to the original metal. Rust can be restored to iron by providing electrons.

$$Fe(OH)_2(aq) + 2e^- \rightarrow Fe(s) + 2OH^-(aq)$$

Copper ions, silver ions and other metal ions can be reduced to the metal.

$$CuCl(s) + e^- \rightarrow Cu(s) + Cl^-$$

$$Ag_2S(s) + 2e^- \rightarrow 2Ag(s) + S^{2-}(aq)$$

Electrolysis, therefore, can be used to restore artefacts without abrading them and losing some of the original metal. This makes it an effective technique. It is also a more efficient method of cleaning artefacts than desalination.

CHAPTER 6

Option Topic

Forensic Chemistry

Past HSC Questions

Question 1 (25 marks) **Marks**

(a) (i) Define *organic compounds*. **1**

(ii) For ONE class of organic compound, describe a chemical test that identifies this class. **2**

(b)* The table shows fatty acid composition of some common oils and fats.

	Fatty acids present (% by weight)				
Fat or oil hydrolysed	Lauric	Palmitic	Stearic	Oleic	Linoleic
Butter	2–3	23–26	10–13	30–40	4–5
Lard	< 1	28–30	12–18	41–48	6–7
Tallow	< 1	24–32	14–32	35–38	2–4
Coconut	45–51	4–10	1–5	2–10	0–2

An oily sample was hydrolysed and the fatty acids analysed.

	Fatty acids present (% by weight)				
	Lauric	Palmitic	Stearic	Oleic	Linoleic
Oil sample	< 1	29	28	36	4

(i) Which fat or oil has been identified? **1**

(ii) Explain the solubility in water of fatty acids, in terms of their structure. **3**

(c) Assess the usefulness of mass spectrometry in providing forensic evidence. **5**

Question 1 continues

* The HSC Chemistry syllabus has been modified, and this question may no longer be within the syllabus.

Question 1 (continued) **Marks**

(d) (i) Name ONE technique used by forensic chemists to separate a mixture of organic compounds. **1**

(ii) In your study of Forensic Chemistry, you performed a first-hand investigation to separate a mixture of organic materials. Outline the procedure. **2**

(iii) Describe ways in which accuracy and reliability could be improved in the procedure described in part (ii). **3**

(e) Discuss the uses of DNA analysis in forensic chemistry. **7**

Question 2 (25 marks)

(a)* (i) Identify the functional group in glycerol. **1**

(ii) Compare the reactions of both glycerol and 1-propanol when they react with cold dilute $KMnO_4$. **3**

(b)* Discuss the value of electron spectroscopy and scanning tunnelling microscopy in the analysis of small samples in forensic chemistry. **4**

(c) (i) What class of compounds is used to break proteins into fragments of different lengths? **1**

(ii) Describe the processes of electrophoresis and chromatography in separating organic compounds. **4**

(d) During your practical work you performed a first-hand investigation to describe the emission spectrum of sodium.

(i) Name the piece of equipment you used to analyse the emission spectrum of sodium in the laboratory. **1**

(ii) Outline the procedure that you used in this investigation. **2**

(iii) Explain how the emission spectrum was produced. **3**

(e) Discuss the uses of DNA analysis in forensic chemistry. **6**

---

* The HSC Chemistry syllabus has been modified, and this question may no longer be within the syllabus.

Question 3 (25 marks) **Marks**

(a) (i) Identify the general class of compounds represented by the formula $C_x(H_2O)_y$. **1**

(ii) Describe TWO tests that can be used to distinguish between some of the following classes of organic compounds: alkanes, alkenes, alkanols and alkanoic acids. **3**

(b) During your practical work you performed a first-hand investigation to distinguish between reducing and non-reducing sugars.

(i) Describe ONE precaution you took to minimise hazards, or to dispose of reactants and products safely. **1**

(ii) Outline the procedure you used to distinguish between reducing and non-reducing sugars. **3**

(c) Analyse how emission spectra of elements assist in the identification of the origins of a mixture. **5**

Question 3 continues

Question 3 (continued) **Marks**

(d) The diagram shows the results of an investigation to identify the parents of a child. The DNA fingerprints from the mother (M) and child (C) are labelled. Also shown are the DNA fingerprints from two possible fathers (F1) and (F2).

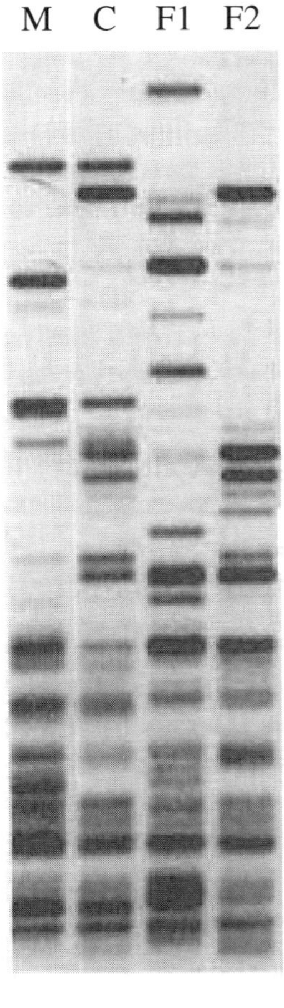

(i) Identify the more probable father. **1**

(ii) Outline the structure and composition of DNA. **2**

(iii) Describe how DNA fingerprints are produced, and explain why they can be used to show that two people belong to the same family. **3**

(e) Evaluate how the development of chromatographic methods has advanced forensic science. **6**

Option Topic

Forensic Chemistry

Worked Answers

Question 1

(a) (i) Organic compounds are those containing carbon and other elements such as hydrogen, and also smaller amounts of oxygen, nitrogen and sulfur. Organic compounds are commonly found in or derived from living things.

(ii) One class of organic compound is alkanoic acids. A chemical test for these is to add a solution of sodium hydrogencarbonate and warm. An acid will release bubbles of carbon dioxide gas.

(b) (i) Tallow.

(ii) Fatty acids have a polar functional group (–COOH). The polar end group attracts polar water molecules and thus mixes with them because of the hydrogen bonding and polar attraction.

As the hydrocarbon tail increases in length, the influence of the polar end-group is less. Longer chain fatty acids are thus less and less soluble in water.

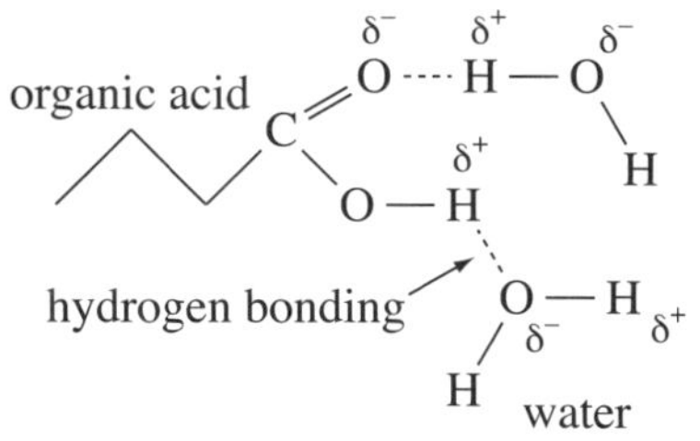

organic acid
hydrogen bonding
water

(c) Mass spectroscopy (MS) is useful in identifying a wide range of organic compounds.

In MS, the molecule is converted to a positively charged ion (the parent or molecular ion) as are numerous fragments of the molecule. Each produces a peak in the resulting charge/mass ratio spectrum.

Thus MS is often combined with gas chromatography (GC) to make more accurate determinations when a mixture of compounds is to be analysed. The combined GC-MS 'fingerprint' allows a mixture of organic compounds to be analysed, for example, in detecting drugs in sport, or matching the known 'fingerprint' of an oil source with an oil spill to determine which vessel was guilty of leaking oil at sea. Only small samples are required. Results are compared with known standards for positive identification.

(d) (i) Gel electrophoresis

(ii) Paper chromatography can be used to separate the various colours used in the coating of M&M sweets. Outline of procedure used:

The sweet was wet with water

The coloured sweet coating was painted about 1 cm above the bottom of a rectangular piece of filter paper.

The paper was placed in a jar in about 0.5 cm of water.

As the water rose up the paper, the colours were carried up to different heights and separated into different coloured bands.

(iii) Accuracy could be improved by using chromatography paper instead of filter paper. This would have given us a better separation of the colours. Accuracy could also be improved by ensuring that adequate amounts of dye coating are transferred to the chromatography paper. If it is too faint, it is hard to identify small amounts of certain colours such as light blue or yellow. Different solvent mixtures should also be investigated to ensure separation of components.

Reliability could be improved by repeating the procedure five or more times ensuring at all times that variables are controlled and that no contamination occurs.

(e) DNA 'fingerprinting' can be used to identify individuals from the traces of their nuclear cell material left at the scene of a crime. This is because portions of DNA are as unique to individuals as their fingerprints — with the exception of identical twins.

For example, skin scrapings under a victim's fingernails can be matched with the DNA of a suspect who has been observed with scratches. The DNA of semen from a rape victim can be matched with the DNA of a suspect. DNA may also be used to establish the innocence of a suspect or of someone wrongly convicted of an offence.

DNA fingerprinting is being used extensively in paternity cases to establish the real father of a child. However, since the DNA of identical twins is the same, it cannot be used to distinguish between them. DNA may be fairly similar between brothers, and a brother may be falsely accused of paternity if both brothers are not tested, or if the test is not carried out in the standard fashion.

In some countries, DNA data banks of known offenders are being established to assist the police. This can be controversial, as some civil liberty groups feel that if a criminal serves his time in jail, his record should be wiped clean.

There are other ethical problems with DNA fingerprinting in respect to insurance matters. Since this technique can also show genetic diseases such as Huntington's, there is a fear that insurance companies may insist that symptomless but likely persons have a test prior to insuring them. Some people who will develop a genetic disorder later on may wish to know about it and make plans. Some people would prefer not to know, especially if they cannot do anything about the oncoming disease.

Question 2

(a) (i) The –OH group

(ii) Both propanol and glycerol will decolourise $KMnO_4$. Glycerol has three OH groups while propanol has only one. This causes glycerol to react faster with the $KMnO_4$ and to use up more of the $KMnO_4$.

(This dot point has been removed from the current syllabus.)

(b) Both ES and STM are useful in analysing small samples as they are non-destructive tests. They both provide data on the surface of the sample. ES determines the elements present in the surface by analysing the energy of the electrons emitted when electrons are focused on the surface.

STM maps the surface features when a probe moves over the surface and a constant current is maintained between the surface and the probe.

(This dot point has been removed from the current syllabus.)

(c) (i) enzymes

(ii) Electrophoresis separates organic compounds based on their mass and charge. An electric field is set across a gel on which the sample is placed. Samples with a higher charge:mass ratio move faster toward the oppositely charged terminal.

Chromatography separates compounds based on their differing attraction to (or solubility in) the stationary and mobile phases used. Compounds more attracted to the mobile phase move faster through the stationary phase to separate from slower moving compounds.

(d) (i) Spectroscope

(ii) A sodium salt was heated in a Bunsen flame and was viewed with the spectroscope. Bands of light were seen. The yellow doublet lines are characteristic of sodium's emission spectrum.

(iii) Electrons in the sodium atom absorbed energy and moved to higher energy levels. When they returned to lower energy levels they released photons of light with a particular frequency. These different frequencies form the line spectrum pattern that is characteristic of each element.

(e) DNA analysis is a more accurate testing procedure than previously used forensic tests like blood group testing.

DNA tests can identify an individual rather than a group of people by matching their DNA to the DNA samples collected at a crime scene. Each person has a unique pattern of bases in the intron or non-coding area.

DNA can also be used in paternity investigations to prove family relationships as close family members have very similar intron sequences, for example a child has the same number of short tandem repeats as either the mother or father at each of the sites tested.

There are however ethical issues concerning the collection and storage of DNA information in databases due to problems of privacy and misuse of the personal information.

Question 3

(a) (i) Carbohydrate

(ii) For each of the two tests you must specify *what reagent* you would add, *the reaction with* your named group, and *the reaction with* another named group.

1. To distinguish an alkanoic acid from all other compounds listed add a few drops of Na_2CO_3 solution. If bubbles of $CO_2(g)$ form (detected by reaction with limewater) then the compound is an acid. None of the other organic compounds listed will react.

AND

2. To distinguish an alkene from all other compounds listed, add few drops of brown $Br_2(aq)$ and carefully shake. The brown bromine colour will decolourise very quickly. No other compound group listed reacts in this way.

OR

3. To distinguish an alkene from an alkane, add few drops of brown $Br_2(aq)$ and carefully shake. The brown bromine colour will decolourise quickly with the alkene. With an alkane, the brown colour will preferentially dissolve in the upper alkane layer and will slowly decolourise in a reaction accelerated by UV light.

OR

4. To distinguish an alkene from an alkane, add a few drops of purple $KMnO_4/H^+(aq)$ and carefully shake. The purple permanganate colour will decolourise quickly with the alkene, changing to a pale colour with brown $MnO_2(s)$. With an alkane, there is no reaction.

(b) (i) Describe ONE of the following

Precaution to minimise hazards

- Use minimal quantities of reagent
- Wear goggles to protect eyes from alkali splashes in Benedict's solution.

Disposal of chemicals

Put left over chemicals in inorganic waste bottle in fume cupboard for later safe and appropriate disposal.

(ii) Place 1 mL glucose solution into a clean test tube, add 10 drops of Benedict's solution and mix. Place the test tube in a hot water bath and wait (up to 5 minutes) until a red-brown precipitate of Cu_2O forms. This is a positive test for a reducing sugar.

Repeat this procedure but using a sucrose solution. *No* red brown precipitate forms, which indicates a non-reducing sugar.

(c) An emission spectrum occurs when a sample of an element is excited in a glass discharge tube and the light emitted by the excited electrons falling back from higher energy levels to their normal level is examined through a spectroscope (prism). A series of lines of coloured light of differing wavelengths are observed on a black background. Each element has a unique emission spectrum which is like a fingerprint of that element. The differences in emission spectra between elements can be used to monitor their concentrations, particularly metals in samples from for example water supplies. When used to test for the toxic metals Hg and Co in industrial waste, the concentration of these metals can be monitored upstream and will be highest in wastewater being discharged from the pipe outfall.

Emission spectroscopy is used for soil analysis because only very small quantities of the sample are needed to detect the element. For example, a sample of soil or mud on the boot of a suspect may be found to contain higher-than-normal levels of lead or other metals using emission spectra. The level of lead and range of elements detected may indicate that the origin of the soil was a high-lead environment such as a mine or paint factory.

Emission spectroscopy is very sensitive. It can detect small amounts of arsenic in food or 'medicine' given to a victim. Arsenic remains behind in a corpse, especially in the hair roots enabling the origin of the victim's illness and death to be determined. Older chemical tests required larger amounts of arsenic for detection. Unfortunately emission spectroscopy cannot detect the nature of compounds as compounds are decomposed to gaseous atoms. The analysis can be done quickly using computer analysis of the spectrum.

(d) (i) F2

(ii) DNA is deoxyribonucleic acid. It is a polymer consisting of a double helix. The molecule consists of two strands composed of a phosphate and deoxyribose backbone. It has one of four nucleobases or nitrogen bases (adenine, guanine, thymine, cytosine) attached to each deoxyribose group. Hydrogen bonding exists between corresponding bases on the opposing strands of the helix (only A–T and G–C can form this bond).

A nucleotide is formed from deoxyribose sugar plus a base.

(iii) Collect a sample of DNA from any cell that contains a nucleus, break the cell membrane, and centrifuge to separate. If it is a small sample such as a blood droplet, amplify the DNA using PCR (polymerase chain reaction) in which the DNA is heated to separate the strands and enzymes and free nucleotides are added to cause the strands to produce the complementary 'other half'.

Then, isolate the DNA, and use restriction enzymes to cut the DNA into fragments (at specific base sequences) – those used in DNA fingerprinting are very short, only 7–15 base pairs.

Use gel electrophoresis to separate fragments. Place DNA into a well on the gel, apply an electric charge across the gel and wait for the fragments to separate (shorter move faster, longer slower). The DNA is split into single strands and the band pattern is transferred to a nylon sheet. The sheet is immersed in a bath of DNA probes such as luminous markers or radioactive sources. The nylon sheet is exposed to X-ray film. The radioactive probes attached to the sorted fragments appear as dark bands on the film. The spacing and banding of the pattern is a DNA fingerprint. The pattern of one person is compared to members of the same family. Close family members will have more similarities in their DNA than complete strangers but will *not* be identical unless the samples came from identical twins. Similar bands line up as at same position on the sheet.

(e) Chromatographic methods have improved and become more advanced over time and as technology has improved. This is particularly due to the use of computers, the decrease in the cost of technology, and the increase in the range of types of chromatography. Chromatography is the name used for separating substances based upon their differential distribution between two phases, one stationary (such as solid or liquid) and one mobile (either liquid or gas). It was originally used where coloured organic molecules could be separated and visibly detected. It has advanced forensic science as a greater range of organic molecules can now be detected, many of which are colourless. The recorder may show the output as peaks on a readout.

If the stationary phase is a solid then separation is based on adsorption of the substance to paper or another solid – some adsorb strongly others weakly. Those that absorb weakly move faster such as yellow ink compared to red ink. This can allow the separation of the ink colours in a pen and allow forensic scientists to then compare them with a specific manufacturer's make.

Chromatographic techniques now include paper, column, thin-layer, gas, high performance liquid (HPLC) and ion exchange chromatography. HPLC has been very costly but is reducing in cost. It is a very important machine as it readily separates small molecules eg. carbohydrates. HPLC can identify colourless organic molecules by running a known sample marker through the column at the same time. Drugs in samples of urine and blood, or mixed with inert carriers, can be detected. Low concentrations (as little as parts per billion) and tiny samples can be used. They are recorded as peaks on a graph.

To determine blood alcohol reading, gas chromatography is used – hooked up to a detector and computer analyser. This is very useful as it is cheaper than HPLC, faster and a readout shows different peaks corresponding to time of separation and relative measurements. Chromatography has advanced forensic science because more detailed and precise measurements can be made using smaller quantities of organic material. Furthermore the cost of these tests is becoming lower and the time to get an accurate result is going down. However, forensic chemists do not use chromatographic techniques in isolation and use a vast array of other modern analytical methods including infrared detection (IR), nuclear magnetic resonance (NMR), mass spectroscopy and electrophoresis.

CHAPTER 7

BOARD OF STUDIES
NEW SOUTH WALES

2009
HIGHER SCHOOL CERTIFICATE
EXAMINATION

Chemistry

General Instructions

- Reading time – 5 minutes
- Working time – 3 hours
- Write using black or blue pen
- Draw diagrams using pencil
- Board-approved calculators may be used
- A data sheet and a Periodic Table are provided at the back of this paper
- Write your Centre Number and Student Number where required

Total marks – 100

Section I

75 marks

This section has two parts, Part A and Part B

Part A – 15 marks

- Attempt Questions 1–15
- Allow about 30 minutes for this part

Part B – 60 marks

- Attempt Questions 16–26
- Allow about 1 hour and 45 minutes for this part

Section II

25 marks

- Attempt ONE question from Questions 27–31
- Allow about 45 minutes for this section

Section I
75 marks

Part A – 15 marks
Attempt Questions 1–15
Allow about 30 minutes for this part

Use the multiple-choice answer sheet for Questions 1–15.

---

1 Which of the following is an important factor in predicting the nuclear stability of an isotope?

(A) Atomic radius

(B) Nuclear radius

(C) The ratio of neutrons to protons

(D) The ratio of electrons to protons

2 Unpolluted rain water in New South Wales is slightly acidic.

Which substance is the major contributor to this acidity?

(A) Ozone

(B) Sulfur dioxide

(C) Carbon dioxide

(D) Nitrogen dioxide

3 Which of the following groups contains ONLY acidic substances?

(A) Antacid tablets, baking soda, laundry detergents

(B) Blood, oven cleaner, seawater

(C) Milk, tea, drain cleaner

(D) Vinegar, wine, aspirin

4 What flame colour is produced by barium ions in a flame test?

(A) Red

(B) Blue

(C) Green

(D) Orange

5 The apparatus shown is used in a first-hand investigation to determine and compare the heat of combustion of three different liquid alkanols.

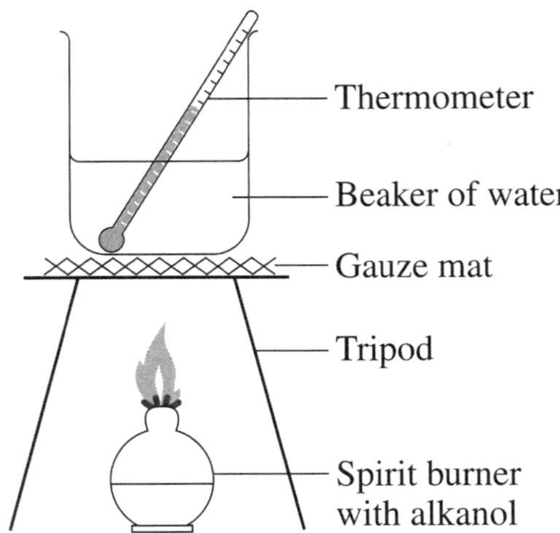

Which is the independent variable?

(A) Type of alkanol used

(B) Amount of water used

(C) Amount of alkanol used

(D) Temperature change in the water

6 Bromine, Br_2, dissolves in unsaturated hydrocarbons and reacts immediately.

Which of the following is the best description of this process?

(A) Bromine is polar and reacts by adding bromine atoms across the double bond.

(B) Bromine is polar and reacts by substituting hydrogen atoms with bromine atoms.

(C) Bromine is non-polar and reacts by substituting hydrogen atoms with bromine atoms.

(D) Bromine is non-polar and reacts by adding bromine atoms across the double bond.

7 What is the conjugate base of HSO_4^-?

(A) SO_3^{2-}

(B) SO_4^{2-}

(C) H_2SO_4

(D) HSO_3^-

8 Three separate colourless solutions each contain one cation, Na^+, Pb^{2+} or Ca^{2+}.

Which of the following would be an appropriate reagent to unambiguously identify the solution containing Pb^{2+}?

(A) KI

(B) K_2CO_3

(C) K_3PO_4

(D) $AgNO_3$

9 One test used for random breath testing in NSW involved crystals of potassium dichromate reacting with ethanol. In this reaction the orange dichromate ion, $Cr_2O_7^{2-}$, changes to the green chromium ion, Cr^{3+}.

Which statement is true for this reaction?

(A) Chromium has lost electrons and reached a lower oxidation state.

(B) Chromium has lost electrons and reached a higher oxidation state.

(C) Chromium has gained electrons and reached a lower oxidation state.

(D) Chromium has gained electrons and reached a higher oxidation state.

10 Which of the following is the main organic product resulting from the reaction of butanoic acid and pentanol?

(A) $H_3C—H_2C—H_2C—C(=O)—O—CH_2—CH_2—CH_2—CH_3$

(B) $H_3C—H_2C—H_2C—C(=O)—O—CH_2—CH_2—CH_2—CH_2—CH_3$

(C) $H_3C—H_2C—H_2C—H_2C—C(=O)—O—CH_2—CH_2—CH_2—CH_2—CH_3$

(D) $H_3C—CH_2—CH_2—CH_2—CH_2—C(=O)—O—CH_2—CH_2—CH_2—CH_3$

11 The following process is used to purify water for drinking.

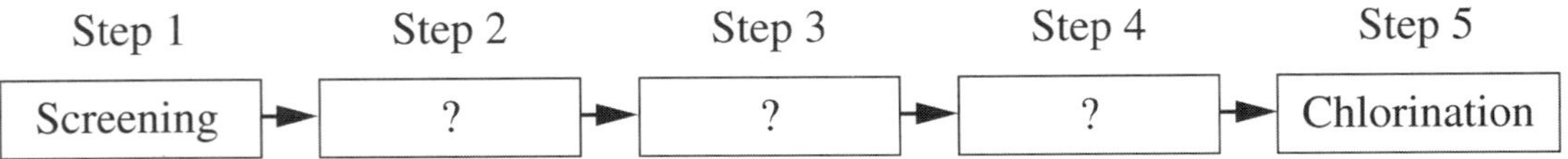

Which sequence represents the correct order of Steps 2, 3 and 4?

	Step 2	*Step 3*	*Step 4*
(A)	Flocculation	pH adjustment	Settling
(B)	pH adjustment	Flocculation	Settling
(C)	pH adjustment	Settling	Flocculation
(D)	Flocculation	Settling	pH adjustment

12 What is the IUPAC name of the following compound?

$$\begin{array}{ccccccccccc} & & CH_3 & & H & & Cl & & F & & \\ & & | & & | & & | & & | & & \\ H & - & C & - & C & - & C & - & C & - & CH_3 \\ & & | & & | & & | & & | & & \\ & & H & & Cl & & H & & H & & \end{array}$$

(A) 3,4–dichloro–2–fluorohexane

(B) 3,4–dichloro–5–fluorohexane

(C) 2–fluoro–3,4–dichlorohexane

(D) 5–fluoro–3,4–dichlorohexane

13 In a fermentation experiment 6.50 g of glucose was completely converted to ethanol and carbon dioxide.

What is the mass of carbon dioxide produced?

(A) 1.59 g

(B) 3.18 g

(C) 9.53 g

(D) 13.0 g

14 Citric acid, the predominant acid in lemon juice, is a triprotic acid. A student titrated 25.0 mL samples of lemon juice with 0.550 mol L^{-1} NaOH. The mean titration volume was 29.50 mL. The molar mass of citric acid is 192.12 g mol^{-1}.

What was the concentration of citric acid in the lemon juice?

(A) 1.04 g L^{-1}

(B) 41.6 g L^{-1}

(C) 125 g L^{-1}

(D) 374 g L^{-1}

15 The graph shows the maximum dissolved oxygen concentration in water as a function of temperature at normal atmospheric pressure.

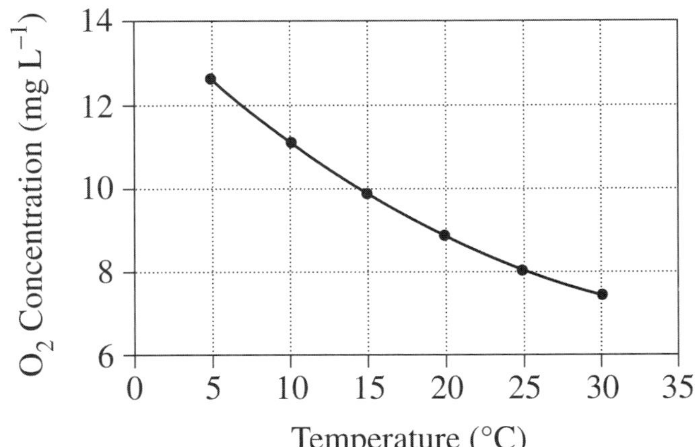

What is the volume of O_2 that can dissolve in 10.0 L of water at 25°C and normal atmospheric pressure?

(A) 62.0 mL

(B) 63.5 mL

(C) 80.0 mL

(D) 124 mL

2009 HIGHER SCHOOL CERTIFICATE EXAMINATION

Chemistry

Centre Number

Section I (continued)

Student Number

Part B – 60 marks
Attempt Questions 16–26
Allow about 1 hour and 45 minutes for this part

Answer the questions in the spaces provided.

Show all relevant working in questions involving calculations.

Question 16 (3 marks)

Describe how to prepare an ester in the school laboratory. Include a specific safety precaution in your answer. **3**

....

....

....

....

....

....

....

....

....

....

Question 17 (4 marks)

Water and ethanol are both used as solvents. **4**

Explain the differences and similarities in their solvent behaviour in terms of their molecular structures. Include a diagram in your answer.

....

....

....

....

....

....

....

....

....

....

2009 HIGHER SCHOOL CERTIFICATE EXAMINATION

Chemistry

Centre Number

Section I – Part B (continued)

Student Number

Question 18 (5 marks)

There has been an increase in the concentration of the oxides of nitrogen in the atmosphere as a result of combustion. **5**

Assess both the evidence to support this statement and the need to monitor these oxides.

Question 19 (6 marks)

Outline the chemical and physical processes involved in the production of ONE of the following from a natural raw material: 6

- a polyethylene bottle
- a polyvinyl chloride pipe
- a polystyrene cup.

Include relevant chemical equations in your answer.

2009 HIGHER SCHOOL CERTIFICATE EXAMINATION

Chemistry

Section I – Part B (continued)

Centre Number

Student Number

Question 20 (4 marks)

(a) Calculate the mass of ethanol that must be burnt to increase the temperature of 210 g of water by 65°C, if exactly half of the heat released by this combustion is lost to the surroundings. **3**

The heat of combustion of ethanol is 1367 kJ mol^{-1}.

(b) What are TWO ways to limit heat loss from the apparatus when performing a first-hand investigation to determine and compare heat of combustion of different liquid alkanols? **1**

Question 21 (6 marks)

The graph shows changes in pH for the titrations of equal volumes of solutions of two monoprotic acids, *Acid 1* and *Acid 2*.

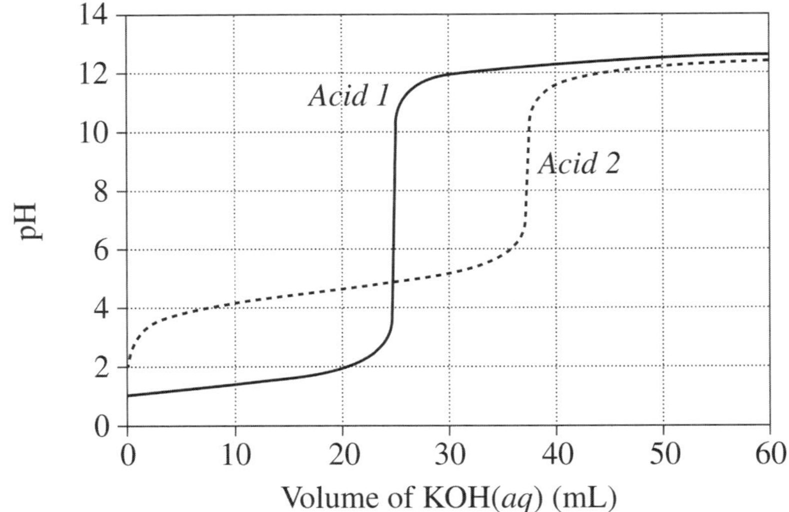

(a) Explain the differences between *Acid 1* and *Acid 2* in terms of their relative strengths and concentrations. **3**

....

....

....

....

....

....

(b) Name the salt produced by the reaction of an acid of the same type as *Acid 2* with KOH(*aq*). **1**

....

(c) Calculate the concentration of hydrogen ions when 20 mL of KOH(*aq*) has been added to *Acid 1*. **1**

....

....

(d) Why would phenolphthalein be a suitable indicator for both titrations? **1**

....

....

2009 HIGHER SCHOOL CERTIFICATE EXAMINATION

Chemistry

Centre Number

Section I – Part B (continued)

Student Number

Question 22 (7 marks)

The nitrogen content of bread was determined using the following procedure:

- A sample of bread weighing 2.80 g was analysed.
- The nitrogen in the sample was converted into ammonia.
- The ammonia was collected in 50.0 mL of 0.125 mol L^{-1} hydrochloric acid. All of the ammonia was neutralised, leaving an excess of hydrochloric acid.
- The excess hydrochloric acid was titrated with 23.30 mL of 0.116 mol L^{-1} sodium hydroxide solution.

(a) Write balanced equations for the TWO reactions involving hydrochloric acid. **2**

....

....

(b) Calculate the moles of excess hydrochloric acid. **1**

....

....

....

(c) Calculate the moles of ammonia. **2**

....

....

....

....

....

(d) Calculate the percentage by mass of nitrogen in the bread. **2**

....

....

....

....

Question 23 (6 marks)

The graph shows the variation in concentration of reactant and products as a function of time for the following system. **6**

$$COCl_2(g) \rightleftharpoons Cl_2(g) + CO(g) \quad \Delta H = +108 \text{ kJ}$$

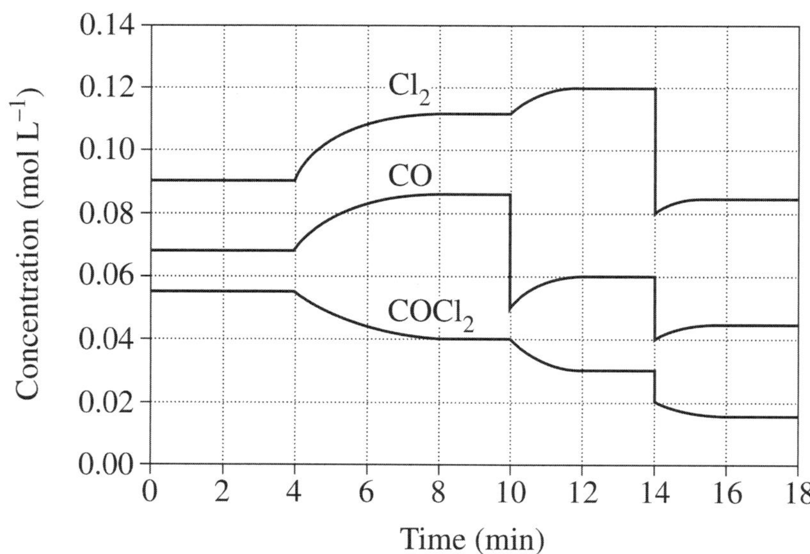

Identify and explain each of the changes in conditions that have shaped the curves during the time the system was observed.

....

....

....

....

....

....

....

....

....

....

....

....

....

2009 HIGHER SCHOOL CERTIFICATE EXAMINATION

Chemistry

Centre Number

Section I – Part B (continued)

Student Number

Question 24 (5 marks)

Describe the principle of atomic absorption spectroscopy and its application in environmental monitoring. Include a diagram in your answer. **5**

Question 25 (7 marks)

An analytical chemist determined the phosphate concentration of water samples from three local streams.

(a) Using the absorbance values in the table and graph, determine the mean absorbance and mean phosphate concentration for each stream and complete the table. **2**

Stream	*Absorbances measured*	*Mean absorbance*	*Mean phosphate concentration* (mg L^{-1})
1	0.090, 0.092, 0.088		
2	0.513, 0.511, 0.514		
3	0.234, 0.237, 0.234		

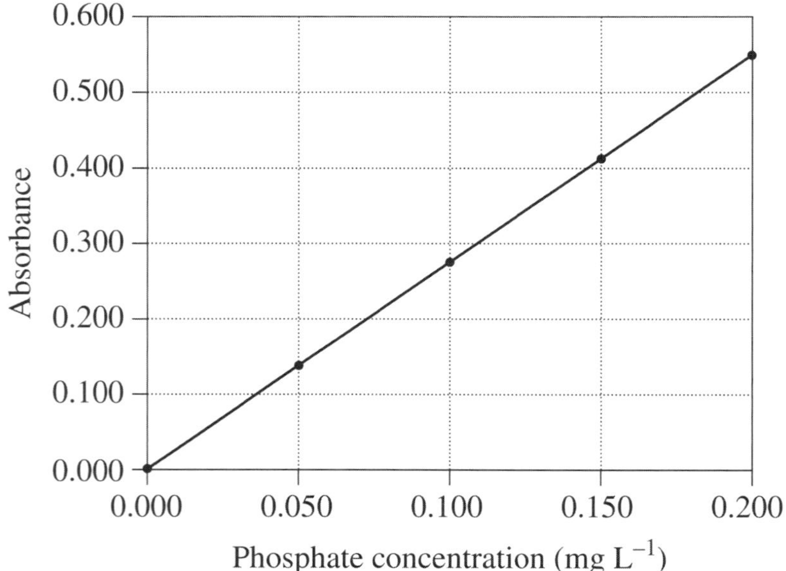

Question 25 continues

Question 25 (continued)

(b) The recommended maximum level of phosphate in streams is 0.100 mg L^{-1}.

With reference to the recommended maximum level of phosphate for stream water, explain why there are differences between the three streams. **3**

....

....

....

....

....

....

....

....

(c) Why is phosphate concentration a water quality issue? **2**

....

....

....

....

End of Question 25

Question 26 (7 marks)

An electrochemical cell is constructed using two half cells. One half cell consists of an inert platinum electrode and a solution of Fe^{2+} and Fe^{3+}. The other half cell consists of a lead electrode and a solution of Pb^{2+}.

Current will flow from one electrode to the other electrode when the cell is completed using a voltmeter and a salt bridge.

(a) Write relevant half equations and a balanced net ionic equation for the overall cell reaction. **2**

....

....

....

....

(b) Calculate the standard cell potential ($E^{\ominus}$). **1**

....

....

(c) Identify the anode, cathode, metals and ions by labelling the following diagram. **3**

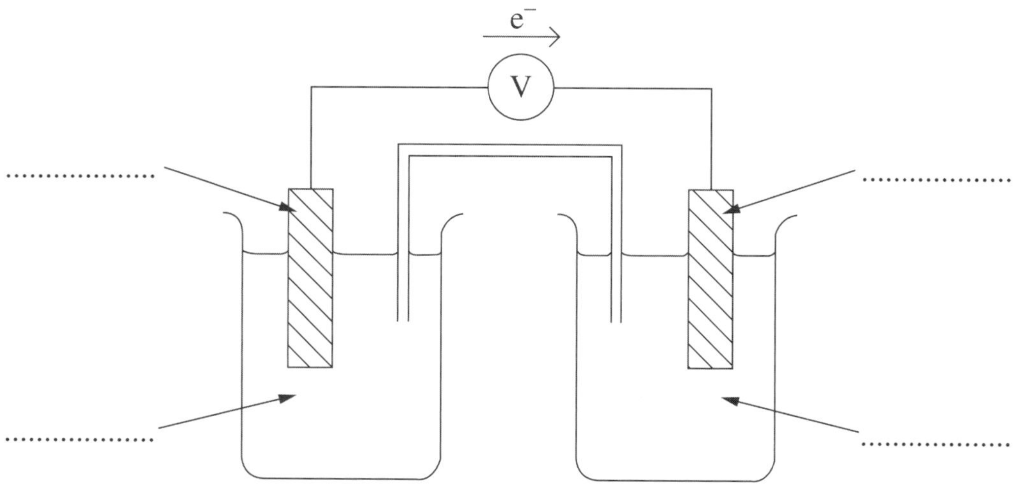

(d) Identify an appropriate electrolyte to use in the salt bridge. **1**

....

2009 HIGHER SCHOOL CERTIFICATE EXAMINATION

Chemistry

Section II

25 marks
Attempt ONE question from Questions 27–31
Allow about 45 minutes for this section

Answer the question in a writing booklet. Extra writing booklets are available.

Show all relevant working in questions involving calculations.

Question 27	Industrial Chemistry
Question 28	Shipwrecks, Corrosion and Conservation
Question 29	The Biochemistry of Movement (*Not included in this reproduction*)
Question 30	The Chemistry of Art (*Not included in this reproduction*)
Question 31	Forensic Chemistry

Question 27 — Industrial Chemistry (25 marks)

(a) Sulfuric acid is one of the world's most significant industrial chemicals because of the variety and importance of its uses.

(i) Identify the major use of sulfuric acid. **1**

(ii) Outline the industrial process for the manufacture of sulfuric acid from its raw materials. **3**

(iii) Account for the safety precautions associated with the industrial transport of sulfuric acid. **2**

(b) At a particular temperature, iodine trichloride dissociates into iodine gas and chlorine gas according to the following equation:

$$2ICl_3(g) \rightleftharpoons I_2(g) + 3Cl_2(g) \quad \Delta H = 240 \text{ kJ}$$

Initially 0.35 mol of $ICl_3(g)$ was introduced into a 1.0 L container and allowed to come to equilibrium. At equilibrium there was 0.45 mol L^{-1} of $Cl_2(g)$.

(i) Write the equilibrium constant expression for this reaction. **1**

(ii) Calculate the value of K at this temperature. **3**

(iii) What are TWO consequences of increasing the temperature of the mixture at equilibrium? **2**

Question 27 continues

Question 27 (continued)

(c) Account for a use of an emulsion in terms of its properties. **2**

(d) (i) Explain the cleaning action of soap in terms of its molecular structure. **2**

(ii) Soap is one product of saponification. Name the other product and draw its structural formula. **2**

(e) The flowchart summarises the fundamental criteria that must be considered in order to find a suitable location for an industrial plant. **7**

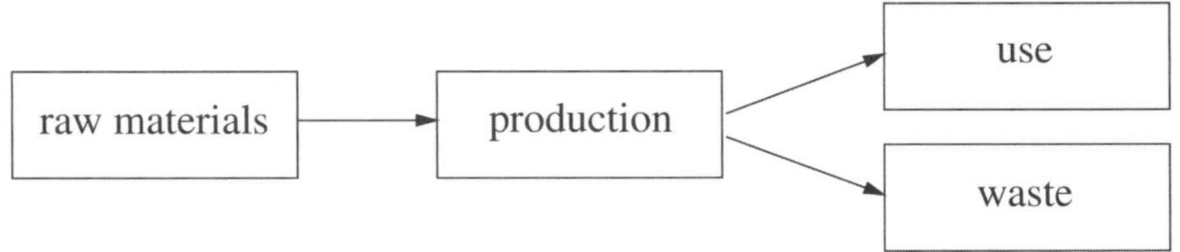

With reference to the flowchart, explain the significance of each criterion to determine a suitable location for an industrial plant to manufacture sodium carbonate.

End of Question 27

Question 28 — Shipwrecks, Corrosion and Conservation (25 marks)

(a) Tins and cans are used to store a range of foods, beverages and general household chemicals.

(i) Identify the passivating metal used in drink cans in Australia. **1**

(ii) Compare the use and effectiveness in preventing corrosion of different protective coatings on containers, such as those shown. **3**

(b) Galvanising is used outside the home on water tanks and fencing. Explain the protection provided to iron by galvanising. **2**

(c) The roof guttering on a garage has rusted through.

(i) Explain, using chemical equations, the cause of this problem. **2**

(ii) The owners have decided to replace the guttering. They have steel screws and a choice of aluminium or copper guttering. **4**

Justify the course of action the owners should take, based on chemical principles.

Question 28 continues

Question 28 (continued)

(d) (i) The diagram shows an electrolytic cell used to conserve an iron artefact recovered from the wreck of a ship that sank in the early 1800s. **3**

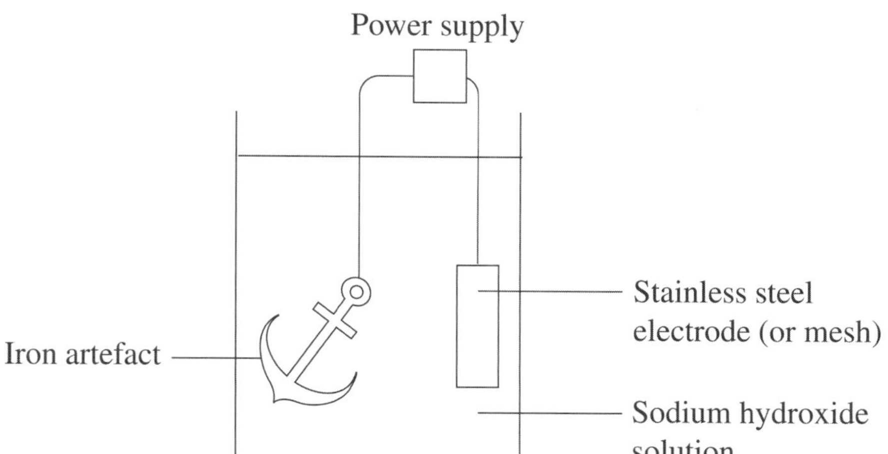

Explain how this process is used to conserve this iron artefact.

(ii) Describe, with the aid of a diagram, how a nickel spoon could be silver-plated using equipment available in a school laboratory. **3**

(e) The flowchart summarises the historical development of the understanding of electron transfer reactions. **7**

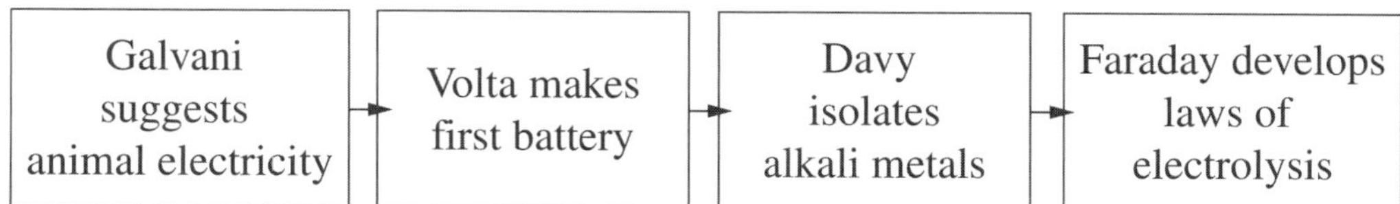

With reference to the flowchart, describe how the work of the scientists led to a better understanding of electron transfer reactions.

End of Question 28

Question 31 — Forensic Chemistry (25 marks)

(a) (i) This structure represents a biological molecule.

```
                           Bond
      H   H   O   |     H   O
      |   |   ||  v     |   ||
  H — N — C — C — N — C — C — O — H
          |       |   |
         CH3      H  CH2
                      |
                      SH

      Alanine     Cysteine
```

Name the specific type of covalent bond indicated. **1**

(ii) Distinguish between the primary, secondary and tertiary structures of proteins. **3**

(iii) How could electrophoresis be used to identify the origins of proteins in an investigation of adulterated food? **2**

Question 31 continues

Question 31 (continued)

(b) An oil spill occurred at a shipping port. To determine which ship was responsible, a sample of the oil was collected and analysed by gas chromatography. Samples of bunker oil were collected from three ships in port at the time and analysed by gas chromatography.

The chromatograms of the samples collected are shown.

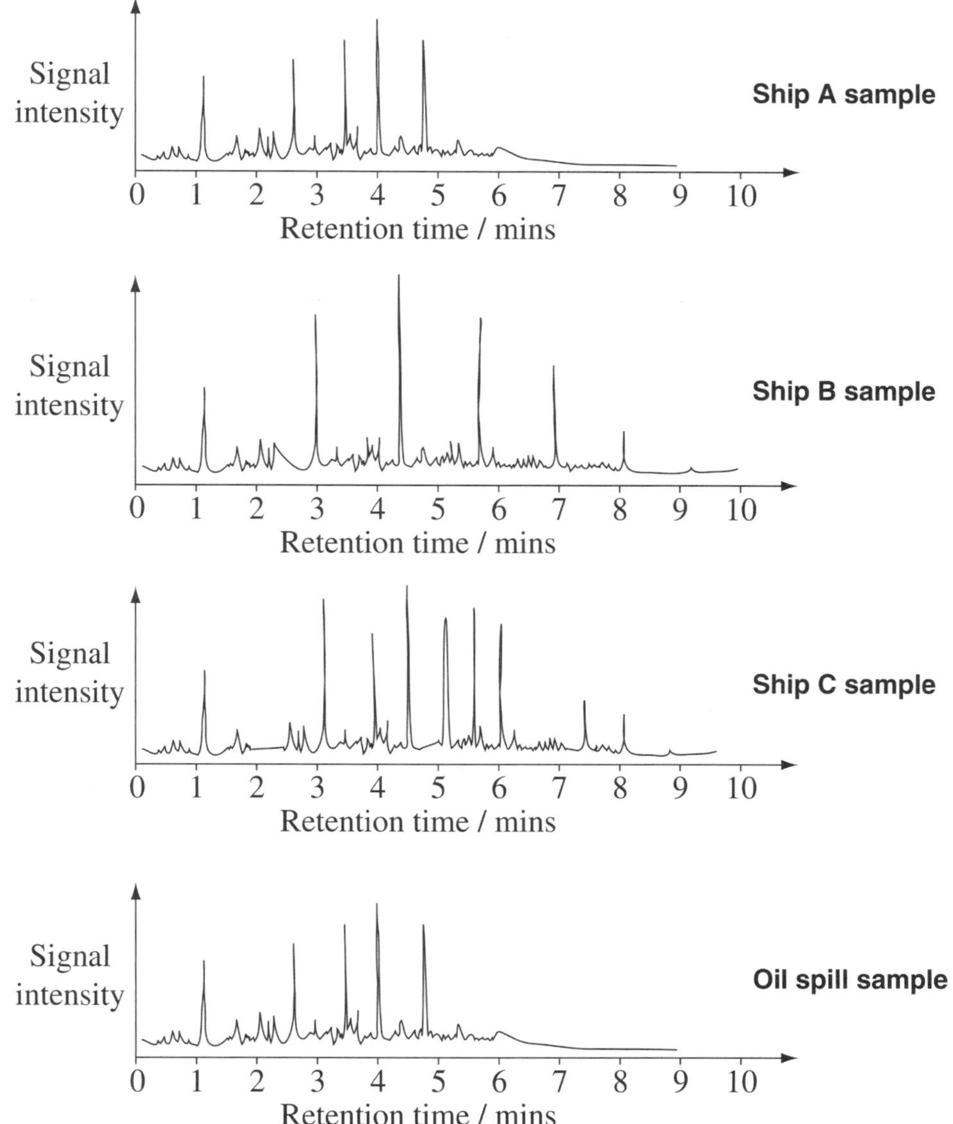

(i) By comparing the chromatograms identify whether the oil spill originated from one of these vessels. **1**

(ii) Explain what the peaks in the chromatograms represent. **2**

(c) Describe the features of instrumental chromatography (GLC or HPLC) that allow the analysis of small samples. **3**

Question 31 continues

Question 31 (continued)

(d) (i) Describe how a first-hand investigation in the school laboratory could be used to separate and identify the components of a mixed food dye. **3**

(ii) Describe the chemical tests that could be used to distinguish between alkenes, alkanols and alkanoic acids. **3**

(e) The flowchart summarises the steps involved in sample processing and presentation of results for a forensic investigation. **7**

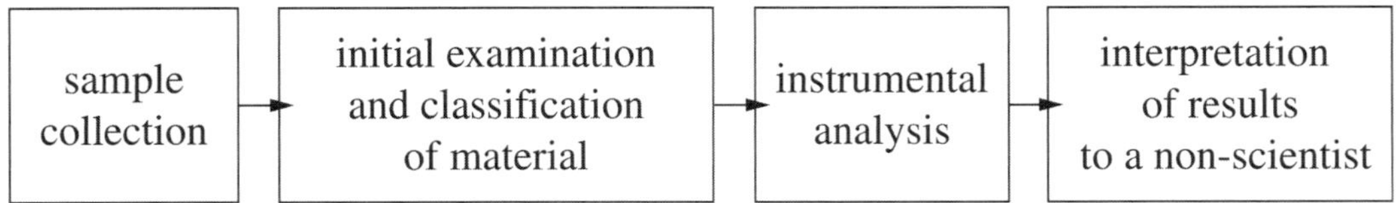

You are the forensic chemist investigating a crime scene and collecting samples. With reference to the flowchart, describe how you would process the sample and present the results. Illustrate your answer with relevant tests and appropriate analyses.

End of paper

2009 HSC Examination Paper

Sample Answers

Section I Part A

1 C In elements with a small mass, the stable ratio of neutrons to protons is 1:1. In large mass elements it is closer to 1.5:1.

2 C CO_2 is found in unpolluted air and dissolves in water forming a weak acid, H_2CO_3. The other gases listed are found in polluted air.

3 D Aspirin contains salicylic acid. In the other options, antacid, oven cleaner and drain cleaner are basic.

4 C Apple green.

5 A As the water is heated and its temperature is raised, the temperature becomes the dependent variable as it is dependent upon the fuel used.

The type of alkanol fuel used is the only other variable which is not held constant and is defined as the independent variable as it is the one that is manipulated by the researcher.

6 D Br_2 is non-polar as it is a symmetrical molecule and therefore has no net dipole. Unsaturated hydrocarbons are those that have a double or a triple bond between carbons. Substituting hydrogen atoms may occur with saturated hydrocarbons (in the presence of UV) but with unsaturated hydrocarbons bromine reacts by adding atoms across double bonds.

7 B HSO_4^- must act as an acid to form a conjugate base. An acid is a proton donor. When HSO_4^- donates an H^+, it becomes SO_4^{2-}.

8 A Pb^{2+} will form a distinctive yellow precipitate with I^- ions, whereas Na^+ and Ca^{2+} will not. Ca^{2+} and Pb^{2+} will both be precipitated by CO_3^{2-} or PO_4^{3-}. NO_3^- will not precipitate any of them.

9 C The oxidation state of chromium in $Cr_2O_7^{2-}$ is +VI while the oxidation state in Cr^{3+} is +III. Reduction results in a drop in the oxidation state and a gain in electrons. Therefore in this example, chromium has been reduced and has gained three electrons to arrive at a lower oxidation state.

10 B Butanoic acid contains four carbon atoms. Pentanol contains five carbon atoms. The ester produced must contain 4 + 5 = 9 carbon atoms. (B) is the only option that contains nine carbon atoms.

11 D Flocculation must occur first to clump suspended particles, and the clumps then settle. The flocculating agent may change the pH of the water. Therefore, the pH may need to be adjusted afterwards.

12 A The longest carbon chain is numbered to give the lowest locant set for the functional groups and in this case the numbering is from right to left (locant set 2,3,4) and not left to right (locant set 3,4,5). The functional groups are then named alphabetically ("chloro" before "fluoro").

13 B $n(C_6H_{12}O_6) = \frac{6.5}{180} = 0.036$ mol

$n(CO_2) = 2 \times n(C_6H_{12}O_6) = 0.072$ mol

$m(CO_2) = 0.072 \times 44.0 = 3.18$ g

14 B $n(NaOH) = 0.550 \times 0.0295 = 0.0162$ mol

$n(H_3X) = \frac{0.0162}{3} = 5.408 \times 10^{-3}$ mol

$c(H_3X) = \frac{5.408 \times 10^{-3}}{0.025} = 0.2163$ mol L^{-1}

$c(H_3X)$ in g $L^{-1} = 0.2163 \times 192.12 = 41.56$ g L^{-1}

15 A At 25°C, 8 mg O_2 will dissolve in 1 L, so 80 mg will dissolve in 10 L.

$n(O_2) = \frac{0.080}{32.0} = 2.5 \times 10^{-3}$ mol

$V(O_2) = 2.5 \times 10^{-3} \times 24.79 = 0.062$ L $= 62.0$ mL

Section I Part B

16 Add 25 mL of an alkanol (eg propan-1-ol) to a similar amount of an alkanoic acid (eg acetic acid) in a reaction flask with a few drops of concentrated sulfuric acid (catalyst).

Set up the apparatus to heat under reflux, using a condenser attached upright.

Heat in a hot-water bath on a hotplate. The hotplate is a safety precaution that avoids the use of a naked flame. This is important as the alkanols are volatile and flammable.

propan-1-ol + acetic acid $\rightleftharpoons$ propyl acetate + water

Boiling chips are added to the flask to avoid "bumping".

Include a labelled diagram of the reflux apparatus.

17 Similarities: Both water and ethanol are polar molecules. Water is more polar than ethanol. The –OH group is polar. Polar solutes are attracted to the polar OH group (dipole–dipole attraction). Ionic solutes are also attracted to the polar OH group. Both liquids are good solvents for polar and ionic substances.

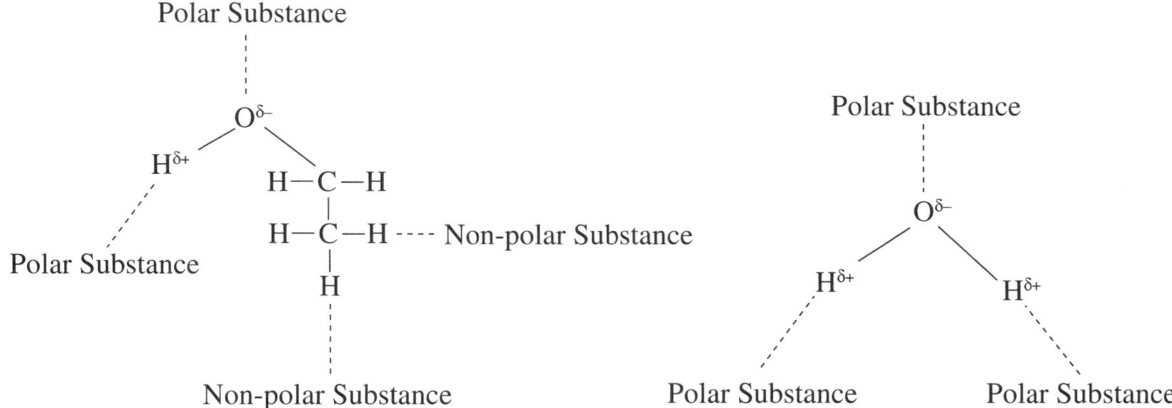

Differences: Ethanol has an additional short carbon chain which is non-polar. This part of the molecule creates dispersion forces with non-polar solutes. Therefore ethanol acts as a solvent for non-polar solutes as well. Water does not have a non-polar section and therefore is a poor solvent for non-polar substances.

18 Evidence:

(Must have two equations.)

Oxides of nitrogen are produced mainly from internal combustion engines and coal-burning power stations:

$$N_2(g) + O_2(g) \rightarrow 2NO(g)$$

$$2NO(g) + O_2(g) \rightarrow 2NO_2(g)$$

The concentrations of these oxides have increased with the rapid increase in car numbers in the world since about 1945. The evidence for this increase comes from both direct measurements of gas concentrations in the atmosphere and indirectly through increases in levels of photochemical smog in cities and increased levels of acid rain. The quantitative measurements are accurate and reliable but the technology has only been available in recent times. The smog and acid rain evidence is less reliable as they could be formed by other oxides and other factors may have led to their increase.

The need to monitor: continued monitoring of nitrogen oxide concentration in the atmosphere is vitally important as only when we record increasing levels of pollution will we act to reduce it. It is important to reduce levels of NO_x as it has a direct negative effect on the population's health, as it can cause respiratory diseases. It also affects the surrounding environment, as acid rain can strip the cuticle off leaves leading to degradation of forests.

$$2NO_2 + H_2O \rightarrow HNO_2 + HNO_3$$

19 The polyethylene bottle is produced from the raw material crude oil, which is fractionally distilled to obtain various fractions. Heavier fractions can undergo thermal or catalytic cracking to produce a variety of products including ethylene. Ethylene is the feedstock for the production of polyethylene.

$$C_{10}H_{22}(l) \rightarrow C_2H_4(g) + C_8H_{18}(l)$$

Ethylene contains a double bond that breaks, allowing the monomers to link together and form the addition polymer, polyethylene. The chemical process involves gaseous ethylene being passed through an alkane solution containing particles of titanium(III) chloride and triethyl aluminium chloride (Ziegler–Natta catalyst) at 60°C and ~200 kPa pressure. This forms unbranched molecules that pack closely together giving a high-density product useful for making bottles.

$$n(C_2H_4) \rightarrow -[CH_2CH_2]_n-$$

The physical process involves heating the polyethylene pellets and melting them, then moulding under high pressure to form bottles.

$$-[CH_2CH_2]_n-(s) \rightarrow -[CH_2CH_2]_n-(l)$$

20 (a) $\Delta H = -mC\Delta T$

$= -210 \times 4.18 \times 65$

$= -57057$ J

$= -57.057$ kJ per x moles of ethanol

If 50% of the heat was lost to the surroundings, then the energy released from the combustion was $57.057 \times 2 = 114.11$ kJ

As one mole of ethanol releases 1367 kJ, then if x moles release 114.11 kJ

$$x = \frac{114.11}{1367} = 0.0835 \text{ mol}$$

$m(CH_3CH_2OH) = 0.0835 \times 46.0$

$= 3.84$ g

$= 3.8$ g (2 sig. fig.)

(b) 1. Make sure the flame from the burning fuel is as close as possible to the container of water being heated to limit heat loss from the flame to the air around it.

2. The base of the water container should be made of a material that conducts heat to transfer as much heat as possible into the water. The sides of the container should be insulated to reduce heat loss. A copper-based container with stainless steel sides would be suitable.

21 (a) Acid 1 is strong as it has an equivalence point at a pH of 7. This occurs because the salt produced from the titration of a strong acid with a strong base is neutral. Acid 2 is weak as its equivalence point is at a pH of 9. This occurs because the conjugate of the weak acid is a strong base.

Acid 2 is more concentrated than Acid 1 because more KOH was required to neutralise it. Given that there are equal volumes of both acids, there must be more moles of Acid 2 present in the volume.

(b) If Acid 2 was acetic acid, the salt would be potassium acetate.

(c) The pH at this point is 2. Therefore $[H^+] = 1 \times 10^{-2}$ mol L^{-1}

(d) Phenolphthalein changes from colourless to pink between pH 8.3 and pH 10.0. Therefore it will change colour at the equivalence points for both acids as the vertical parts of both titration curves include this pH range.

22 (a) $NH_3(aq) + HCl(aq) \rightarrow NH_4Cl(aq)$

$HCl(aq) + NaOH(aq) \rightarrow H_2O(l) + NaCl(aq)$

(b) $n(NaOH)$ reacting $= 0.116 \times 0.02330 = 2.7028 \times 10^{-3}$ mol

$n(HCl)$ excess reacting $= 2.7028 \times 10^{-3}$ mol $= 2.70 \times 10^{-3}$ mol (3 sig. fig.)

(c) $n(HCl)$ initial $= 0.125 \times 0.05 = 6.25 \times 10^{-3}$ mol

$n(HCl)$ reacting with $NH_3 = 6.25 \times 10^{-3} - 2.7028 = 3.547 \times 10^{-3}$ mol

$n(NH_3) = n(HCl)$ reacting $= 3.547 \times 10^{-3}$ mol
$= 3.55 \times 10^{-3}$ mol (3 sig. fig.)

(d) n(N atoms) $= 3.55 \times 10^{-3}$ mol

m(N atoms) $= 3.55 \times 10^{-3} \times 14.01 = 0.049696$ g

$$\% \text{ N by mass} = \frac{0.049696}{2.8} \times 100 = 1.77\%$$

23 Between 0 and 4 minutes the system is in equilibrium. At 4 minutes the system was heated as the forward reaction is endothermic and heating will cause a forward shift in the equilibrium resulting in a gradual increase in the concentration of products and a gradual decrease in concentration of reactants, as evidenced in the graph.

At 10 minutes some CO was removed from the system. This shows in the graph as a sudden drop in its concentration only, followed by a gradual increase as the system shifts forward to counteract the loss of CO. The forward shift also gradually increases the concentration of the other product, Cl_2 and decreases the concentration of the reactant, $COCl_2$.

At 14 minutes the volume of the system was increased. This shows on the graph as a sudden decrease in the concentration of all species present. The system then shifts forward to counteract this change by increasing the total concentration of gas molecules in the system. This shows as a gradual increase in the concentration of products and a gradual decrease in the concentration of the reactant.

(You must identify an equilibrium state, e.g. 0–4 or 8–10.)

24 AAS relies on the principle that electrons exist in discrete energy levels around an atom's nucleus. Once a gas has been atomised, an input of energy can excite electrons out of their normal ground state energy level into a higher level. When they fall back, energy of particular wavelengths is emitted. Each wavelength corresponds to the energy required to excite an electron in the atom from its ground state to a higher state. Each element has its own unique absorption spectrum. In AAS, one unique wavelength of light is directed into a gaseous sample containing a particular element. By measuring the fraction of the light absorbed of a particular wavelength, the concentration of that element present in a sample can be determined.

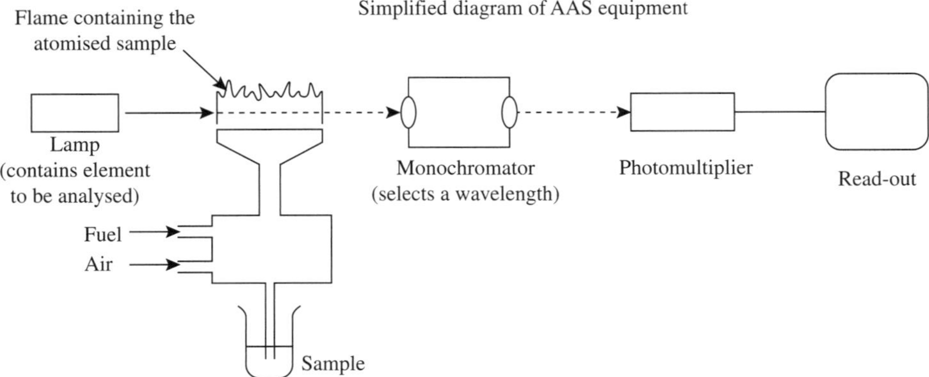

AAS can be used to determine very small concentrations of trace elements in soils accurately. Elements such as chromium, copper and zinc are nutrient requirements of plants for proper metabolism. If a soil is lacking in these trace elements, plants exhibit deficiency diseases. The roles of these elements were not known until AAS was developed but now AAS is used to measure their concentration and if low they can be added to soils.

AAS can also be used to measure the concentration of heavy metals in waterways accurately to very low concentrations.

25 (a)

Stream	*Absorbances measured*	*Mean absorbance*	*Mean phosphate concentration* (mg L^{-1})
1	0.090, 0.092, 0.088	0.090	0.030
2	0.513, 0.511, 0.514	0.513	0.180
3	0.234, 0.237, 0.234	0.235	0.085

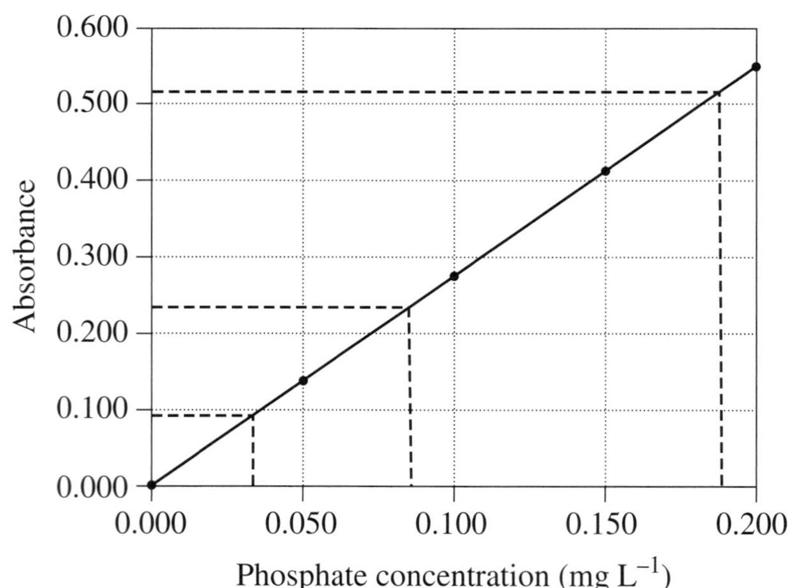

(b) Stream 2's mean phosphate concentration of 0.180 mg L^{-1} is higher than the recommended maximum level of 0.100 mg L^{-1}. This may be because the measurement point in Stream 2 is downstream from a farm where phosphate fertiliser was used and escaped into the stream through run-off or via ground water. Streams 1 and 3 have lower phosphate concentrations than the recommended maximum level. Stream 3 has a higher level than stream 1 and this may be because it is downstream from a town where phosphate detergents have been used and some has found its way into the waterway but still within acceptable levels. Stream 1 has the lowest levels of phosphate and may be up-stream from farmland or in a more pristine environment away from farms and streams.

(c) Phosphate acts as fertiliser for water plants such as algae. High levels of phosphate may lead to an algal bloom, which in turn can lead to eutrophication, in which the algae use up the dissolved oxygen in the water, causing the death of other living things in that waterway.

26 (a) $Pb(s) \leftrightarrow Pb^{2+}(aq) + 2e^-$ $E° = 0.13$ V

$Fe^{3+}(aq) + e^- \leftrightarrow Fe^{2+}(aq)$ $E° = 0.77$ V

$Pb(s) + 2Fe^{3+}(aq) \leftrightarrow Pb^{2+}(aq) + 2Fe^{2+}(aq)$

(b) $E° = 0.13 + 0.77 = 0.90$ V

(c)

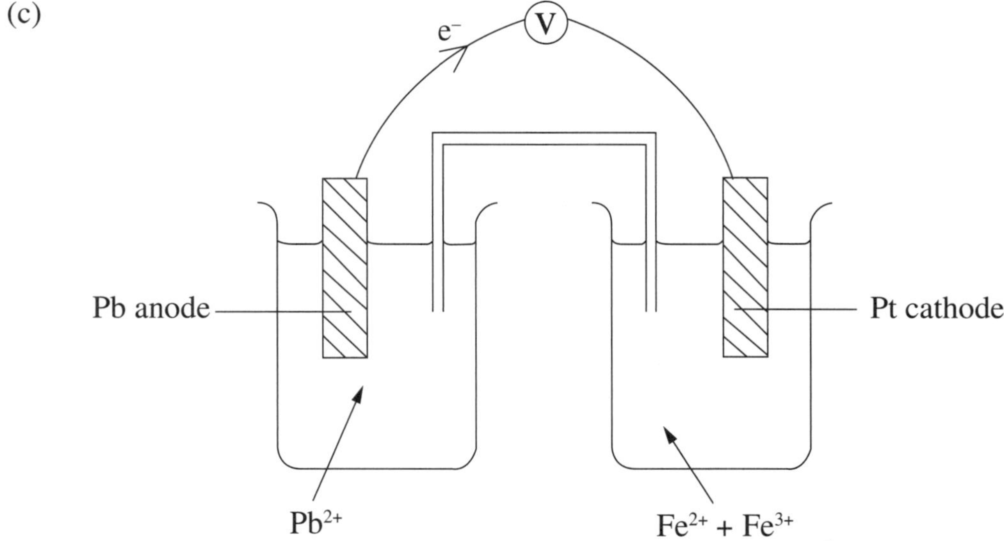

(d) $KNO_3(aq)$

Section II—Options

Question 27—Industrial Chemistry

(a) (i) To make fertiliser.

(ii) Sulfuric acid is prepared by the Contact process. Sulfur is melted and sprayed into air at high temperature, where it reacts with oxygen:

$$S(l) + O_2(g) \rightarrow SO_2(g)$$

The SO_2 is further reacted with O_2 to make SO_3. This is done at moderate temperatures, using three successively cooler V_2O_5 catalyst beds. The successively lower temperatures maximise the yield of SO_3 and the catalyst beds increase the reaction rate.

$$2SO_2(g) + O_2(g) \leftrightarrow 2SO_3(g)$$

The SO_3 gas is then bubbled into concentrated H_2SO_4 to make oleum:

$$SO_3(g) + H_2SO_4(l) \rightarrow H_2S_2O_7(l)$$

The oleum is then very carefully diluted with water, which produces concentrated sulfuric acid:

$$H_2S_2O_7(l) + H_2O(l) \rightarrow 2H_2SO_4(l).$$

(iii) Sulfuric acid is transported as concentrated acid using reinforced steel tanks (by road, sea or rail), or in reinforced steel pipelines. Reinforced steel tanks are used because of their added strength, to minimise the chances of a spill. Hazardous chemical codes are affixed to the trucks to identify the acid, and to provide instructions for emergency personnel in case of an emergency.

(b) (i) $K = \dfrac{[Cl_2]^3[I_2]}{[ICl_3]^2}$

(ii) $2ICl_3 \leftrightarrow I_2 + 3Cl_2$

Initial concentrations (mol/L): $[ICl_3] = 0.35$; $[I_2] = 0.0$; $[Cl_2] = 0.0$

Final concentrations (mol/L):

$$[Cl_2] = 0.45$$

$$[ICl_3] = 0.35 - (\frac{2}{3} \times 0.45) = 0.05;$$

$$[I_2] = \frac{0.45}{3} = 0.15;$$

$$K = \frac{0.45^3 \times 0.15}{0.05^2} = 5.5.$$

(iii) Two consequences of increasing the temperature of the equilibrium mixture are (since the forward reaction is endothermic):

- the equilibrium position will shift to the right, increasing the concentrations of I_2 and Cl_2, and decreasing the concentration of ICl_3;
- the rates of the forward and reverse reactions will increase.

(c) Cold cream is an example of an emulsion. It is used as a makeup remover. Two of its properties make it useful for this. First, it is slightly acidic, consistent with the pH of skin, which helps prevent infection. Second, it is miscible with oil, which makes it suitable for dissolving the oily substances in makeup, and it is miscible with water, so that it can be washed off the face with water.

(d) (i) The molecular structure of soap consists of a long hydrophobic hydrocarbon "tail", and a hydrophilic, anionic "head". The hydrophobic end dissolves in oil and grease droplets (on clothing for example) due to dispersion forces, and the hydrophilic end dissolves in water, due to ion–dipole interactions. The interaction of soap molecules, grease or oil droplets and water results in micelle formation, which solubilizes the grease or oil droplets, allowing them to be washed away.

(ii) Apart from soap, saponification also produces glycerol (1,2,3-propanetriol). Its structure is shown below:

```
       H    OH    H
       |    |     |
HO  —  C  — C  —  C  — OH
       |    |     |
       H    H     H
```

(e) The Solvay process to produce sodium carbonate occurs as follows. Solid calcium carbonate is heated to produce CO_2 and CaO. The CaO is mixed with water to produce $Ca(OH)_2$. The CO_2 is bubbled into purified brine, to which ammonia has been added (ammoniacal brine). When cooled, this produces a precipitate of sodium hydrogen carbonate and a solution of ammonium chloride. The sodium hydrogen carbonate is heated to produce solid sodium carbonate, and CO_2 which is recycled. The ammonium chloride reacts with the $Ca(OH)_2$ produced above, to regenerate the ammonia and water (which are recycled), and calcium chloride. The overall reaction is:

$$CaCO_3(s) + 2NaCl(aq) \rightarrow Na_2CO_3(s) + CaCl_2(s)$$

The flowchart illustrates the important considerations in the location of a Solvay plant. For example, proximity to raw materials is important. The raw materials are calcium carbonate and brine. Thus locating a plant at or near sources of sodium chloride (solid or aqueous) or calcium carbonate reduces transportation costs (and possibly purchasing costs) associated with obtaining these raw materials. Production costs are also an important consideration in locating a plant. These include labour costs and energy costs. The Solvay process does not require a large energy input, and labour costs may be fairly constant throughout a region or country. Thus, production costs are not as important as other considerations in this process.

Sodium carbonate is used primarily in the production of glass, but also has a range of other applications, including the production of fine chemicals and soaps and detergents, paper recycling, and in the food and beverage industries. Location of the plant relative to shipping and transportation infrastructure can be an important consideration in producing a commercially competitive product.

Waste products are another crucial consideration in the location of a plant. In the Solvay process $CaCl_2(s)$ is produced as a waste. Some is sold, for example as a concrete additive, as a road de-icer, and as a drying agent, but production far exceeds demand. The remaining salt must not be disposed of into rivers or lakes, where it would drastically alter the dissolved ion concentrations, but can be safely disposed of into the ocean, or by burial in special leak-proof holding containers. The process also produces large amounts of heat, and thus large volumes of water are required for cooling. Heated water must be stored for cooling before discharge back into the environment. Similarly, excess calcium hydroxide is produced, which must be neutralised with HCl and stored before careful release. These waste disposal issues require either large tracts of land, or close proximity to the ocean, where the huge volumes of water into which wastes are discharged effectively dilute them. How an industrial plant deals with its waste products has become increasingly important for communities and government, and thus this criterion is likely to become the most important consideration for choosing a location for a Solvay plant.

Question 28—Shipwrecks, Corrosion and Conservation

(a) (i) Aluminium.

(ii) Steel cans used to store food are coated on the inside with tin. Tin is less reactive than steel so it forms a physical barrier to oxidants touching the steel. However, if the tin is scratched the underlining steel will readily corrode.

Steel cans are often coated on the outside with a plastic label. This also forms a physical barrier to oxidants, however if scratched or torn it will also allow the steel to readily corrode. Tin is a more effective physical barrier than plastic as it binds strongly to the steel and provides a more complete coverage.

Drink cans are made from aluminium. Aluminium forms a coating of aluminium oxide which offers an impervious layer to oxidants. This is a more effective coating than tin or plastic as it does not allow oxidants to touch the underlying metal and if scratched, reforms instantly.

(b) Galvanising is fixing a complete coating of zinc to the iron. This offers protection to the iron against corrosion in two ways. First, the zinc offers a physical barrier to oxidants coming in contact with the iron. Second, if the zinc is scratched, it will act as a sacrificial anode because zinc is more reactive than iron and will oxidise in preference to iron. The iron will become cathodic and hence will not corrode.

(i) Rusting refers to the corrosion of iron and steel. It is an oxidation–reduction reaction involving a transfer of electrons. Rusting requires the presence of both oxygen and water.

The iron roof guttering is the anode because oxidation occurs on its surface:

$$Fe(s) \rightarrow Fe^{2+}(aq) + 2e^-$$

The reduction equation is:

$$O_2(g) + 2H_2O(l) + 4e^- \rightarrow 4OH^-(aq)$$

The overall equation is:

$$2Fe(s) + O_2(g) + 2H_2O(l) \rightarrow 2Fe^{2+}(aq) + 4OH^-(aq)$$

The iron ion migrates toward the hydroxide ion and they combine to form insoluble iron(II) hydroxide.

$$Fe^{2+}(aq) + 2OH^-(aq) \rightarrow Fe(OH)_2(s)$$

The iron is further oxidised to Fe^{3+} in the form of hydrated iron(III) oxide.

Iron(III) oxide, ($Fe_2O_3.xH_2O$) is known as rust. O_2 is required in this step as well.

$$4Fe(OH)_2(s) + O_2(g) \rightarrow 2Fe_2O_3.H_2O(s) + 2H_2O(l)$$

The overall reaction for the formation of rust can therefore be summarised as:

$$4Fe(s) + 3O_2(g) + 4H_2O(l) \rightarrow 2Fe_2O_3.2H_2O(s)$$

(ii) The owners should choose the aluminium guttering and use the steel screws to attach it. Copper is less reactive than steel, so the steel screws would act as the anode if screwed into copper and as such would rapidly corrode. Copper is also significantly more expensive than aluminium, making it less desirable. Although aluminium is more reactive than steel, it is a passivating metal, which slows its corrosion. The aluminium will slowly corrode around the screw but the reaction will be much slower than that with the copper–steel combination. In either case, the steel screws should be physically separated from the metal of the guttering so that a galvanic cell is not established with an anode and a cathode. Plastic washers can be used to keep the metals apart.

(d) (i) By making the artefact the cathode and driving electrons into it, some of the iron ions are returned to iron metal. This also prevents any further corrosion, thereby conserving the artefact. Making the artefact cathodic and negatively charged helps to remove ions from small cracks throughout the metal surface by repelling anions such as Cl^- and SO_4^{2-}. The stainless steel mesh acts as the anode and water is oxidised to oxygen gas on its surface. The bubbles of oxygen provide a soft abrasion that helps to clean the surface of the artefact of any loose material without damaging the original metal.

Sodium hydroxide solution acts as the electrolyte, providing a conducting solution and also preventing the surrounding solution from becoming acidic and accelerating corrosion.

(ii) A nickel spoon can be electroplated by driving a low voltage current into it, making it the cathode. Positive silver ions from a dilute silver nitrate electrolyte will move toward the negatively charged spoon and reduce on its surface to solid silver:

$$Ag^{+}(aq) + e^{-} \rightarrow Ag(s)$$

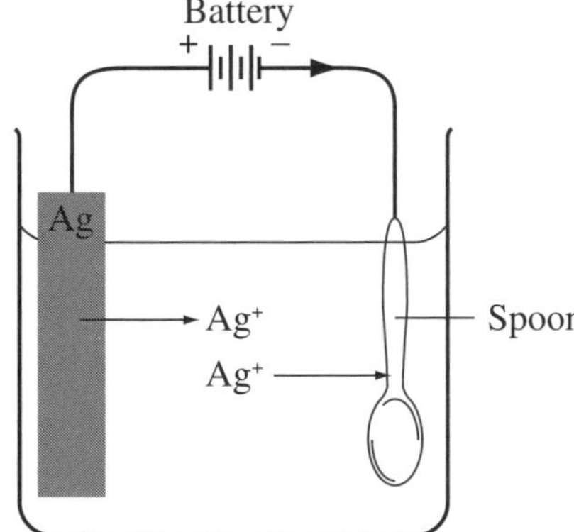

(e) The work of these four scientists in the late eighteenth and early nineteenth centuries helped to increase our understanding of electron transfer reactions.

Galvani was an Italian physician and physicist who investigated the nature and effects of what he considered the electricity in animal tissue.

He pressed a copper hook into a frog's spinal cord and hung the hook on an iron rail. The metallic contact between the leg muscle and spinal cord led to twitching. He concluded that animal tissue contained a vital force which he termed "animal electricity" as mentioned in the flow chart, which activated nerve and muscle when spanned by metal probes. He considered the brain to be the most important organ for the secretion of this "electric fluid" and the nerves to be conductors of the fluid to muscle. Galvani was correct in attributing muscular contractions to electrical stimulus but wrong in identifying it as "animal electricity".

Galvani's work was refuted by Alessandro Volta, but provided a major stimulus for his work.

Volta was an Italian physicist born in Como. He recognised that Galvani's conclusion that the twitching of muscles of dead frogs was evidence of animal electricity was incorrect. Through experimentation he realised that the two different metal objects holding the frog's leg might be the source of the action. Over a period of several years he worked out that the wet muscle tissue conducted a current between the two different types of metal. He modified this effect to produce the first continuous flow of electric current.

Volta observed the electrical interaction between two different metals submerged near each other in an acidic solution. Based on this principle, he constructed the first battery, as indicated in the flow chart, consisting of a series of alternating copper and zinc rings in an acid electrolyte. He called his invention a column battery, although it later became known as the Volta battery or Voltaic cell.

In 1800, Volta made a "voltaic pile" from a stack of alternating silver and zinc discs separated by felt pads soaked in brine (salt solution). When the bottom zinc disc was joined to the upper silver disc current would flow through attached wires. This was the prototype of electrical batteries.

Davy then reasoned that a chemical combination occurred between substances of opposite charge and that chemical action was the cause of the electricity produced in simple electrolytic cells. From this he concluded that electrolysis could be used to separate compounds into their constituent elements. His discoveries led him to successfully isolate the elements sodium and potassium, as mentioned in the flow chart, by electrolysis. He also discovered other group I and II elements.

Michael Faraday was Davy's lab assistant and he continued Davy's work on electrolysis. Faraday quantified electrochemistry through his laws of electrolysis, as mentioned in the flow chart.

His first law states that *the mass of metal produced (deposited on the cathode) in an electrolytic cell is directly proportional to the quantity of electricity passing through the cell*. By using his formulas the number of moles of electrons flowing through a circuit can be calculated as long as current and time are known. This can then be used to calculate the number of moles of product forming at electrodes.

Faraday's second law can be outlined in simple terms as *the masses of different substances produced in electrolytic cells connected in series are directly proportional to the molar masses of the substances and inversely proportional to the number of electrons in each half-equation*.

Through his laws of electrolysis and other research in electrolysis, Faraday was able to relate the quantity of electricity to the mass of metals deposited on the cathode in electrolysis.

Faraday introduced terminology still used today in electrolysis such as electrolyte, cation and anion.

Question 31—Forensic Chemistry

(a) (i) Peptide bond.

(ii) The *primary structure* of proteins is the number and sequence of the amino acids in the polypeptide chain. Peptide bonds link the amino acids.

The *secondary structure* is the shape of the protein molecule caused by hydrogen bonding between –C=O and –N—H groups within the chain. The two main shapes are:

- alpha helix: the chains are twisted into coils. The type of coil depends partly on the number of hydrogen bonds formed, e.g. wool fibres.
- beta sheet: the polypeptide chains are aligned side by side and are connected by hydrogen bonds between all the peptide links, e.g. silk.

The *tertiary structure* consists of a polypeptide chain, coiled and folded asymmetrically in the protein molecule caused by the interactions of the R groups. This interaction may be a result of hydrogen bonding, dipole–dipole interactions, covalent bonding or ionic bonding, depending on the polarity of the R groups.

(iii) Electrophoresis can be used to identify the origins of proteins. First, the mixture of proteins is applied to a gel and an electric field from a DC source is imposed. The separation is obtained through the movement of the negatively charged protein molecules through the gel. Standards of known molecular weight (called markers) are run alongside the sample and are used to distinguish proteins in the unknown sample. The higher the charge on the ion, the faster it moves between the electrodes, and the greater the mass of the ion, the slower it moves between the electrodes. After the proteins are separated they are stained so they can be located and identified.

(b) (i) The oil spill originated from ship A, because the retention times for four of the peaks on the chromatogram for the ship A sample match the retention time of the peaks of the oil spill sample.

(ii) The peaks in the chromatograms represent the various components present in the oil mixture (qualitative analysis) and the area of the peaks represents how much of each component is present in the mixture (quantitative analysis).

(c) The features of gas–liquid chromatography (GLC) that allow for the analysis of small samples are:

- the degree of solubility is determined by the polarity of the substance and the solvent;
- the technique is extremely sensitive and can detect minute quantities of a compound at the nanogram level;
- GLC allows for the rapid separation of a mixture into individual components and it is both quantitative and qualitative.

(d) (i)

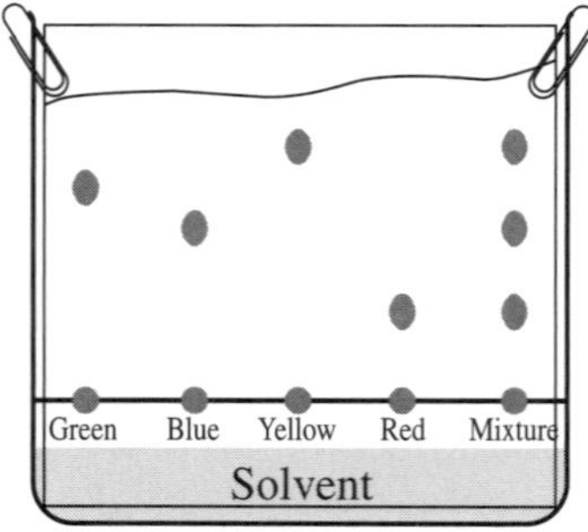

A paper chromatography procedure was used to separate a mixture of food dyes into components. A baseline was drawn 2 cm from the bottom of a piece of chromatography paper using a pencil. Using a dropper, a drop of the food dye mixture and four standard food dye colours: red, blue, green and yellow were placed on the baseline of the chromatography paper. The paper was then placed in a beaker containing a sodium chloride solution 1 cm deep and secured at the top of the beaker with a clip. When the solvent had almost reached the top, the paper was removed and allowed to dry. The paper showed a series of three spots corresponding to the different components in the food dye mixture (red, yellow and blue) and four spots corresponding to the standard colours. The distance travelled by each component in the mixture was measured and compared to the distances travelled by the standard colours. If the distances of the components in the mixture (red, yellow and blue) match the distances of the standard coloured spots, then the coloured standard dyes (red, yellow and blue) are present in the mixture.

(ii) A chemical test that can be used to distinguish an alkene from an alkanoic acid or an alcohol is to place some of the mixture in a test tube and add bromine water dropwise. If the bromine water changes from red-brown to colourless then an alkene is present.

To test for an alkanoic acid a small amount of sodium carbonate is added to a small sample of the mixture and bubbles of carbon dioxide gas are liberated. For an alcohol, acidified potassium dichromate is added to a small sample of the mixture and if the acidified potassium dichromate changes from an orange colour to a green colour, then an alcohol is present.

(e) A sample residue can be taken from a crime scene, e.g. a small chip of blue paint from a car at a murder scene, and analysed by a forensic chemist to determine its chemical composition. Chemical tests will be used to determine whether the sample is organic or inorganic. If it is organic, what hydrocarbons and functional groups are present? If it is inorganic, further chemical tests can reveal if it is acidic, basic or a neutral salt.

Samples are collected by using sterile forceps or latex gloves and the sample must be placed into correctly labelled sterile sealable plastic bags or containers. The evidence must be examined objectively and all scientific analysis must be accurate and reproducible.

The sample can be placed into a high vacuum chamber of a mass spectrometer then slowly heated until some of the sample vaporises. The vapour then passes through an ionization chamber where the molecules or atoms are ionized (electrons removed) to acquire a positive charge. These positively charged molecules or ions are very unstable and break up into numerous smaller fragments. They are then accelerated and pass through a magnetic field where they are deflected. The higher the mass of the ion the less it will be deflected and the higher the charge, the more the ion is deflected. A detector identifies the mass of each particle.

The data are recorded on a spectrogram, each line representing a peak for an ion of the molecule or a region of the molecule that may have broken apart when ionization took place. The height of a peak indicates the relative abundance of each fragment. Computers can store thousands of known spectra and compare them with the spectra produced by unknown samples, but there are specific mass to charge ratio values that are common to many molecules and classes of compounds and these are easily identifiable by a computer. These results can be interpreted by a non-scientist, for example to identify the brand of car that was involved at the murder scene.

CHAPTER 8

BOARD OF STUDIES
NEW SOUTH WALES

2010

HIGHER SCHOOL CERTIFICATE EXAMINATION

Chemistry

General Instructions

- Reading time – 5 minutes
- Working time – 3 hours
- Write using black or blue pen
- Draw diagrams using pencil
- Board-approved calculators may be used
- A data sheet and a Periodic Table are provided at the back of this paper
- Write your Centre Number and Student Number where required

Total marks – 100

Section I

75 marks

This section has two parts, Part A and Part B

Part A – 20 marks
- Attempt Questions 1–20
- Allow about 35 minutes for this part

Part B – 55 marks
- Attempt Questions 21–31
- Allow about 1 hour and 40 minutes for this part

Section II

25 marks

- Attempt ONE question from Questions 32–36
- Allow about 45 minutes for this section

Section I
75 marks

Part A – 20 marks
Attempt Questions 1–20
Allow about 35 minutes for this part

Use the multiple-choice answer sheet for Questions 1–20.

---

1 Water is released during a polymerisation reaction.

Which monomer is likely to have been involved in the reaction?

(A) Ethene

(B) Glucose

(C) Styrene

(D) Vinyl chloride

2 Which of the following is an example of a transuranic element?

(A) C–14

(B) Co–60

(C) U–238

(D) Cm–249

3 Which substance shows the correct indicator colour?

	Substance	*pH*	*Indicator*	*Colour*
(A)	Stomach acid	2	Methyl orange	Yellow
(B)	Lemon juice	3	Phenolphthalein	Pink
(C)	Soda water	4	Phenolphthalein	Pink
(D)	Seawater	8	Methyl orange	Yellow

4 The diagram shows the structural formula of a gas.

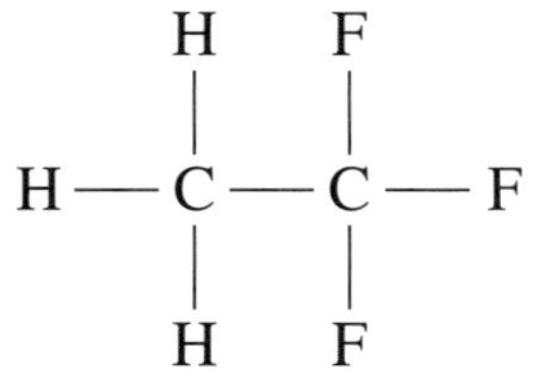

$$\begin{array}{ccccccc} & & H & & F & & \\ & & | & & | & & \\ H & — & C & — & C & — & F \\ & & | & & | & & \\ & & H & & F & & \end{array}$$

How many isomers does this compound have?

(A) 1

(B) 2

(C) 3

(D) 4

5 An imbalance of which two substances causes the eutrophication of waterways?

(A) H^+ and OH^-

(B) Mg^{2+} and Ca^{2+}

(C) Oxygen and ozone

(D) Phosphorus and nitrogen

6 The diagram shows a section of a polymer.

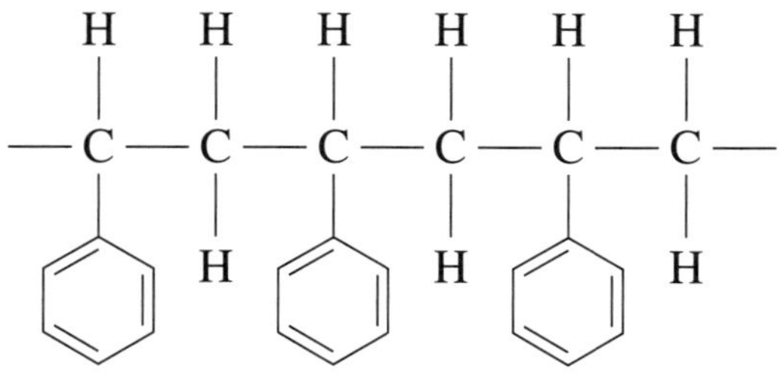

$$\begin{array}{c} H \quad H \quad H \quad H \quad H \quad H \\ | \quad\; | \quad\; | \quad\; | \quad\; | \quad\; | \\ —C—C—C—C—C—C— \\ | \quad\; | \quad\; | \quad\; | \quad\; | \quad\; | \\ C_6H_5 \; H \; C_6H_5 \; H \; C_6H_5 \; H \end{array}$$

What is the systematic name of the monomer?

(A) Polybenzene

(B) Benzylethene

(C) Ethylbenzene

(D) Ethenylbenzene

7 Equal volumes of four 0.1 mol L^{-1} acids were titrated with the same sodium hydroxide solution.

Which one requires the greatest volume of base to change the colour of the indicator?

(A) Citric acid

(B) Acetic acid

(C) Sulfuric acid

(D) Hydrochloric acid

8 In a research report a student wrote, 'Acids are compounds that contain hydrogen and can dissolve in water to release hydrogen ions into solution.'

Who originally stated this theory of acids?

(A) Arrhenius

(B) Brönsted–Lowry

(C) Davy

(D) Lavoisier

9 What types of reaction occur in the Haber process during the production of ammonia?

(A) Redox and synthesis

(B) Hydration and redox

(C) Decomposition and oxidation

(D) Reduction and decomposition

10 A sample of water from a stream, suspected to be contaminated with metal ions, was analysed.

The results of some tests on the water are recorded in the table.

Test	*Result*
Add dilute HCl	No change
Add Na_2SO_4 solution	White precipitate formed
Flame test	Pale green colour

What is the most likely contaminant in the water?

(A) Ba^{2+}

(B) Ca^{2+}

(C) Cu^{2+}

(D) Fe^{3+}

11 An organic liquid, when reacted with concentrated sulfuric acid, produces a compound that decolourises bromine water.

What is the formula of the organic liquid?

(A) C_6H_{12}

(B) C_6H_{14}

(C) $C_6H_{11}OH$

(D) $C_5H_{11}COOH$

12 In which of the following reactions does the metal atom show the greatest change in oxidation state?

(A) MnO_4^- to Mn^{2+}

(B) MnO_2 to $Mn(OH)_3$

(C) PbO_2 to $PbSO_4$

(D) VO_2^+ to VO^{2+}

13 The diagram shows a galvanic cell.

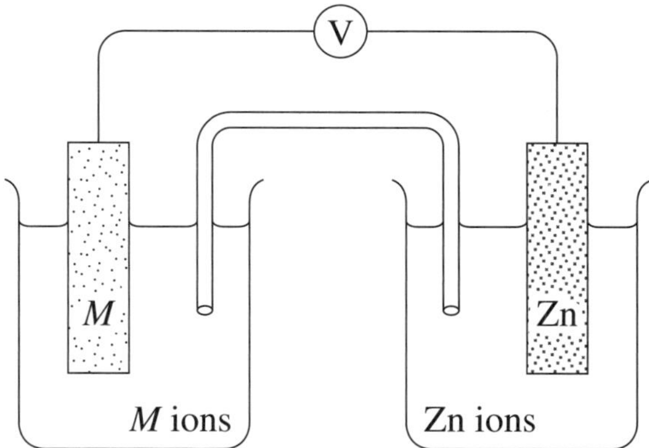

Which of the following metals (*M*) acting as an anode would produce the lowest theoretical potential for the cell?

(A) Calcium

(B) Copper

(C) Iron

(D) Manganese

14 The table shows information about three carbon compounds.

Compound	*Structural formula*	*Molecular weight*	*Boiling point*
X	H H H | | | H—C—C—C—O—H | | | H H H	60	97°C
Y	H | //O H—C—C | \O—H H	60	118°C
Z	H O | || H—C—O—C—H | H	60	?

What is the best estimate for the boiling point of compound *Z*?

(A) 31°C

(B) 101°C

(C) 114°C

(D) 156°C

15 What mass of ethanol is obtained when 5.68 g of carbon dioxide is produced during fermentation, at 25°C and 100 kPa?

(A) 2.95 g

(B) 5.95 g

(C) 33.6 g

(D) 147.2 g

16 Which of the following Lewis structures does NOT contain a coordinate covalent bond?

(A) $\left[\begin{array}{c} \text{H} \; :\ddot{\text{O}}: \; \text{H} \\ \text{H} \end{array}\right]^+$

(B) $\left[\begin{array}{c} \text{H} \\ \text{H} \; :\ddot{\text{N}}: \; \text{H} \\ \text{H} \end{array}\right]^+$

(C) $:\underset{\cdot\cdot}{\text{O}}::\text{C}::\underset{\cdot\cdot}{\ddot{\text{O}}}$

(D) O_3 (ozone Lewis structure)

17 A student completed an experiment to determine the amount of energy absorbed by a volume of water.

The following data were recorded.

Mass of beaker	215.6 g
Mass of beaker plus water	336.1 g
Final temperature of water	71.0°C
Energy absorbed	21.2 kJ

What was the initial temperature of the water?

(A) 15°C

(B) 25°C

(C) 29°C

(D) 42°C

18 Chromate and dichromate ions form an equilibrium according to the following equation.

$$2CrO_4^{2-}(aq) + 2H^+(aq) \rightleftharpoons Cr_2O_7^{2-}(aq) + H_2O(l)$$

Which solution would increase the concentration of the chromate ion (CrO_4^{2-}) when added to the equilibrium mixture?

(A) Sodium nitrate

(B) Sodium chloride

(C) Sodium acetate

(D) Ammonium chloride

19 Sodium azide is used in automobile airbags to provide a source of nitrogen gas for rapid inflation in an accident.The equation shows the production of nitrogen gas from sodium azide.

$$2NaN_3(s) \rightarrow 2Na(s) + 3N_2(g)$$

What mass of sodium azide will produce 40 L of N_2 at 100 kPa and 0°C?

(A) 70 g

(B) 76 g

(C) 114 g

(D) 172 g

20 Solutions containing copper ions were analysed by AAS. A standard solution of 10 ppm copper had an AAS absorbance of 0.400. A second solution of unknown concentration was found to have an absorbance of 0.500.

100 mL of this second solution was reacted with excess sodium carbonate solution. The precipitate was then dried and weighed.

What mass of precipitate was formed?

(A) 1.25×10^{-3} g

(B) 2.43×10^{-3} g

(C) 1.54 g

(D) 2.43 g

2010 HIGHER SCHOOL CERTIFICATE EXAMINATION

Chemistry

Centre Number

Section I (continued)

Student Number

Part B – 55 marks
Attempt Questions 21–31
Allow about 1 hour and 40 minutes for this part

Answer the questions in the spaces provided. These spaces provide guidance for the expected length of response.

Show all relevant working in questions involving calculations.

Question 21 (3 marks)

A 0.001 mol L^{-1} solution of hydrochloric acid and a 0.056 mol L^{-1} solution of ethanoic acid both have a pH of 3.0. **3**

Why do both solutions have the same pH?

....

....

....

....

....

....

Question 22 (6 marks)

A student prepared the compound methyl propanoate in a school laboratory.

(a) Give a common use for the class of compounds to which methyl propanoate belongs. **1**

....

....

(b) In the preparation of this compound a few drops of concentrated sulfuric acid were added to the starting materials. The mixture was then refluxed for a period of time. **2**

Why was it necessary to reflux the mixture?

....

....

....

....

(c) Name the TWO reactants used in preparing the methyl propanoate and draw their structural formulae. **3**

....

....

....

....

....

....

2010 HIGHER SCHOOL CERTIFICATE EXAMINATION

Chemistry

Section I – Part B (continued)

Centre Number

Student Number

Question 23 (3 marks)

(a) Write a balanced chemical equation for the complete combustion of 1-butanol. **1**

....

....

(b) A student measured the heat of combustion of three different fuels. The results are shown in the table. **2**

Fuel	*Heat of combustion* (kJ g^{-1})
A	–48
B	–38
C	–28

The published value for the heat of combustion of 1-butanol is 2676 kJ mol^{-1}.

Which fuel from the table is likely to be 1-butanol? Justify your answer.

....

....

....

....

....

....

Question 24 (4 marks)

In the margarine industry, alkenes are often hydrogenated to convert unsaturated oils into solid fats that have a greater proportion of saturated molecules.

(a) Using ethene as an example, write an equation for this reaction and state the type of reaction this represents. **2**

....

....

....

(b) Describe a test that could be used to confirm that all the ethene has been converted. **2**

....

....

....

....

2010 HIGHER SCHOOL CERTIFICATE EXAMINATION

Chemistry

Centre Number

Section I – Part B (continued)

Student Number

Question 25 (5 marks)

What is the relationship between dissolved oxygen and biochemical oxygen demand and why is it important to monitor both in natural waterways? **5**

....

....

....

....

....

....

....

....

....

....

....

....

2010 HIGHER SCHOOL CERTIFICATE EXAMINATION

Chemistry

Section I – Part B (continued)

Centre Number

Student Number

Question 26 (4 marks)

A gas is produced when 10.0 g of zinc is placed in 0.50 L of 0.20 mol L^{-1} nitric acid. **4**

Calculate the volume of gas produced at 25°C and 100 kPa. Include a balanced chemical equation in your answer.

....

....

....

....

....

....

....

....

Question 27 (2 marks)

The diagram shows a particular cell with relevant half equations. **2**

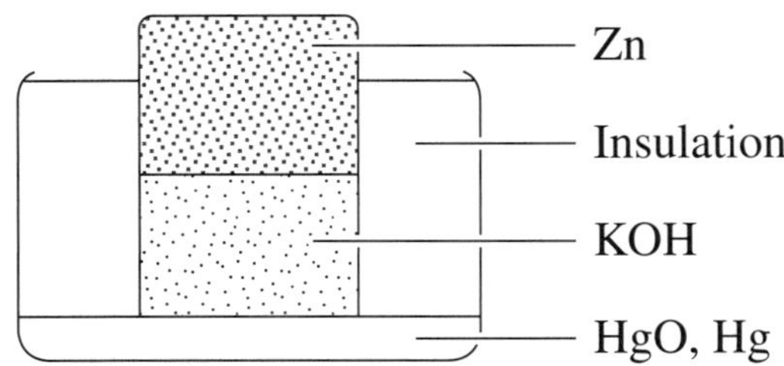

$$Zn(s) + 2OH^-(aq) \rightarrow ZnO(s) + H_2O(l) + 2e^-$$

$$HgO(s) + H_2O(l) + 2e^- \rightarrow Hg(l) + 2OH^-(aq)$$

Identify the anode, cathode and electrolyte for this cell.

....

....

....

....

2010 HIGHER SCHOOL CERTIFICATE EXAMINATION

Chemistry

Centre Number

Section I – Part B (continued)

Student Number

Question 28 (8 marks)

The flowchart shown outlines the sequence of steps used to determine the concentration of an unknown hydrochloric acid solution. **8**

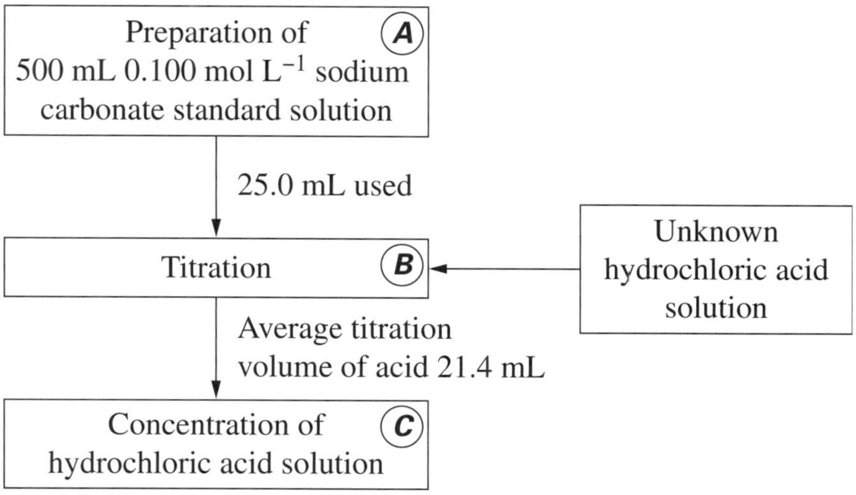

Preparation of 500 mL 0.100 mol L^{-1} sodium carbonate standard solution (A)

↓ 25.0 mL used

Titration (B) ← Unknown hydrochloric acid solution

↓ Average titration volume of acid 21.4 mL

Concentration of hydrochloric acid solution (C)

Describe steps ***A***, ***B*** and ***C*** including correct techniques, equipment and appropriate calculations. Determine the concentration of the hydrochloric acid.

....

....

....

....

....

....

....

....

....

....

Question 28 continues

Question 28 (continued)

....

....

....

....

....

....

....

....

....

....

....

....

....

....

End of Question 28

Question 29 (6 marks)

The flowchart shown outlines the process used to determine the amount of sulfate present in a sample of lawn fertiliser.

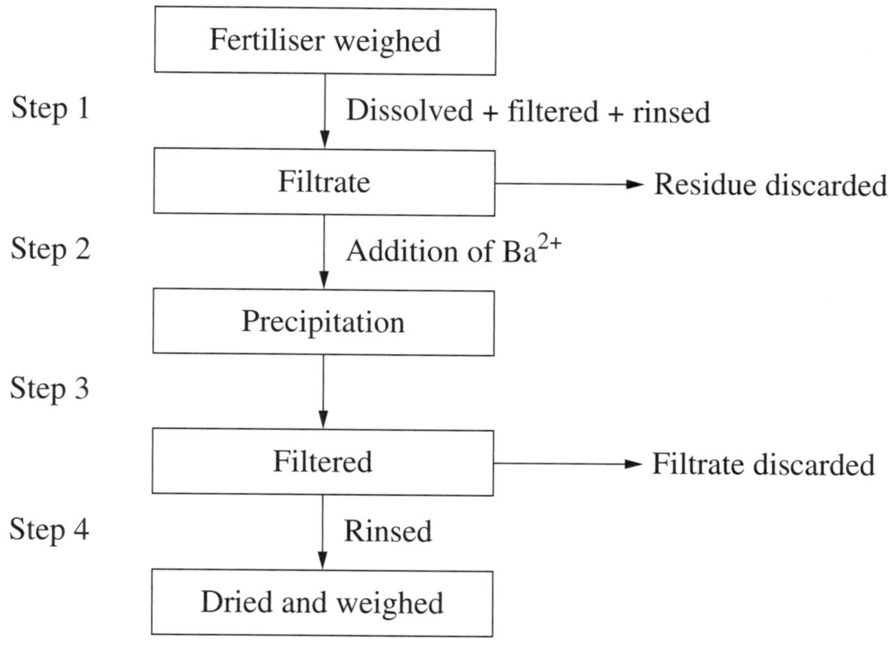

(a) What assumptions were made and how do these affect the validity of this process? **3**

....

....

....

....

....

....

(b) It was found that 4.25 g had a sulfate content of 35%. **3**

What is the mass of the dried precipitate at Step 4? Include a chemical equation in your answer.

....

....

....

....

....

....

2010 HIGHER SCHOOL CERTIFICATE EXAMINATION

Chemistry

Centre Number

Section I – Part B (continued)

Student Number

Question 30 (8 marks)

(a) Compare the process of polymerisation of ethylene and glucose. Include relevant chemical equations in your answer. **3**

....

....

....

....

....

....

....

....

Question 30 continues

Question 30 (continued)

(b) Explain the relationship between the structures and properties of THREE different polymers from ethylene and glucose, and their uses. **5**

....

....

....

....

....

....

....

....

....

....

....

....

End of Question 30

2010 HIGHER SCHOOL CERTIFICATE EXAMINATION

Chemistry

Centre Number

Section I – Part B (continued)

Student Number

Question 31 (6 marks)

(a) A student collected a 250 mL sample of water from a local dam for analysis. The data collected are shown in the table.

Mass of filter paper	0.23 g
Mass of filter paper and solid	0.47 g
Mass of evaporating basin	43.53 g
Mass of basin and solid remaining	44.67 g

(i) The water was filtered and the filtrate evaporated to dryness. **2**

Calculate the percentage of the total dissolved solids in the dam sample.

....

....

....

....

(ii) It is suspected that the water in the dam has a high concentration of chloride ions. **2**

Describe a chemical test that could be carried out on the water sample to determine the presence of chloride ions. Include an equation in your answer.

....

....

....

....

Question 31 continues

Question 31 (continued)

(b) Name an ion other than chloride that commonly pollutes waterways, and identify its source and the effect of its presence on water quality. **2**

....

....

....

....

End of Question 31

2010 HIGHER SCHOOL CERTIFICATE EXAMINATION

Chemistry

Section II

25 marks
Attempt ONE question from Questions 32–36
Allow about 45 minutes for this section

Answer parts (a)–(c) of the question in a writing booklet. Answer parts (d)–(e) of the question in a SEPARATE writing booklet.

Show all relevant working in questions involving calculations.

Question 32	Industrial Chemistry
Question 33	Shipwrecks, Corrosion and Conservation
Question 34	The Biochemistry of Movement *(Not included in this reproduction)*
Question 35	The Chemistry of Art *(Not included in this reproduction)*
Question 36	Forensic Chemistry

Question 32 — Industrial Chemistry (25 marks)

Answer parts (a)–(c) in a writing booklet.

(a) Identify the type of cell shown and outline the process used in the extraction of sodium hydroxide. **3**

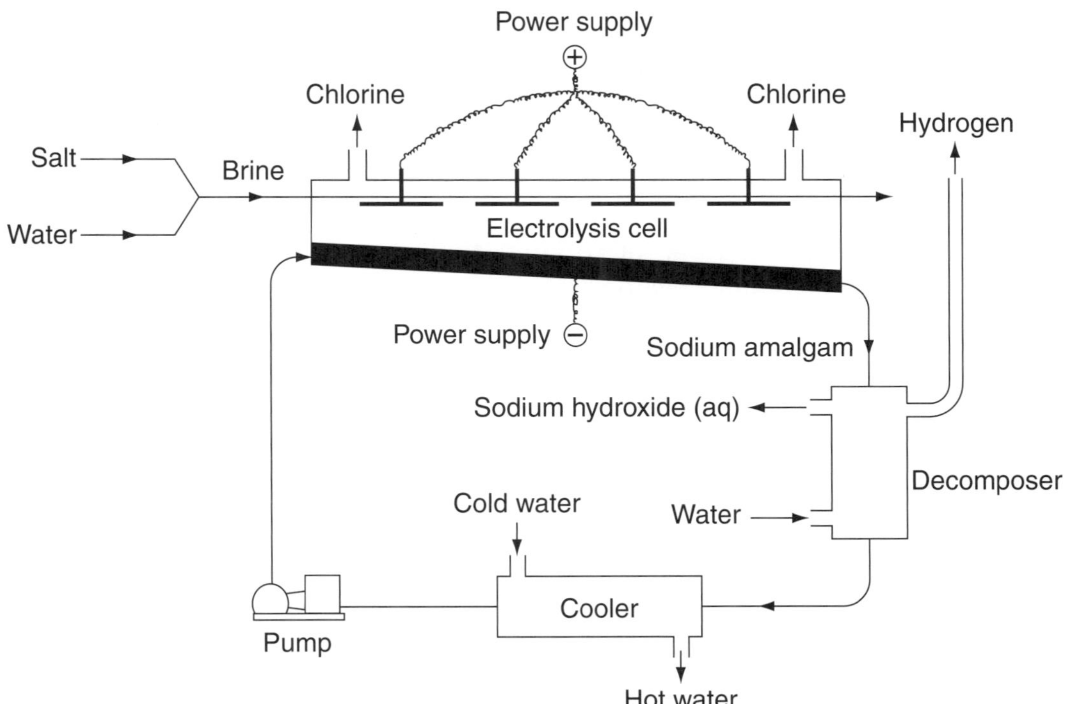

(b) Compare the electrolysis of molten sodium chloride and aqueous sodium chloride. Write the relevant half equations and overall reaction for each process. **5**

(c) At room temperature 0.80 moles of SO_2 and 0.40 moles of O_2 were introduced into a sealed 10 L vessel and allowed to come to equilibrium.

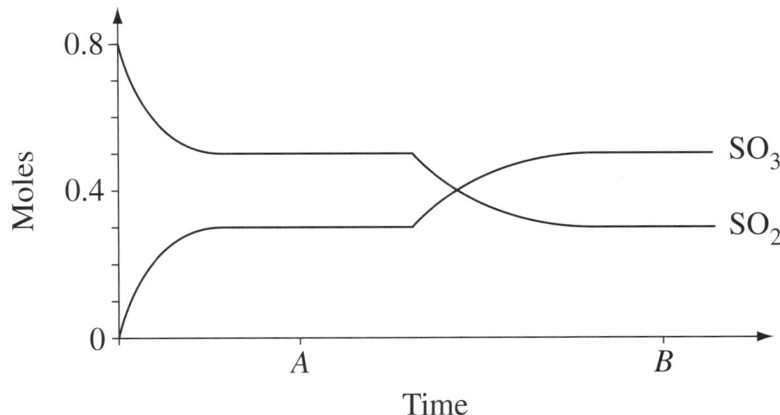

(i) Write the equilibrium constant expression and calculate the value for the equilibrium constant at time *A*. **3**

(ii) Explain why a new equilibrium position was established at time *B*. **2**

Question 32 continues

Question 32 (continued)

Answer parts (d)–(e) in a SEPARATE writing booklet.

(d) The equation represents a reaction that can be performed in a school laboratory.

$$\text{Oil} + 3\,\boxed{\text{A}} \rightarrow 3\text{KO}\overset{\overset{\text{O}}{\|}}{\text{C}}(\text{CH}_2)_{14}\text{CH}_3 + \text{Glycerol}$$

(i) Identify both this type of reaction and the reactant *A*. **2**

(ii) Describe how this type of reaction could be carried out in a school laboratory including specific safety precautions for this process. **3**

(e) Assess both the importance and resulting environmental impacts of using limestone in the Solvay Process. **7**

End of Question 32

Question 33 — Shipwrecks, Corrosion and Conservation (25 marks)

Answer parts (a)–(c) in a writing booklet.

(a) The following artefact was retrieved from a ship that sank 150 years ago off the coast of New South Wales. **3**

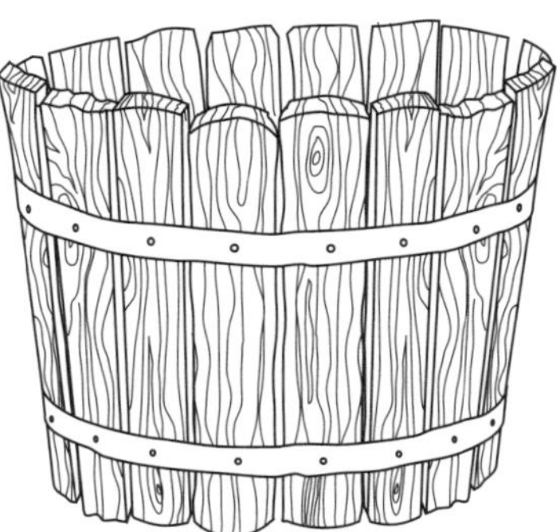

Outline the effect that the marine environment would have had on the artefact.

(b) (i) Use a fully labelled diagram to show the electrolysis of an aqueous solution of potassium chloride. Write the relevant half equations and the overall reaction for the cell. **4**

(ii) How would the cathode be identified? **1**

(c) The following table shows the composition of four types of steel. **5**

Steel	*Composition*
1	99.8% Fe, 0.2% C
2	98.5% Fe, 1.5% C
3	94% Fe, 4% C, 1% Mn, 1% Si
4	75% Fe, 15% Cr, 10% Ni

Explain how the composition of each type of steel determines its properties and uses.

Question 33 continues

Question 33 (continued)

Answer parts (d)–(e) in a SEPARATE writing booklet.

(d) (i) An investigation into environmental factors that affect the rate of corrosion of iron can be performed in a school laboratory. **4**

Describe how you could perform this investigation in relation to THREE environmental factors.

(ii) Explain how the effect of ONE of the factors could be reduced in a marine environment. **1**

(e) Evaluate the suitability of techniques used for restoring and conserving wooden and copper artefacts that have been immersed in salt water for at least 100 years. **7**

End of Question 33

Question 36 — Forensic Chemistry (25 marks)

Answer parts (a)–(c) in a writing booklet.

(a) The diagrams show the arrangement of glucose units in two polysaccharides. **3**

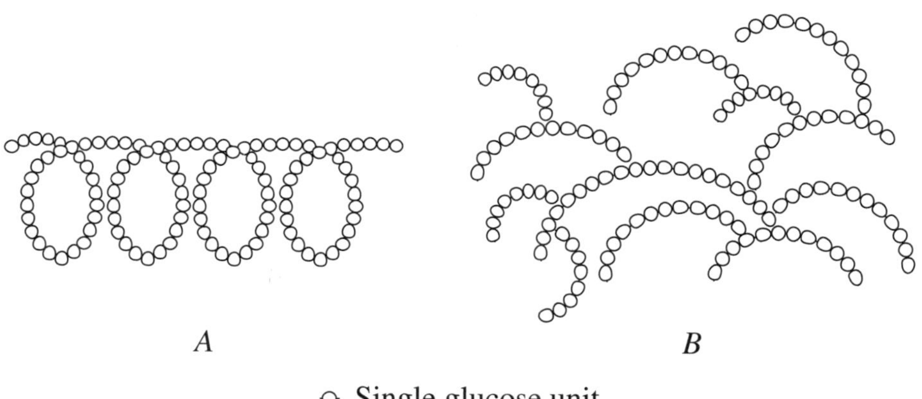

Identify *A* and *B* and outline their differences in structure and origin.

Question 36 continues

Question 36 (continued)

(b) The structures of three amino acids at pH 6 and pH 8 are shown.

<table>
<tr><th></th><th>Glutamic acid</th><th>Lysine</th><th>Valine</th></tr>
<tr><td>pH 6</td><td>O = C – O⁻
H – C – H
H – C – H
H O
⁺H – N – C – C
H H O⁻</td><td>H H H
H – C – C – N – H⁺
H H
H – C – H
H – C – H
H O
⁺H – N – C – C
H O⁻</td><td>H H H H
C H C
H C H
H O
⁺H – N – C – C
H H O⁻</td></tr>
<tr><td>pH 8</td><td>O = C – O⁻
H – C – H
H – C – H
H O
N – C – C
H O⁻</td><td>H H H
H – C – C – N – H⁺
H H
H – C – H
H – C – H
H O
⁺H – N – C – C
H H O⁻</td><td>H H H H
C H C
H C H
H O
N – C – C
H H O⁻</td></tr>
</table>

Samples of each amino acid underwent electrophoresis at pH 6 and pH 8. The results are shown below.

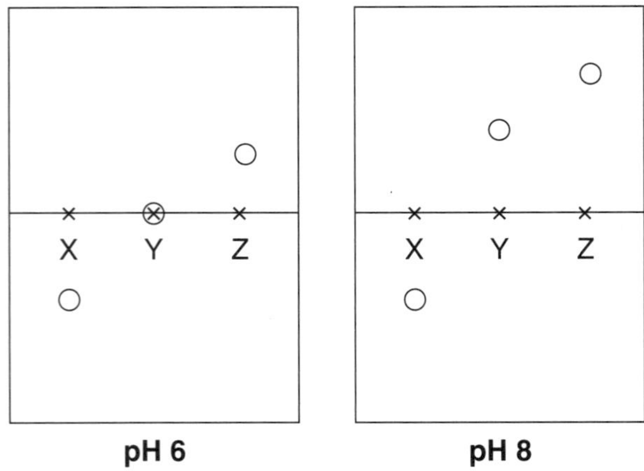

(i) Outline the process of electrophoresis used in identifying amino acids. **2**

(ii) Identify the amino acids **X**, **Y** and **Z** and justify your answer. **3**

Question 36 continues

Question 36 (continued)

(c) The following emission spectra were used to identify the manufacturer of a piece of pottery.

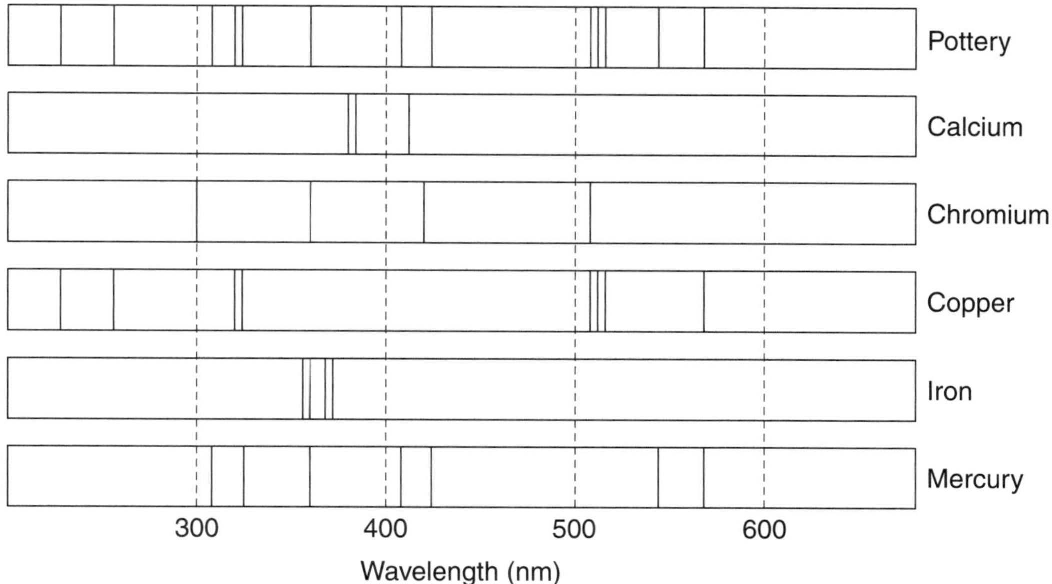

Manufacturer	*Metal atoms in pottery*
A	copper, chromium
B	copper, mercury
C	calcium, chromium

(i) Which manufacturer made the pottery – A, B or C? **1**

(ii) Explain the production and use of emission spectra in identifying the manufacturer of this pottery. **4**

Answer parts (d)–(e) in a SEPARATE writing booklet.

(d) A forensic chemist analyses soil from a pair of shoes worn by a person suspected of committing a crime.

(i) Identify FOUR properties of soil a forensic chemist would investigate. **2**

(ii) Describe both an organic and an inorganic test that could be performed on the soil that could match it to soil at the crime scene. **3**

(e) Describe the techniques used to analyse DNA and the applications of these techniques in forensic analysis. **7**

End of paper

2010 HSC Examination Paper

Sample Answers

Section I Part A
(Total 20 marks)

1 B The polymerisation of glucose to cellulose is a condensation reaction where water molecules are removed, allowing the monomers to link up. The other three options are all addition polymerisation reactions that do not involve the removal of molecules but rely on adding to a double bond.

2 D Transuranic elements are those beyond uranium. Curium, Cm, is element number 96, which is beyond uranium (element number 92).

3 D Methyl orange indicator is yellow at pHs of 6 and above.

4 B The two isomers are 1,1,1-trifluoroethane and 1,1,2-trifluoroethane.

5 D Eutrophication of waterways results from pollution from excessive amounts of nitrogen and phosphorus nutrients released into the waterways from farms and sewage treatment plants.

6 D

7 A Citric acid is triprotic so that each molecule will require three hydroxide ions to neutralise it. Sulfuric acid is diprotic, requiring two hydroxide ions for neutralisation. Acetic acid and hydrochloric acid are monoprotic, requiring one.

8 A Arrhenius hypothesised that acids ionise in water whereas Davy proposed that it was the presence of hydrogen in acids that gave the acidic properties.

9 A The Haber process involves the reduction of nitrogen and the oxidation of hydrogen to form, or synthesise, ammonia molecules.

10 A Chloride will not precipitate any of the ions; however, barium sulfate is insoluble as a white solid and the barium flame test colour is apple green.

11 C Concentrated sulfuric acid is a dehydrating agent, removing water from a molecule. Cyclohexanol is dehydrated to form cyclohexene. Bromine water is a diagnostic test for alkenes such as cyclohexene and will react with it, decolourising in the process. (Note: candidates should note that the syllabus does not require a knowledge of the formulae and structure of cycloalkenes and cycloalkanols. Answer C can be obtained by elimination of the other three incorrect answers).

12 A Manganese goes from an oxidation state of +VII in MnO_4^- to +II in Mn^{2+}, a change of 5. This is a greater change than in answer B (a change of 1), C (a change of 2) or D (a change of 1).

13 D To act as an anode with zinc as the cathode, the metal would need to have a higher oxidation potential than zinc. Calcium and manganese both do. The lowest potential would be provided by the metal with the lowest oxidation potential of these two, and this is manganese.

14 A Molecules X and Y have hydrogen bonding between molecules and this is a stronger force than the dipole/dipole forces between Z molecules. It takes less energy to break these weaker forces, hence the lower boiling point.

15 B $C_6H_{12}O_6 \rightarrow 2CH_3CH_2OH + 2CO_2$

$$n(CO_2) = \frac{5.68}{44.01} = 0.13 \text{ mol}$$

$$n(CH_3CH_2OH) = n(CO_2) = 0.13 \text{ mol}$$

$$m(CH_3CH_2OH) = 0.13 \times 46.07 = 5.95 \text{ g}$$

16 C The other three alternatives all contain a pair of shared electrons where both electrons originate from the same atom.

17 C

$$\Delta H = -mC\Delta T$$

$$21\,200 \text{ J} = -(336.1 - 215.6) \times 4.18 \times (71 - x)$$

$$71 - x = 42$$

$$x = 71 - 42 = 29$$

18 C Na^+ ions are neutral but acetate is the conjugate base of the weak acid, acetic acid. As a base it will remove H^+ ions from the mixture, reducing their concentration. The equilibrium will shift to the left to partially counteract this change and in the process increase the concentration of chromate.

19 B $n(N_2) = \frac{40}{22.71} = 1.76 \text{ mol}$

$$n(NaN_3) = 1.76 \times \frac{2}{3} = 1.17 \text{ mol}$$

$$m(NaN_3) = 1.17 \times 65 = 76.3 \text{ g}$$

20 B Assuming a linear calibration graph, $[Cu^{2+}] = 10 \text{ ppm} \times \frac{0.5}{0.4} = 12.5 \text{ ppm} = 12.5 \text{ mg/L}$

1.25 mg/100 mL = 0.00125 g of Cu

$$n(Cu) = \frac{0.00125}{65.55} = 1.97 \times 10^{-5} \text{ mol}$$

$$m(CuCO_3) = 1.97 \times 10^{-5} \times 123.55 = 2.43 \times 10^{-3} \text{ g}$$

Section I Part B

21 HCl is a strong acid and ionises completely, producing 0.001 mol L^{-1} H_3O^+.
pH = −log(0.001) = 3.0

CH_3COOH is a weak acid only partially ionising. The more concentrated 0.056 mol L^{-1} solution partially ionises to form the same 0.001 mol L^{-1} concentration of H_3O^+ and therefore the same pH. *(3 marks)*

22 (a) Methyl propanoate is an ester. Esters are commonly used for their fruity smells in foods and cosmetics. *(1 mark)*

(b) The reactants are volatile and require heating for a period of time. Refluxing allows the reactant vapours to condense back to the mixture for heating rather than escaping. Esterification reactions are quite slow and continued heating for a long time is required to reach equilibrium. *(2 marks)*

(c) Methanol and propanoic acid

Methanol

```
     H
     |
H—C—O—H
     |
     H
```

Propanoic acid

```
     H   H      O
     |   |     //
H—C—C—C
     |   |     \
     H   H      O—H
```

(3 marks)

23 (a) $C_4H_9OH(l) + 6O_2(g) \rightarrow 4CO_2(g) + 5H_2O(l)$ *(1 mark)*

(b) Fuel C is most likely to be 1-butanol (butan-1-ol) because the published value for the heat of combustion is 2676 kJ mol^{-1}, which is $\frac{2676}{74.12}$ = 36.1 kJ g^{-1}. When heat of combustion is measured in the school laboratory there is much heat loss to the environment, resulting in a lower value than that published. Fuel C is the only one listed that has a lower value than the published one. *(2 marks)*

(Note: candidates should use the preferred IUPAC nomenclature of butan-1-ol. The examiners have used an older nomenclature, which is accepted in HSC marking.)

24 (a) $C_2H_4(g) + H_2(g) \rightarrow C_2H_6(g)$ Addition reaction *(2 marks)*

(b) Bubble the ethylene through 10 mL of bromine water in a test tube. If the bromine water does not decolourise, all the ethylene has been converted. *(2 marks)*

(Note: this answer has used the preferred IUPAC terminology of ethylene rather than 'ethene' as used by the examiners. The syllabus uses the preferred term.)

25 Dissolved oxygen, DO, refers to the concentration of oxygen dissolved in a water sample and is measured in parts per million. A healthy waterway will have a concentration of 6 ppm or higher. Biochemical oxygen demand, BOD, is a measure of the change in the concentration of DO, usually over a five-day period, in a sample of water that has been stored in the dark, at 20°C in a filled, closed container. Unpolluted waterways have a BOD of less than 5 ppm.

It is important to monitor DO as aquatic organisms require oxygen to respire and it is, therefore, an indicator of a healthy waterway supporting a natural ecosystem. It is important to monitor BOD as a high BOD is due to organic matter such as sewage effluent in the water. Microbes decompose the organic matter, reproduce and use the oxygen in the water to respire themselves. Therefore a high BOD is an indicator of an unhealthy waterway. *(5 marks)*

26 $Zn(s) + 2HNO_3(aq) \rightarrow H_2(g) + Zn(NO_3)_2(aq)$

(Note: zinc reacts with dilute nitric acid forming $NO(g)$ but the intention of the question was that H_2 is formed.)

$n(Zn) = \dfrac{10.0}{65.41} = 0.153$ mol (excess reagent)

$n(HNO_3) = 0.20 \times 0.50 = 0.10$ mol (limiting reagent)

$n(H_2) = \dfrac{0.10}{2} = 0.05$ mol

$V(H_2) = 0.05 \times 24.79 = 1.24$ L

$= 1.2$ L (2 sig fig)

An alternative answer that uses the correct chemical change is:

$3Zn(s) + 8HNO_3(aq) \rightarrow 3Zn(NO_3)_2(aq) + 2NO(g) + 4H_2O(l)$

$n(Zn) = \dfrac{10.0}{65.41} = 0.153$ mol (excess reagent)

$n(HNO_3) = 0.20 \times 0.50 = 0.10$ mol (limiting agent)

$n(NO) = (0.10) \times \dfrac{2}{8} = 0.025$ mol

$V(NO) = 0.025 \times 24.79 = 0.62$ L (2 sig fig) *(4 marks)*

27 Anode = $Zn(s)$ as oxidation occurs on the zinc (oxidation is the loss of electrons)

Cathode = $HgO(s)$ as reduction occurs on the HgO (reduction is the gain of electrons)

Electrolyte = $KOH(aq)$ as this is the conducting solution that allows the current to flow between electrodes. *(2 marks)*

28 *Step A:* Heat a Na_2CO_3 sample in an oven to remove any moisture. Allow to cool in a desiccator. Weigh 5.30 g into a clean, dry 100 mL beaker.

$m(Na_2CO_3) = (0.100 \times 0.500) \times 105.99 = 5.2995$ g

Add enough distilled water to dissolve the solid.

Pour into a clean 500 mL volumetric flask and wash the 100 mL beaker with distilled water into the flask.

Add distilled water to the flask up to the graduated 500 mL mark, cap and invert ten times to mix the solution.

Step B: Rinse a 50 mL burette with distilled water and then with the HCl. Fill the burette to the 0 mL graduated mark with the acid.

Rinse a 25 mL pipette with the standard and pipette 25 mL of the standard into a clean, dry 250 mL conical flask.

Add three drops of methyl orange indicator to the flask and swirl to mix.

Titrate the sodium carbonate with the HCl until the indicator turns from yellow to a permanent red colour. Record the volume of acid needed. Repeat the procedure from the cleaning of the conical flask three times and average the titration volumes. Alternatively, repeat until two readings agree to within 0.05 mL.

Step C: $Na_2CO_3 + 2HCl \rightarrow H_2O + CO_2 + 2NaCl$

$n(Na_2CO_3) = 0.100 \times 0.025 = 0.0025$ mol

$n(HCl) = 0.0025 \times 2 = 0.005$ mol

$$[HCl] = \frac{0.005}{0.0214}$$
$$= 0.2336$$
$$= 0.234 \text{ mol L}^{-1} \text{ (3 sig fig)}$$

(8 marks)

29 (a) We are assuming that all sulfate is dissolved in Step 1. If not, the process is invalid as the amount present in the fertiliser will be an underestimate. We are assuming that all the precipitate is barium sulfate and that all the sulfate is precipitated by the barium. If the barium is not in excess, the process is invalid and the amount calculated will be an underestimate. We are assuming that all the precipitate is captured as residue in Step 3. If some passes through the filter, the process is invalid as it will result in an underestimate of the amount of sulfate in the fertiliser.

As the solution was not acidified initially, the precipitate could contain barium phosphate and barium carbonate. The assumption that the precipitate is barium sulfate only may be incorrect. *(3 marks)*

(b) $m(SO_4^{2-}) = 4.25 \times 0.35 = 1.49$ g

$n(SO_4^{2-}) = 1.49/96.07 = 0.0155$ mol

$Ba^{2+}(aq) + SO_4^{2-}(aq) \rightarrow BaSO_4(s)$

$n(BaSO_4) = 0.0155$ mol

$m(BaSO_4) = 0.0155 \times 233.4$
$= 3.6199$ g

= 3.62 g (3 sig fig) *(3 marks)*

30 (a) Polymerisation of ethylene is addition polymerisation where a double bond opens out in the ethylene monomers, forming single bonds with neighbouring monomers until a very long chain of polyethylene forms. There are two processes used to form two different types of polyethylene. One uses an organic peroxide initiator to form a highly branched, low-density polymer (LDPE) and the other uses a catalyst to form a more crystalline, high-density form (HDPE).

$$n(C_2H_4) \rightarrow (C_2H_4)_n$$

However, the polymerisation of glucose to cellulose, or starch, involves condensation polymerisation. This involves the removal of water molecules from pairs of glucose monomers, allowing monomers to link together forming the long chain polymer.

$$n(C_6H_{12}O_6) \rightarrow (C_6H_{10}O_5)_n + (n - 1)H_2O$$

Both are single monomer polymers. *(3 marks)*

(Note: the wording of this question was ambiguous. The intent of the question was for students to describe the two different ways of polymerising ethylene and the way in which glucose monomers could be polymerised using condensation polymerisation.)

(b) *Low density polyethylene* (LDPE), a polymer formed from the monomer ethylene, is a long chain polymer with significant branching so that the chains cannot pack closely together and produce a low density product that is soft and flexible, owing to weak dispersion forces between chains, and transparent, owing to the limited scattering of light as it passes through the plastic. It is therefore used for making cling-wrap, plastic bags and electrical insulation.

High density polyethylene (HDPE), a polymer formed from ethylene, consists of closely packed chains with dispersion forces between chains holding them together tightly. This results in a more rigid and stronger plastic that consists of crystalline regions that refract light. The polymer is therefore translucent. It is used to make freezer bags, pipes, buckets and garbage bins as it is also water resistant.

Cellulose is a condensation polymer formed from the beta-glucose monomer and consists of long chains of unbranched glucose with every second one inverted. OH groups on one molecule form hydrogen bonds with OH on neighbouring chains, leaving few OH exposed and available to attract water molecules. Cellulose is therefore insoluble, fibrous and strong and has been used to make linen and cotton products such as clothing and paper. *(5 marks)*

31 (a) (i) $m(\text{dissolved solid}) = 44.67 - 43.53 = 1.14$ g

$$\%(\text{dissolved solid}) = 1.14 \times \frac{100}{250}$$

$$= 0.46\% \text{ w/v}$$

(2 marks)

(ii) Chloride is precipitated by silver ions. Add a few drops of 1.0 mol L^{-1} $AgNO_3$. A white solid that darkens in sunlight is a positive test for chloride.

$$Ag^+(aq) + Cl^-(aq) \rightarrow AgCl(s)$$ *(2 marks)*

(b) Phosphate commonly pollutes waterways. It is derived from detergents, where it is added as phosphate builders, and finds its way into waterways in household effluent. Phosphates are also present in run-off from fertiliser spread over farmers' fields. As phosphate is a nutrient, it causes the algae in waterways to rapidly increase in numbers, causing an algal bloom. The algae cut off sunlight to the waterway and use up available dissolved oxygen, leading to their demise. Bacterial decomposer numbers increase as they thrive on dead plant matter and these release toxins, causing the death of the waterway in this process called eutrophication. *(2 marks)*

Section II—Options

Question 32—Industrial Chemistry

(a) The cell is a mercury cell and it was used to produce sodium hydroxide hydrogen, and chlorine from saturated sodium chloride (brine) (eq. 1). Brine is introduced into the electrolysis cell, and chloride ions are oxidised at the anode (eq. 2) to produce chlorine gas. Sodium ions are reduced at the cathode (eq. 3), which is made of flowing mercury. The resulting sodium metal is dissolved in the cathode, to produce an amalgam, which is subsequently sprayed into water. The sodium reacts with the water to produce sodium hydroxide and hydrogen gas (eq. 4). The mercury is pumped back (recycled) into the electrolysis cell.

Equation 1: $2NaCl(aq) + 2H_2O(l) \rightarrow 2NaOH(aq) + Cl_2(g) + H_2(g)$

Equation 2: $2Cl^- \rightarrow Cl_2 + 2e^-$

Equation 3: $Na^+ + e^- \rightarrow Na(s)$

Equation 4: $2Na(s) + 2H_2O(l) \rightarrow 2NaOH(aq) + H_2(g)$ *(3 marks)*

(b) The electrolysis of molten sodium chloride in the Downs cell produces only sodium and chlorine. Sodium ions are reduced at the cathode: $Na^+ + e^- \rightarrow Na$. Chloride ions are oxidised at the anode: $2Cl^- \rightarrow Cl_2 + 2e^-$. The overall reaction is:

$$2NaCl(l) \rightarrow 2Na(s) + Cl_2(g).$$

The products of the electrolysis of aqueous sodium chloride depend on the concentration of the solution. In $NaCl(aq)$ the following reactions are, in principle, possible and have the associated E° values shown:

Oxidations (at the anode): $2Cl^- \rightarrow Cl_2(g) + 2e^-$ $E^o = -1.36\text{ V}$

$2H_2O(l) \rightarrow 4H^+ + O_2(g) + 4e^-$ $E^o = -1.23\text{ V}$

Reductions (at the cathode): $Na^+ + e^- \rightarrow Na$ $E^o = -2.71\text{ V}$

$2H_2O(l) + 2e^- \rightarrow H_2(g) + 2OH^-$ $E^o = -0.83\text{ V}$

In dilute solution, as expected from the E° values, water is electrolysed, producing oxygen gas at the anode and hydrogen gas at the cathode. The overall reaction is:

$$2H_2O(l) \rightarrow 2H_2(g) + O_2(g).$$

In more concentrated solutions, chlorine gas is preferentially produced at the anode. The overall reaction in this case is:

$$2NaCl(aq) + 2H_2O(l) \rightarrow 2NaOH(aq) + Cl_2(g) + H_2(g).$$ *(5 marks)*

(c) (i) The equilibrium reaction occurring is $2SO_2(g) + O_2(g) \rightleftharpoons 2SO_3(g)$.

The equilibrium constant expression is $K = \frac{[SO_3]^2}{[O_2][SO_2]^2}$

At time = *A*:

$n(O_2) = 0.40 - 0.15 = 0.25$ mol, and $[O_2] = \frac{0.25}{10} = 0.025$ mol/L

$n(SO_2) = 0.80 - 0.30 = 0.50$ mol, and $[SO_2] = \frac{0.50}{10} = 0.05$ mol/L

$n(SO_3) = 0.30$, and $[SO_3] = \frac{0.30}{10} = 0.03$ mol/L

$K = \frac{(0.03)^2}{(0.05)^2(0.025)} = 14.4$ *(3 marks)*

(ii) In the production of $SO_3(g)$ from $SO_2(g)$ and $O_2(g)$, the forward reaction is exothermic. Shortly after time A the temperature of the system was decreased, resulting in a shift to the right of the equilibrium position, as shown in the graph. *(2 marks)*

(Note: this question is not a random equilibrium calculation question. It relates to a key step in the production of sulfuric acid, and students are expected to know that the production of SO_3 from SO_2 and O_2 is an equilibrium reaction, and that the forward reaction is exothermic.)

(d) (i) This type of reaction is known as saponification—the production of soap by the hydrolysis of a triglyceride. The reactant *A* is KOH (potassium hydroxide). *(2 marks)*

(ii) This reaction is carried out in a school laboratory as follows:

- Place 20 mL of cooking oil in a 250 mL beaker (cover with a clock glass).
- Add 20 mL of concentrated NaOH(*aq*) to the beaker.
- Heat the mixture on an electric hotplate for 10–15 minutes.
- Allow the reaction mixture to cool.
- Add 20 mL of cooled, concentrated NaCl(*aq*).
- Obtain the solid soap product by filtration and wash it three times with ice-cold water.
- Allow the soap to dry in the filter paper.

Safety precautions that should be followed are:

- Concentrated NaOH is highly corrosive. Avoid skin contact by wearing gloves. Avoid eye contact by wearing safety glasses.
- Concentrated NaCl can be irritating to the eyes and skin. The same precautions apply as for NaOH.
- Using an electric hotplate can result in burns or fire. Prevent these by ensuring safe and careful handing, and by removing flammable objects from the vicinity of the experiment. Allow hot materials to cool before handling. *(3 marks)*

(e) The Solvay Process is the method of production of Na_2CO_3 from calcium carbonate and brine. The overall reaction for the process is:

$$CaCO_3(s) + 2NaCl(aq) \rightarrow Na_2CO_3(s) + CaCl_2(s)$$

This reaction is the reverse of the reaction that would spontaneously occur if solutions of sodium carbonate and calcium chloride were mixed. The use of solid calcium carbonate is very important in ensuring that the overall process proceeds.

The solid calcium carbonate is heated strongly to produce $CO_2(g)$ and $CaO(s)$:

$$CaCO_3(s) \rightarrow CaO(s) + CO_2(g).$$

Simultaneously in the Solvay plant, ammonia is bubbled into brine to produce ammoniacal brine. When the $CO_2(g)$ is bubbled into this solution the following ions are present: HCO_3^-, NH_4^+, Na^+ and Cl^-. When this mixture is cooled, $NaHCO_3$ precipitates, as it is the least soluble compound. $NaHCO_3$ is thus able to be separated by the simple and inexpensive process of filtration and, when heated, it produces $Na_2CO_3(s)$ of high purity:

$$2NaHCO_3(s) \rightarrow Na_2CO_3(s) + CO_2(g).$$

The CO_2 is recycled back into the ammoniacal brine solution.

The generation of CO_2 from $CaCO_3$ also produces CaO. This is used to regenerate the ammonia, so that it too can be recycled, to make the ammoniacal brine. The CaO is reacted with the NH_4Cl solution to make $CaCl_2$, H_2O and NH_3. The $CaCl_2$ is a waste product, some of which is sold but most of which is disposed of.

Thus, in this overall process, the use of solid $CaCO_3$ is very important for a number of reasons, including its environmental impact:

- The $CaCO_3$ produces CO_2, which generates the hydrogen carbonate ion. It is this ion that allows Na_2CO_3 to be produced after a very simple and inexpensive separation process, followed by heating.
- The $CaCO_3$ also produces CaO, which is used to regenerate NH_3. Rather than needing to dispose of an ammonium salt, the NH_3 is recycled. This reduces the environmental impact of the process and also saves money for plant operators, who do not have to continuously buy ammonia.
- $CaCO_3$ is an abundant and inexpensive solid and is also easily transported. This reduces the operating costs of a Solvay plant.
- The waste product resulting from $CaCO_3$ use is $CaCl_2$. This has minimal environmental impact (it can be disposed of in the ocean, where it has negligible impact on overall ion concentration), or it can be buried. Furthermore, some of it can be sold to de-ice roads, for example.

On the other hand:

- A large amount of energy is required to heat $CaCO_3^-$ to convert it to CO_2 and CaO. This heat energy is produced by fossil fuels and this makes CO_2, which is released into the atmosphere, contributing to the enhanced greenhouse effect.

- If the $CaCl_2$ were released into rivers or lakes, where there would be relatively little dilution effect, the concentration of dissolved ions would be too high to support life in those environments.
- Mining $CaCO_3$ may lead to habitat loss and also requires energy, the production of which contributes to the enhanced greenhouse effect.

Overall, while the process is not without environmental impacts, the benefits of production outweigh the costs, and the use of $CaCO_3$ is very important in this process. *(7 marks)*

Question 33—Shipwrecks, Corrosion and Conservation

(a) The timber would likely be riddled with marine boring worms and permeated by salt water, softening the timber. The steel collars would be highly corroded. The iron would have been oxidised to iron(III) oxide, causing the original lattice to be replaced with porous rust coating. The timber and iron would be encrusted with shellfish, corals and other calcareous organisms. *(3 marks)*

(b) (i)

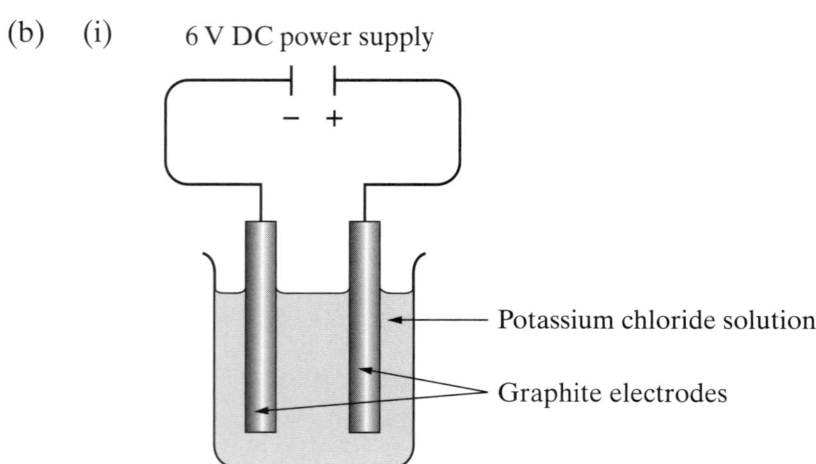

At the anode: $H_2O(l) \rightarrow \frac{1}{2}O_2(g) + 2H^+(aq) + 2e^-$

At the cathode: $H_2O(l) + e^- \rightarrow \frac{1}{2}H_2(g) + OH^-(aq)$

Overall equation for reaction: $H_2O(l) \rightarrow \frac{1}{2}O_2(g) + H_2(g)$ *(4 marks)*

(ii) The cathode is the place where reduction occurs. Hydrogen gas bubbles are produced at the cathode. It is the negative electrode. *(1 mark)*

(c) Steel 1 contains a high percentage of iron with a very small percentage of carbon, giving a soft and malleable steel that is ductile and can be used to make car bodies, ships' hulls, nails and wire.

Steel 2 contains less iron and a high enough percentage of carbon to give a harder and less ductile steel that corrodes readily. It is therefore used to make train tracks, girders and beams.

Steel 3 contains 4% carbon and as such would be very brittle. This mixture is known as pig iron and is too brittle to be used for anything. It is converted into more useful steels with lower carbon contents.

Steel 4 is stainless steel. Owing to its 15% chromium content it has high corrosion resistance and owing to its 10% nickel it is strong and hard. It is therefore used to make surgical instruments, cutlery and kitchen fixtures such as taps and sinks. *(5 marks)*

(d) (i) The three factors being tested are pH, temperature and oxygen concentration.

- pH: three test tubes were set up, each half full of water. The first had a pH of 7, the second a pH of 5 and the third a pH of 3.
- Temperature: three test tubes were set up, each half full of water. The first was half filled with water at 4°C and left in the fridge, the second half filled with water at 20°C and kept on a lab bench and the third half filled with water at 35°C and kept in an incubator.
- Oxygen: three test tubes were set up, each full of water. The first was filled with highly aerated water from a water tap, the second with still water that had been left to settle and the third with water that had been boiled to remove oxygen and left to cool. All were sealed with 1 mm of oil so that oxygen could not enter or leave the water.

A small piece of iron was placed in each tube so that it was half in and half out of the solution, or completely submerged in the oxygen tests. The volume of water, surface area of metal and other factors that could vary were held the same for all tubes. Every day for seven days observations were made of evidence of corrosion on the metal, such as a rust-coloured precipitate, and precipitate in the solution. The result was recorded on a 1 to 5 scale, 1 being very little evidence and 5 being extensive evidence of corrosion and a large amount of rusty precipitate. *(4 marks)*

(ii) As oxygen is the oxidant in the corrosion of iron in a marine environment, one way to reduce its effect is to create a physical barrier between the iron and the oxygen. For ships, this physical barrier can take the form of a polymer-based paint that is impervious to oxygen and water. *(1 mark)*

(e) *Wooden artefacts* are first cleaned of silt and mud by washing in cold water. The artefact is then soaked in distilled water so that salts leach out of the cells. The water is replaced periodically so that the concentration gradient remains high and salts continue to diffuse out. This is a suitable technique as it does no damage to the artefact and when it is eventually removed from water, there will not be salt crystallising within the cells and causing the wood to lose its structure.

The artefact is then soaked in a polyethylene glycol (PEG) solution in gradually increasing concentrations. This is a suitable technique as the PEG diffuses into the wood cells and replaces the water with a stronger material that adds stability to the wood.

The artefact is then dried and coated on the outside with PEG or a wood lacquer. This is a suitable technique as it protects the surface from further damage and prevents organisms attacking the wood at a later stage.

Copper artefacts corrode much less than iron and do not become encrusted with concretions from organisms as copper is poisonous to living things. When copper is oxidised it forms copper(I) chloride and copper(II) hydroxychloride compounds on its surface. Copper(II) sulphide may also form at depth. These surface deposits can be dissolved by soaking in dilute weak acids such as 5% citric acid containing 1% thiourea as a corrosion inhibitor for days to weeks. This is a suitable technique as it removes the surface deposit without affecting the underlying copper.

The artefact is then soaked in a sodium hydrogen carbonate solution to initiate the leaching of chloride ions from small cracks in the structure. Electrolysis is then used to drive out any remaining chlorides and to reverse the corrosion process. Cu^{+} ions will be reduced to solid copper.

$$CuCl(s) + e^{-} \rightarrow Cu(s) + Cl^{-}(aq)$$

Electrolysis is a suitable technique as it returns some of the lost copper back to its elemental form without any damage to the existing copper, as well as driving chlorides out of microscopic cracks in the artefact. If this did not occur, the salt crystals could grow in these cracks as the artefact is dried and cause structural damage.

The artefact is finally dried and coated in microcrystalline wax or a clear lacquer. This is a suitable technique as it forms a physical barrier between the restored copper and oxidising agents that may cause further corrosion. *(7 marks)*

Question 36—Forensic Chemistry

(a) Polysaccharide *A* is amylose, which is a polymer of several hundred to a few thousand alpha glucose rings all joined the same way by alpha(1-4) glycosidic bonds. The resulting molecule is unbranched and wound into a helix. The origin of amylose is in the starch granules stored in the cytoplasm of plant cells.

Polysaccharide *B* is glycogen, which is a highly branched chain structure. The branch points in glycogen involve the first atom on one glucose molecule and the sixth carbon atom in a glucose molecule in the backbone. The origin of glycogen is in the liver and muscles of animals. *(3 marks)*

(b) (i) Electrophoresis is a method of separating charged substances based upon the different sign of their charges and their size (molecular mass). Electrophoresis uses an electrolyte containing a suitable buffer and when a voltage is applied a sample of amino acids can be separated on a porous paper or agar gel. A smaller ion will move more rapidly than a larger ion and an ion that has a higher charge will move more rapidly than an ion that has a lower charge. The movement of the amino acids depends on their structure.

At low pHs amino acids have a positive charge and migrate towards the negative electrode, while at high pHs amino acids have a negative charge and move towards the positive electrode. Known standard amino acids can be run through a buffered electrolyte on an agar gel when voltage is applied. A mixture of amino acid is also run on an agar gel and the distance of each component of the amino acid mixture can be matched against the standard amino acids to identify the amino acids in the mixture. *(2 marks)*

(ii) At pH 8 amino acids glutamic acid and valine have a negative charge. Since the glutamic acid has a greater charge than valine it will move faster and further up towards the positive electrode on the electrophoresis paper, and since lysine has a positive charge it will move in an opposite direction to glutamic acid and valine on the electrophoresis paper, down towards the negative electrode. Hence at pH 8 X is lysine, Y is valine and Z is glutamic acid.

The electrophoresis paper on the left (pH 6) has a stationary amino acid Y, which is valine since it is neutral overall (isoelectric point). At pH 6 the other two amino acids are oppositely charged and will move in opposite directions. Since Z moves up towards the positive electrode, it is glutamic acid since it has an overall negative charge and X is lysine since it moves down towards the negative electrode due to its overall positive charge. *(3 marks)*

(c) (i) Manufacturer B *(1 mark)*

(ii) Electrons can exist only in discrete quantised energy states. Each time an electron drops to a lower state, a quantum of energy is released in the form of a photon of light.

Emission spectra can be used to identify elements present in the pottery. A minute piece of the pottery sample is ground to a powder and heated to vaporise it and excite the electrons, in either a flame or a carbon arc (electric discharge across a pair of carbon electrodes). Emissions from the excited electrons move back to lower energy states in the sample and emit light of a characteristic frequency, which is then concentrated with a lens and passed through a monochromator such as a prism, which separates the emitted radiation into its different wavelengths. The various wavelengths are detected by a photomultiplier. A computer is then used to record the emission spectrum of the pottery sample, producing coloured lines on a black background.

The elements in the pottery can be identified by emission spectroscopy and used to identify the metal atoms present in order to work out the manufacturer. The emission spectra of many elements can be produced, in this case calcium, chromium, copper, iron and mercury, and the wavelength of the lines of the emission spectrum of the pottery can be matched up with the position of the lines of each of the emission spectra of the different elements to identify which elements are present in the pottery. *(4 marks)*

(d) (i) Four properties of a soil a forensic chemist would investigate are:

- particle size—some particles are larger than others
- colour—soils differ in colour from red to brown to black
- texture—depends on the amount of sand, clay and humus
- pH of the soil. *(2 marks)*

(ii) An organic test that could be performed on the soil from the pair of shoes and from the crime scene is to determine the humus content of the soil by ignition loss, by heating oven-dried soil of the same mass at about 700°C for 30 minutes. The humus is burnt off and the composition by mass is determined. If the humus compositions of the soils are identical, then the soil from the shoes will match the soil from the crime scene.

An inorganic test that could be performed on the soil from the pair of shoes and from the crime scene is a pH test. The same amount of soil from the shoes and crime scene is added to each of two beakers containing the same volume of water, stirred and then filtered. To each filtrate a few drops of universal indicator are added and the change in colour is matched to a pH chart to give a pH reading, or a pH meter can be used to record the pH of the filtrates. If the pHs of the soils are identical, then the soil from the shoes will match the soil from the crime scene. *(3 marks)*

(e) The non-coding sequences along the DNA strand, called introns, vary significantly from one person to another and it is these regions of the DNA molecules that are used for DNA analysis in forensic analysis. If people are related, then their introns show some similarity. Fifty per cent of a person's DNA comes from each parent.

One technique used is the polymerase chain reaction (PCR). In this the double stranded DNA is separated into single strands by incubation. Short pieces of purified DNA, called primers, are then added. When the temperature is lowered, copies of primers bond to the DNA. A heat stable DNA polymerase is added and then the sample is warmed, causing the primers to synthesise complementary strands of each of the single strands. This process is repeated many times to amplify the original DNA sequence. The amplified DNA is then digested by adding restriction enzymes. These cut the DNA into a series of fragments of various sizes.

The DNA fragments are separated according to chain length (i.e. weight) using gel electrophoresis. Each sample is placed in a separate well of a polyacrylamide-agrose gel. The longer the fragment, the more slowly it moves through the gel. A standard sample containing polynucleotide samples of known molecular weight are also run at the same time in other wells and the number of base pairs in the synthesised fragments is obtained by comparing the positions of the DNA fragments in the sample with those in the standard. After electrophoresis the DNA is transferred to a nylon membrane. Radioactive probes are used so that the DNA sequences to which they become attached can be tracked. An X-ray is placed next to the membrane and is later developed and the bands become visible. This barcode pattern is the DNA fingerprint. The final step is the band pattern comparison.

Genetic differences between individuals will be identified by differences in the location and distribution of the band patterns.

The application of this technique in forensic analysis is to identify individuals and relationships between people because DNA in itself is unique to each individual. Relationships can be identified since portions of DNA from both parents are present in all individuals.

Other uses include:

- In criminal cases DNA helps to convict criminals by identifying the person who produced a biological sample at a crime scene.
- Medical scientists use DNA analysis to determine the structure of genes, locating changes to genes that lead to genetic disorders.
- DNA analysis is used in archaeology in dating as DNA is destroyed as organisms become fossilised. In some cases it can be extracted from buried remains and analysed.

(7 marks)

CHAPTER 9

BOARD OF STUDIES
NEW SOUTH WALES

2011

HIGHER SCHOOL CERTIFICATE EXAMINATION

Chemistry

General Instructions

- Reading time – 5 minutes
- Working time – 3 hours
- Write using black or blue pen
  Black pen is preferred
- Draw diagrams using pencil
- Board-approved calculators may be used
- A data sheet and a Periodic Table are provided at the back of this paper
- Write your Centre Number and Student Number where required

Total marks – 100

Section I

75 marks

This section has two parts, Part A and Part B

Part A – 20 marks

- Attempt Questions 1–20
- Allow about 35 minutes for this part

Part B – 55 marks

- Attempt Questions 21–32
- Allow about 1 hour and 40 minutes for this part

Section II

25 marks

- Attempt ONE question from Questions 33–37
- Allow about 45 minutes for this section

Section I
75 marks

Part A – 20 marks
Attempt Questions 1–20
Allow about 35 minutes for this part

Use the multiple-choice answer sheet for Questions 1–20.

---

1 Which of the following industrial processes is used to produce ethanol from ethylene?

(A) Hydration
(B) Dehydration
(C) Addition polymerisation
(D) Condensation polymerisation

2 Which of the following shows two products that result from the fermentation of glucose?

(A) Cellulose and water
(B) Ethanol and oxygen
(C) Carbon dioxide and water
(D) Ethanol and carbon dioxide

3 Which of the following lists contains ONLY basic substances?

(A) Oven cleaner, urine, vinegar
(B) Lemonade, drain cleaner, blood
(C) Baking soda, ammonia, sea water
(D) Antacid, dishwashing detergent, lemon juice

4 Which of the following gases can cause major depletion of the ozone layer?

(A) O_2
(B) NO_2
(C) CO_2
(D) CCl_3F

5 Iron (III) chloride and aluminium sulphate are two chemicals that can be used in the purification of town water supplies. What is the role of these chemicals?

(A) To disinfect water by removing bacteria

(B) To remove particulate material by flocculation

(C) To control the concentration of total dissolved solids

(D) To control the pH of the water within the required range

6 Which property would be most useful in distinguishing between butan-1-ol and propan-1-ol?

(A) Boiling point

(B) Colour

(C) Conductivity

(D) Density

7 Which of the following lists contains ONLY unstable isotopes?

(A) $^{207}_{82}Pb$, $^{99}_{43}Tc$, $^{12}_{7}N$

(B) $^{214}_{82}Pb$, $^{46}_{20}Ca$, $^{99}_{43}Tc$

(C) $^{238}_{92}U$, $^{40}_{20}Ca$, $^{12}_{7}N$

(D) $^{238}_{92}U$, $^{40}_{20}Ca$, $^{99}_{43}Tc$

8 What is the systematic name of the molecule shown?

$$H_3C—CH_2—C(=O)—O—CH_2—CH_2—CH_2—CH_3$$

(A) Butyl butanoate

(B) Propyl butanoate

(C) Butyl propanoate

(D) Propyl propanoate

9 What property of O_3 makes it more soluble in water than O_2 in water?

(A) O_3 is a polar molecule.

(B) O_3 has a resonance structure.

(C) O_3 is a highly reactive molecule.

(D) O_3 has a coordinate covalent bond.

10 An aqueous sample containing the following anions is analysed.

Cl^- CO_3^{2-} SO_4^{2-}

In which order should the reagents be added to determine the amount of chloride in the sample?

	Reagent 1	*Reagent 2*	*Reagent 3*
(A)	$AgNO_3$	H_2SO_4	$BaSO_4$
(B)	HCl	$Pb(NO_3)_2$	$AgNO_3$
(C)	HNO_3	$Ba(NO_3)_2$	$AgNO_3$
(D)	$Ba(NO_3)_2$	$AgNO_3$	CH_3COOH

11 Which compound can form when bromine water reacts with propene?

(A) 1-bromopropane

(B) 2-bromopropane

(C) 1,1-dibromopropane

(D) 1,2-dibromopropane

12 A table of redox couples and their standard reduction potentials is shown.

Redox couple	$E^{\ominus}$
Ag^+/Ag	0.80 V
Cd^{2+}/Cd	−0.40 V
Pd^{2+}/Pd	0.92 V
Ni^{2+}/Ni	−0.24 V

Which of the following ranks the metals in decreasing order of their electrochemical activity?

(A) Ni > Cd > Ag > Pd

(B) Pd > Ag > Cd > Ni

(C) Pd > Ag > Ni > Cd

(D) Cd > Ni > Ag > Pd

13 When chloride ions are added to a solution containing $Co(H_2O)_6^{2+}(aq)$, the following equilibrium is established.

$$\underset{\text{Pink}}{Co(H_2O)_6^{2+}(aq)} + 4Cl^-(aq) \rightleftharpoons \underset{\text{Blue}}{CoCl_4^{2-}(aq)} + 6H_2O(\ell)$$

Which of the following statements about the colour of the solution is true?

(A) Diluting the solution with water will make it turn blue.

(B) If the reaction is exothermic, heating the solution will make it turn blue.

(C) If the reaction is endothermic, cooling the solution will make it turn pink.

(D) Adding a large amount of solid potassium chloride to the solution will make it turn pink.

14 How many isomers are there for C_3H_6BrCl?

(A) 3

(B) 4

(C) 5

(D) 6

Use the information provided to answer Questions 15 and 16.

> Using 0.100 mol L^{-1} NaOH, a student titrated 25.0 mL of a 0.100 mol L^{-1} weak monoprotic acid, and separately titrated 25.0 mL of a 0.100 mol L^{-1} strong monoprotic acid.

15 Which statement about the volume of base required to reach the equivalence point is correct?

(A) The weak acid will require the same volume of base as the strong acid.

(B) The weak acid will require a larger volume of base than the strong acid.

(C) The weak acid will require a smaller volume of base than the strong acid.

(D) The volume of base required will depend on the molar mass of the acid used.

16 Which statement correctly describes the pH at each titration equivalence point?

(A) The pH of both solutions will be the same.

(B) One of the solutions will be neutral while the other will have a pH higher than 7.

(C) One of the solutions will be neutral while the other will have a pH lower than 7.

(D) One of the solutions will have a pH higher than 7 while the other will have a pH lower than 7.

17 The molar heat of combustion of pentan-1-ol is 2800 kJ mol^{-1}. A quantity of pentan-1-ol was combusted, generating 108 kJ of heat.

What mass of pentan-1-ol was combusted?

(A) 2.29 g

(B) 2.86 g

(C) 3.32 g

(D) 3.40 g

18 A household cleaning agent contains a weak base with the formula NaX. 1.00 g of this compound was dissolved in water to give 100.0 mL of solution. A 20.0 mL sample of the solution was titrated with 0.100 mol L^{-1} hydrochloric acid, and required 24.4 mL of the acid for neutralisation.

What is the molar mass of the weak base?

(A) 82.0 g

(B) 84.0 g

(C) 122 g

(D) 410 g

19 All of the carbon dioxide in a soft drink with an initial mass of 381.04 g was carefully extracted and collected as a gas. The final mass of the drink was 380.41 g.

What volume would the carbon dioxide occupy at 100 kPa and 25°C?

(A) 0.33 L

(B) 0.35 L

(C) 0.56 L

(D) 0.63 L

20 When charcoal reacts in the presence of oxygen, carbon monoxide and carbon dioxide are produced according to the following chemical reactions.

$$C(s) + \tfrac{1}{2}O_2(g) \rightarrow CO(g)$$

$$C(s) + O_2(g) \rightarrow CO_2(g)$$

What would be the total mass of gas produced when 400 g of charcoal is reacted, assuming equal amounts are consumed in each reaction?

(A) 0.93 kg

(B) 1.2 kg

(C) 1.5 kg

(D) 2.5 kg

2011 HIGHER SCHOOL CERTIFICATE EXAMINATION

Chemistry

Centre Number

Section I (continued)

Student Number

Part B – 55 marks
Attempt Questions 21–32
Allow about 1 hour and 40 minutes for this part

Answer the questions in the spaces provided. These spaces provide guidance for the expected length of response.

Show all relevant working in questions involving calculations.

Question 21 (4 marks)

What features of the molecular structure of ethanol account for its extensive use as a solvent? Include a diagram in your answer. **4**

....

....

....

....

....

....

....

....

....

....

Question 22 (4 marks)

(a) Use chemical equations to show how ozone is depleted in the stratosphere. **2**

....

....

....

....

(b) Outline ONE method used to monitor stratospheric ozone. **2**

....

....

....

....

2011 HIGHER SCHOOL CERTIFICATE EXAMINATION

Chemistry

Section I – Part B (continued)

Centre Number

Student Number

Question 23 (3 marks)

(a) Element 112 was first synthesised in 1996 and officially named in 2009 as copernicium, Cn. **1**

Explain why the transuranic isotope $^{278}_{112}Cn$ is unstable.

....

....

(b) Describe a method by which transuranic elements can be synthesised. **2**

....

....

....

....

Question 24 (7 marks)

A galvanic cell was constructed as shown in the diagram.

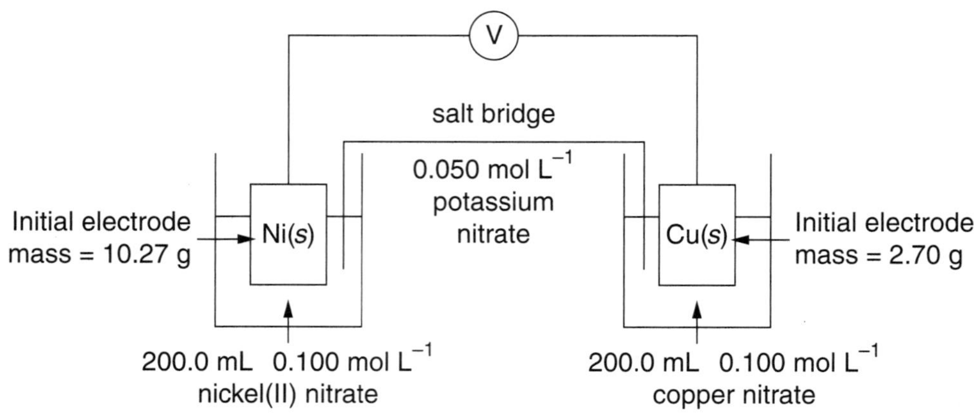

(a) Calculate the standard cell potential ($E^{\ominus}$). In your answer, include a net ionic equation for the overall cell reaction. **2**

....

....

....

....

(b) After a period of time, a solid deposit that had formed on the copper electrode was removed and dried. The deposit had a mass of 0.395 g. **3**

(i) Calculate the final mass of the nickel electrode.

....

....

....

....

....

(ii) Calculate the final concentration of the nickel(II) nitrate solution. **2**

....

....

....

....

2011 HIGHER SCHOOL CERTIFICATE EXAMINATION

Chemistry

Section I – Part B (continued)

Centre Number

Student Number

Question 25 (3 marks)

Explain the role of the conjugate acid/base pair, $H_2PO_4^-/HPO_4^{2-}$, in maintaining the pH of living cells. Include chemical equations in your answer. **3**

Question 26 (6 marks)

A manufacturer makes lemon cordial by mixing flavouring, sugar syrup and citric acid. The concentration of the citric acid is determined by titration with NaOH.

The sodium hydroxide solution is prepared by dissolving 4.000 g of NaOH pellets in water to give 1.000 L of solution. This solution is standardised by titrating 25.00 mL with a 0.1011 mol L^{-1} standardised solution of HCl. The average titration volume is found to be 24.10 mL.

To analyse the lemon cordial 50.00 mL of the cordial is diluted to 500.0 mL. Then 25.00 mL of the diluted solution is titrated with the NaOH solution to the phenolphthalein endpoint.

The following data were collected during one of the analysis runs of the lemon cordial.

Titration #1 volume	26.55 mL
Titration #2 volume	27.25 mL
Titration #3 volume	27.30 mL
Titration #4 volume	27.20 mL

(a) Why is the calculated concentration of the standardised NaOH solution different from the concentration calculated using the mass given, assuming no human error occurred? **2**

....

....

....

....

....

(b) Determine the concentration of citric acid in the lemon cordial. **4**

....

....

....

....

....

....

....

....

2011 HIGHER SCHOOL CERTIFICATE EXAMINATION

Chemistry

Centre Number

Section I – Part B (continued)

Student Number

Question 27 (5 marks)

The following extract was taken from a blog about environmental issues. **5**

> ... the use of long-lasting polymers for short-lived applications can cause problems for the preservation of living systems ... Plastic debris has a costly impact on waste management for municipalities.

Assess the uses of polystyrene and a named biopolymer in terms of their properties, with reference to the statements made in this blog.

2011 HIGHER SCHOOL CERTIFICATE EXAMINATION

Chemistry

Centre Number

Section I – Part B (continued)

Student Number

Question 28 (4 marks)

A student investigating the water quality of stormwater in a drain near the school collected samples for testing in the school laboratory. The student conducted the following tests to measure the quality of the stormwater. **4**

- Hardness
- Phosphate level
- Biochemical oxygen demand
- Total dissolved solids
- Turbidity
- Nitrate level

For TWO of these tests, outline the chemical or physical principle involved and the procedure followed in a school laboratory.

Question 29 (4 marks)

(a) Justify the continued use of the Arrhenius definition of acids and bases, despite the development of the more sophisticated Brönsted–Lowry definition. **3**

....

....

....

....

....

....

....

....

(b) Why does the neutralisation of any strong acid in an aqueous solution by any strong base always result in a heat of reaction of approximately –57 kJ mol^{-1}? **1**

....

....

Question 30 (6 marks)

The flowchart outlines the sequence of steps in the Ostwald process for the manufacture of nitric acid. 6

Step 1: $4NH_3(g) + 5O_2(g) \overset{900°C}{\rightleftharpoons} 4NO(g) + 6H_2O(g) \quad \Delta H = -950 \text{ kJ}$

↓

Step 2: $2NO(g) + O_2(g) \rightleftharpoons 2NO_2(g) \quad \Delta H = -114 \text{ kJ}$

↓

Step 3: $3NO_2(g) + H_2O(\ell) \rightarrow 2HNO_3(aq) + NO(g) \quad \Delta H = -117 \text{ kJ}$

Explain the reaction conditions required at each step of the Ostwald process to maximise the yield and production rate of nitric acid.

....

....

....

....

....

....

....

....

....

....

....

....

....

....

....

....

2011 HIGHER SCHOOL CERTIFICATE EXAMINATION

Chemistry

Section I – Part B (continued)

Centre Number

Student Number

Question 31 (4 marks)

A council monitors the water quality of a local river daily. The records for six days are shown in the table. During this period contamination from unknown sources was detected. **4**

	Mon	*Tue*	*Wed*	*Thu*	*Fri*	*Sat*
pH	7.2	7.3	7.2	7.0	7.2	7.3
Turbidity (NTU)	5	67	5	25	10	8
Total dissolved solids (ppm)	340	436	342	380	370	360
Temperature (°C)	15.6	16.0	15.8	15.8	16.0	16.2
Dissolved oxygen (ppm)	9.7	9.6	9.7	4.2	4.3	8.0
Biochemical oxygen demand (ppm)	2.6	4.0	3.2	52	20	7.8
Faecal coliforms (cfu/100 mL)	10	10	12	514	60	18

Propose possible sources of the contamination, justifying your answer with reference to the data provided.

....

....

....

....

....

....

....

....

Question 32 (5 marks)

To determine the pH of garden soil, a sample was first saturated with distilled water in a petri dish. Barium sulfate powder was added to the surface of the sample, and drops of the three indicators listed below were added to separate parts of the sample. The colours observed are shown in the table.

Experimental results

Indicator	Methyl yellow	Methyl red	Phenolphthalein
Colour observed	Yellow	Red	Colourless

Indicator colour ranges

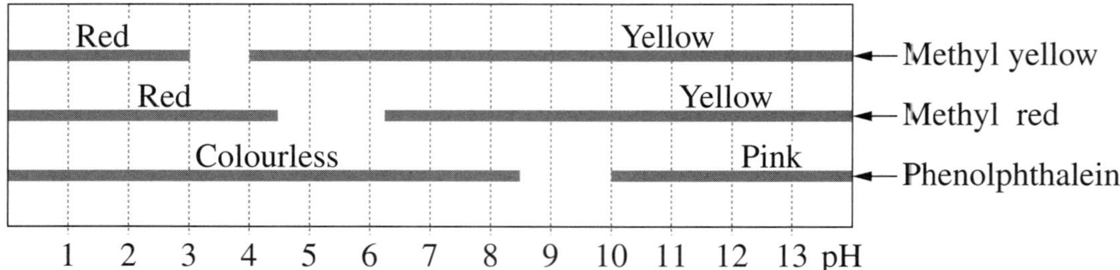

Plant response

Plant	*soil pH range for optimal growth*
Carrot	5.5 – 6.8
Chrysanthemum	6.0 – 6.3
Hydrangea Blue	4.0 – 5.0
Hydrangea White	6.5 – 8.0
Potato	5.0 – 5.7

(a) Why is barium sulfate powder added when testing soil pH? **1**

....

....

(b) Using the information given, select the plant that will grow well at the current soil pH, and justify your selection. **2**

....

....

....

....

(c) Outline the method you would use to test a natural indicator that has been prepared in the school laboratory. **2**

....

....

....

2011 HIGHER SCHOOL CERTIFICATE EXAMINATION

Chemistry

Section II

25 marks
Attempt ONE question from Questions 33–37
Allow about 45 minutes for this section

Answer parts (a)–(c) of the question in Section II Answer Booklet 1.
Answer parts (d)–(e) of the question in Section II Answer Booklet 2.
Extra writing booklets are available.

Show all relevant working in questions involving calculations.

Question 33	Industrial Chemistry
Question 34	Shipwrecks, Corrosion and Conservation
Question 35	The Biochemistry of Movement *(Not included in this reproduction)*
Question 36	The Chemistry of Art *(Not included in this reproduction)*
Question 37	Forensic Chemistry

Question 33 — Industrial Chemistry (25 marks)

Answer parts (a)–(c) in Section II Answer Booklet 1.

(a) A student set up the following experiment to model a process and observed the colour change of the crystals. **3**

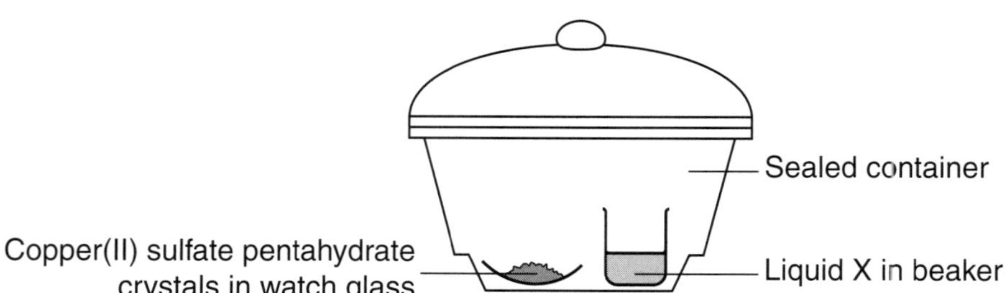

Identify liquid X and explain the colour change of the crystals. Include a chemical equation in your answer.

(b) (i) Why are the large volumes of $CO_2(g)$ produced during the Solvay process of little environmental concern? Include chemical equations in your answer. **3**

(ii) Calcium chloride is also produced during the Solvay process. Calculate the mass of calcium chloride produced per tonne of sodium chloride used in the Solvay process. **2**

(c) A 0.05 mol L^{-1} solution of sodium chloride was electrolysed using graphite electrodes. Separate pieces of litmus paper were dipped into the solution next to each electrode.

The following observations were made.

Polarity of electrode	*Observation*	*Colour of litmus paper*
Positive	Bubbles	Red
Negative	Bubbles	Blue

(i) Account for the observations at the anode and cathode. Include relevant chemical equations in your answer. **3**

(ii) What is the difference between the electrolytic cell described above and a galvanic cell, in terms of energy requirements? **2**

Question 33 continues

Question 33 (continued)

Answer parts (d)–(e) in Section II Answer Booklet 2.

(d) (i) Models are often used to help explain complex concepts. Outline a first-hand investigation that can model an equilibrium reaction. **2**

(ii) Assess the validity of the information that could be collected in this investigation. **3**

(e) Evaluate the impact on society of the environmental issues associated with THREE of the industrial processes that you have studied in this option. **7**

End of Question 33

Question 34 — Shipwrecks, Corrosion and Conservation (25 marks)

Answer parts (a)–(c) in Section II Answer Booklet 1.

(a) The diagram shows a hot water tank. **3**

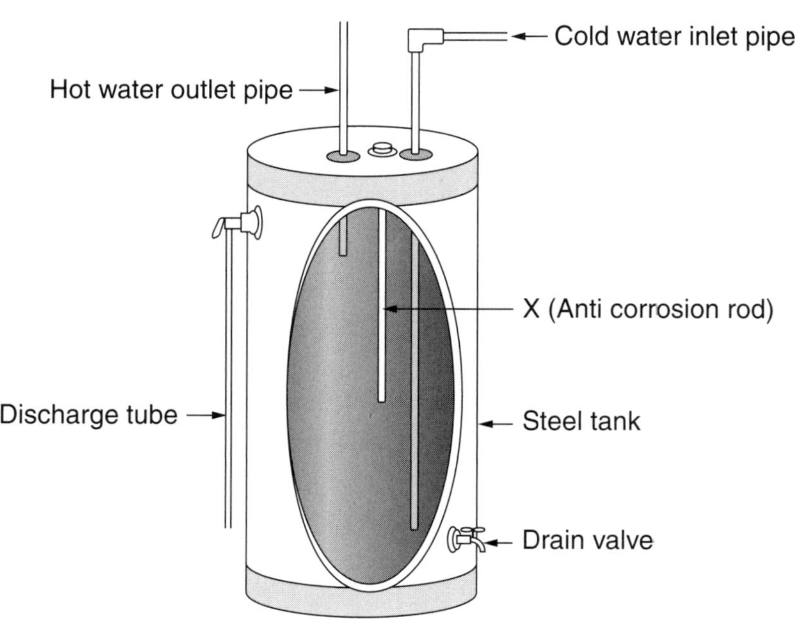

Name an appropriate material from which X can be made, and justify your choice.

(b) There are two common household methods to remove the tarnish on silver cutlery: polishing with a soft cloth and toothpaste or placing the cutlery into an aluminium tray filled with warm sodium hydrogen carbonate solution and leaving it overnight.

(i) Explain, with the use of equations, the chemistry involved in the aluminium tray method. **4**

(ii) Identify an advantage of the aluminium tray method. **1**

Question 34 continues

Question 34 (continued)

(c) A dilute solution of potassium sulfate was electrolysed using graphite electrodes. Separate pieces of litmus paper were dipped into the solution next to each electrode.

The following observations were made.

Polarity of electrode	*Observation*	*Colour of litmus paper*
Positive	Bubbles	Red
Negative	Bubbles	Blue

(i) Draw and label a diagram to represent this cell. **3**

(ii) Identify the products formed at the anode and cathode by writing the equations for each of these reactions. **2**

Answer parts (d)–(e) in Section II Answer Booklet 2.

(d) A student set up the following investigation to compare the rate of corrosion of a variety of metals and alloys, in order to identify those best suited for use in marine vessels. Photos were taken of the experiment on a daily basis for several weeks.

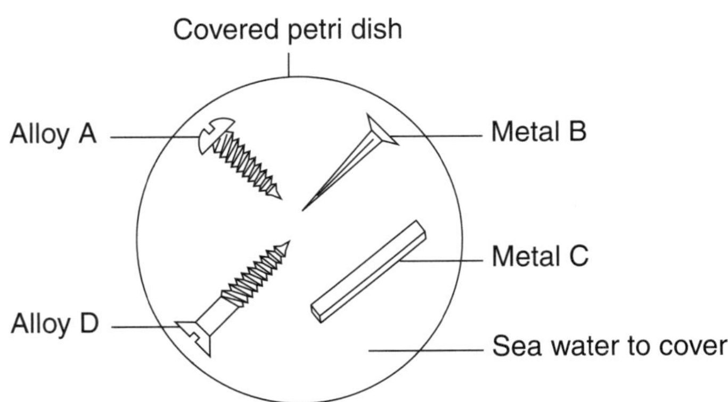

(i) Explain the method the student should have used to prepare the samples for the investigation. **2**

(ii) Assess the validity of the data collected in this experiment. **3**

(e) 'Corrosion at great depth in the ocean was predicted to be slow.' **7**

Evaluate this statement with reference to recent maritime investigations. Include chemical equations in your answer.

End of Question 34

Question 37 — Forensic Chemistry (25 marks)

Answer parts (a)–(c) in Section II Answer Booklet 1.

(a) Identify the structure shown, and justify your answer. **3**

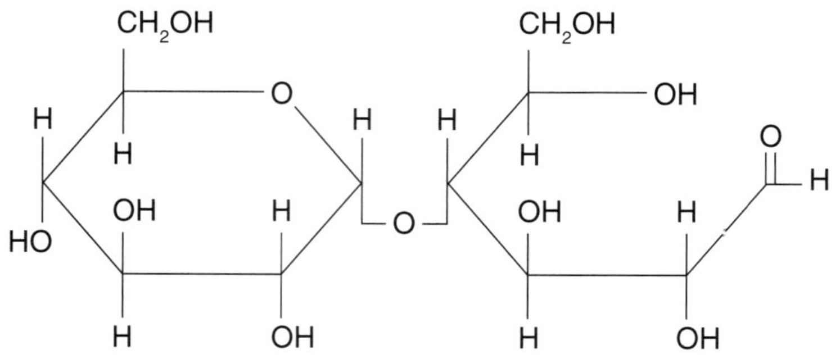

(b) (i) Using the general formula of an amino acid, write an equation to show the formation of a dipeptide. **2**

(ii) Explain the principles of paper chromatography, with reference to the separation of amino acids. **3**

Question 37 continues

Question 37 (continued)

(c) (i) Describe the technique by which DNA is analysed and then used to identify relationships between people. **4**

(ii) The diagram shows the DNA fingerprints (W, X, Y and Z) from a child, the child's biological parents and a family friend. **1**

Identify the DNA fingerprint of the child.

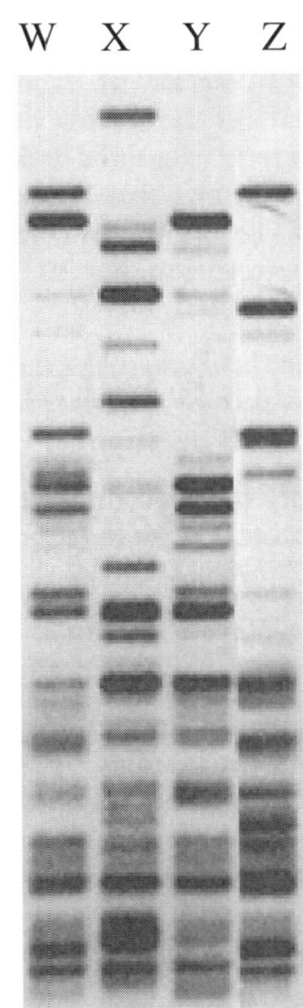

Answer parts (d)–(e) in Section II Answer Booklet 2.

(d) A student has four unlabelled bottles all containing white crystals. The four substances are known to be sucrose, potassium chloride, glycine (amino acid) and acetylsalicylic acid (aspirin).

(i) Explain a method the student could use in the school laboratory to identify the contents of the four unlabelled bottles. **4**

(ii) Identify ONE destructive and ONE non-destructive test the student may have used. **1**

(e) Select THREE instrumental techniques you have studied in this option. Evaluate the optimal use of each technique with reference to the analysis of forensic samples. **7**

End of paper

2011 HSC Examination Paper

Sample Answers

Section I Part A

(Total 20 marks)

1 A Hydration involves the addition of water to the ethylene molecule, one hydrogen added to one carbon and an OH added to the other, forming ethanol.

2 D

3 C As vinegar contains acetic acid, lemonade carbonic acid and lemon juice citric acid, option C is the only one that cannot contain any acids.

4 D Ozone is being decomposed by CFC molecules. CCl_3F is the only CFC option.

5 B The iron(III) and aluminium ions attract suspended matter (particularly colloidal matter), forming larger clumps that settle, clearing the water. This process is known as flocculation.

6 A Butan-1-ol contains one more carbon in its chain than propan-1-ol. This gives it more points where dispersion forces can form. It has a higher boiling point because of the additional dispersion forces and this is an easy property to measure and would, therefore, be most useful in distinguishing these compounds.

7 B For heavy elements, a n:p ratio outside 1.5:1 is unstable and for light elements a ratio outside 1:1. All examples in option B are outside the stable ratios.

8 C

9 A Being polar, O_3 has a net dipole that forms an attraction to water's net dipole. O_2 is non-polar and does not form a dipole-dipole attraction with water and, therefore, has very low solubility in water.

10 C The nitric acid would react with the carbonate, producing carbon dioxide gas. The barium nitrate would then precipitate the sulfate. The sample could then be filtered and silver nitrate added to the filtrate to determine how much chloride is present, as the silver ions will precipitate the chloride.

11 D When bromine reacts with the double bond in propene, the double bond opens up, revealing an unbonded electron on both of the first two carbon atoms. Bromine atoms add across the double bond.

12 D The E° values in the table are reduction potentials for the reduction of the metal ions. The reverse of these potentials is the oxidation potentials of the metals. Cd would have the highest oxidation potential at +0.40 and Pd the lowest at −0.92 V.

13 C If the forward reaction is endothermic, cooling the solution will result in the system shifting in reverse to counteract the cooling and release heat. In the process a reverse shift will produce more of the pink $Co(H_2O)_6^{2+}$ ion.

14 C The four isomers are 1-bromo-1-chloropropane, 1-bromo-2-chloropropane, 2-bromo-1-chloropropane, 1-bromo-3-chloropropane and 2-bromo-2-chloropropane.

15 A The volume of base required to neutralise an acid is not dependent on the strength of the acid. The weak acid will require the same amount of base as its equilibrium shifts to counteract the loss of hydronium ions as they are neutralised.

16 B A strong acid/strong base titration will result in a neutral salt at the equivalence point. A weak acid/strong base titration will result in a basic salt as the conjugate of the weak acid is a strong base.

17 D 108 kJ of heat will be produced by combusting $\frac{108}{2800} = 0.0386$ mol of pentan-1-ol.
Mass $= 0.0386 \times 88.1 = 3.40$ g

18 A $n(\text{HCl}) = 0.100 \times 0.0244 = 0.00244$ mol

$n(\text{base}) = n(\text{acid}) = 0.00244$ mol

$m(\text{base reacting}) = 1.00 \times 20/100 = 0.2$ g

$M(\text{base}) = \frac{0.2}{0.00244} = 82.0$ g/mol

19 B $m(CO_2) = 381.04 - 380.41 = 0.63$ g

$n(CO_2) = \frac{0.63}{44.0} = 0.0143$ mol

$V(CO_2) = 0.0143 \times 24.79 = 0.35$ L

20 B $n(\text{C}) = \frac{400}{12.01} = 33.306$ mol of total charcoal burnt. Half of this carbon is burnt to form CO and the other half to form CO_2.

$n(\text{CO}) = \frac{33.306}{2} = 16.653$ mol

$n(CO_2) = 16.653$ mol

$m(\text{CO}) = 16.653 \times 28.01 = 466.5$ g

$m(CO_2) = 16.653 \times 44.01 = 732.9$ g

total mass of gases $= 466.5 + 732.9 = 1199.4$ g $= 1.2$ kg

Section I Part B

21 Ethanol is a polar molecule. The $-OH$ functional group is polar (due to high electronegativity of the O atom) and the $-C_2H_5$ tail is non-polar.

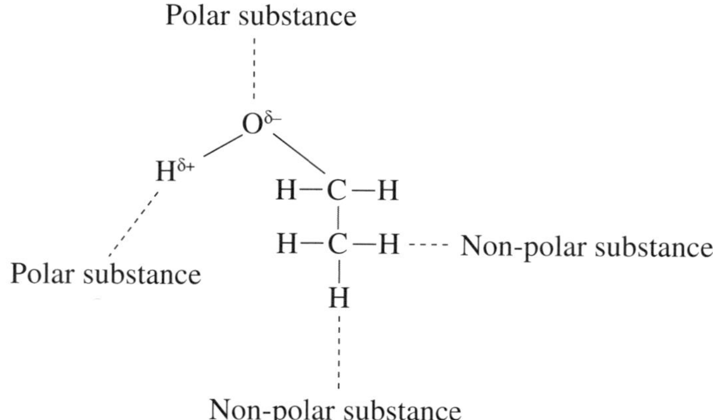

The polar head group interacts with polar solutes and assists their dissolution in the ethanol solvent via dipole-dipole attractive forces. The $-OH$ functional group can also form a hydrogen bond with functional groups such as $-NH_2$ and $-COOH$ present in solutes. The non-polar ethyl group tail can interact with other non-polar groups in a solute via dispersion forces.

Examples:

1. Solute = propanoic acid (CH_3CH_2COOH). Propanoic acid has a polar head group ($-COOH$) that can hydrogen bond with the polar head group of the ethanol solvent. The CH_3CH_2 tail of the propanoic acid can interact with the ethanol tail group via dispersion forces. Thus, this solute is very soluble in ethanol.

2. Solute = iodine (I_2). Iodine is a non-polar molecule. It is more soluble in ethanol than water as a result of dispersion interactions with the non-polar tail of the ethanol. An antiseptic medication called 'tincture' of iodine contains iodine dissolved in a water-ethanol mixed solvent. The water component reduces the iodine solubility a little but enhances the interaction of the solution with human tissue that has a high amount of water. *(4 marks)*

22 (a) CFC molecules decompose in the stratosphere, releasing reactive chlorine radicals. UV light breaks a $C-Cl$ bond:

$$CFCl_3(g) \rightarrow CFCl_2\bullet(g) + Cl\bullet(g)$$

Chlorine radicals that form react with ozone molecules, decomposing them and leading to ozone depletion:

$$O_3(g) + Cl\bullet(g) \rightarrow O_2(g) + ClO\bullet(g)$$

The chlorine radicals are recycled and a single CFC can lead to the decomposition of thousands of ozone molecules:

$ClO\bullet(g) + O\bullet(g) \rightarrow O_2(g) + Cl\bullet(g)$ *(2 marks)*

(b) Pulses of UV light are fired into the stratosphere from lasers on the ground. UV light spectrometers on telescopes measure how much of this light is absorbed, and from this ozone concentrations can be calculated. *(2 marks)*

23 (a) The ratio of neutrons to protons is far greater than the 1.5:1 stable ratio. *(1 mark)*

(b) A beam of ions can be accelerated to near the speed of light in a particle accelerator such as a synchrotron. It then strikes a target nucleus such as lead or another large nucleus and the nuclei combine to form a nucleus larger than uranium. *(2 marks)*

24 (a) $Ni(s) + Cu^{2+}(aq) \rightarrow Ni^{2+}(aq) + Cu(s)$ $E^{o} = 0.24 + 0.34 = 0.58\ V$ *(2 marks)*

(b) (i) $n(Cu) = \dfrac{0.395}{63.55} = 6.2155 \times 10^{-3}$ mol

$n(\text{Ni lost}) = n(\text{Cu gained}) = 6.2155 \times 10^{-3}$ mol

$m(\text{Ni lost}) = 6.2155 \times 10^{-3}\ \text{mol} \times 58.69 = 0.3648$ g

Final mass of Ni $= 10.27 - 0.3648 = 9.91$ g *(3 marks)*

(ii) $n(Ni^{2+})$ formed $= n(Ni)$ lost $= 6.2155 \times 10^{-3}$ mol

Initial $n(Ni^{2+})$ in solution $= 0.100 \times 0.200 = 0.0200$ mol

Final $n(Ni^{2+})$ in solution $= 0.0200 + 6.2155 \times 10^{-3} = 0.0262$ mol

Final $c(Ni^{2+}) = \dfrac{n}{V} = \dfrac{0.0262}{0.200} = 0.131$ mol/L *(2 marks)*

25 The $H_2PO_4^{-}/HPO_4^{2-}$ acid/base pair form one of the buffer systems in cells. The role of the system is to maintain pH over a narrow range in an environment where acidic and basic substances enter cells. The ions form the following equilibrium:

$H_2PO_4^{-}(aq) + H_2O \mid HPO_4^{2-}(aq) + H_3O^{+}(aq)$

When an acid enters the cell the concentration of H_3O^+ increases and the equilibrium counteracts the change by shifting to the left. The strong B/L acid (H_3O^+) reacts with HPO_4^{2-} and converts it into a weak B/L acid ($H_2PO_4^-$) and thus the cell pH does not change very much.

$H_3O^+ + HPO_4^- \rightleftharpoons H_2O + H_2PO_4^-$

When a base enters the cell, OH^- reacts with the weak B/L acid ($H_2PO_4^-$), which is in a high concentration in the buffer and converts the OH^- ions into water and the weak base (HPO_4^{2-}) forms. Thus the pH does not change very much in the cell.

$OH^- + H_2PO_4^- \rightleftharpoons H_2O + HPO_4^{2-}$ *(3 marks)*

26 (a) The calculated concentration of NaOH is $0.1011 \times 0.02410/0.025 = 0.09746$ mol L^{-1}.

The concentration given the mass is $4.000/39.998/1.000 = 0.1000$ mol L^{-1}.

The calculated concentration is lower because NaOH solid absorbs moisture from the air so that the weighed 1.000 g would not be pure NaOH and the real mass would be less than 1.000 g. *(2 marks)*

(b) $n(\text{NaOH}) = 0.09746 \times 0.02725 = 2.656 \times 10^{-3}$ mol

$n(C_7H_8O_6) = 2.656 \times 10^{-3}/3 = 8.8526 \times 10^{-4}$ mol

$C_7H_8O_6 + 3NaOH \rightarrow 3H_2O + Na_3C_6H_5O_7$

$c(\text{diluted } C_7H_8O_6) = 8.8526 \times 10^{-4}/0.025 = 0.03541$ mol L^{-1}

$c(C_7H_8O_6 \text{ in cordial}) = 0.03541 \times 500/50 = 0.3541$ mol L^{-1} *(4 marks)*

27 Polystyrene is used, after gas is blown through it while cooling (to produce Styrofoam), for packaging around electrical and white goods for transport and storage. Styrofoam is lightweight and so reduces transport costs; absorbs shock, thus preventing damage to goods during movement; and is water resistant, preventing goods from coming into contact with water and being water damaged. Styrofoam is disposed of in landfill after its short-lived application, and as it is non-biodegradable it does not decompose readily. This leads to the need to take space from living systems to provide more places for landfill and the high cost in building such areas.

The biopolymer polyhydroxybutanoate, PHB, is used for packaging groceries, for carry bags and for shampoo bottles as it is strong, flexible and water resistant so that foods do not spoil. However, unlike Styrofoam, PHB is biodegradable and will decompose readily and, therefore, it does not put pressure on municipalities to take space from native living systems. Being biodegradable also reduces the cost of disposal, as it breaks down quickly and does not lead to the need to build more landfill waste areas. *(5 marks)*

28 **Turbidity**: a physical principle involved in measuring turbidity is that suspended particles scatter light so that it penetrates less far through turbid water.

A procedure that uses this principle to measure turbidity is a turbidity tube. Water is added to the tube until a mark on the base of the tube is no longer visible. The less water needed, the more turbid the sample.

Total dissolved solids (TDS): a physical principle involved in measuring TDS is that the higher the concentration of dissolved ions, the greater the electrical conductivity.

A procedure that uses this principle is the TDS conductivity probe and data logger. The probe has two stainless steel electrodes and a battery, and once the probe is calibrated it can determine the concentration of dissolved solids to low concentrations in the units microsiemens, which can be converted to mg/L. *(4 marks)*

29 (a) The Arrhenius definition of an acid is a substance that ionises to produce hydrogen ions when it is dissolved in water, while a base produces hydroxide ions when it is dissolved in water. This is a simpler and more easily understood definition than the more sophisticated Brönsted-Lowry definition, which states that an acid is a proton donor and a base is a proton acceptor.

Many chemical reactions (especially those at school level) occur in aqueous solutions and the Arrhenius theory is quite satisfactory in explaining the concept of neutralisation in terms of the reaction between hydronium ions and hydroxide ions for strong and weak acids and bases. It also explains the production of salts during neutralisation. The more sophisticated Brönsted-Lowry theory is, however, required to explain some properties and reactions, such as the behaviour of buffers and why some salts are acidic or basic. It is also required to explain reactions in non-aqueous solvents. *(3 marks)*

(b) No matter what the identity of strong acids and bases, they all react the same way when neutralisation occurs. The neutralisation reaction occurs when the hydrogen ion from the acid reacts with the hydroxide ion from the base, producing water.

$$H^+(aq) + OH^-(aq) \rightarrow H_2O(l)$$

One mole of hydrogen ions reacts with one mole of hydroxide to produce one mole of water and a release of 57 kJ of energy. *(1 mark)*

30 *Step 1:* Low pressure and temperatures favour the production of NO. At low pressure the reaction shifts forward as 9 moles of gas react to form 10 moles and this results in an increase in pressure. At low temperatures the forward reaction is favoured as the forward reaction is exothermic and the reaction would shift forward to counteract the low temperature. However, these conditions reduce the rate of the reaction. The high activation energy means that the reaction cannot be run at a low temperature and a compromise must be chosen. A catalyst (Pt/Rh) would be used so that a moderate temperature can be used but one high enough to overcome the activation energy.

Step 2: A high pressure would be used as 3 moles of gas react to form 2 moles and a high pressure favours the forward reaction as this reduces the pressure. A high pressure also increases the rate as the gas molecules undergo more collisions per second. A low temperature favours the production of NO_2 as the forward reaction is exothermic and the forward reaction counteracts the low temperature by releasing heat. However, a low temperature results in a slow production rate. Therefore, a balance would be chosen of a moderate temperature to raise the rate and yet not reduce the yield too much.

The reaction equilibrium can also be driven to the right by absorbing the NO_2 into water (see step 3).

Step 3: As this step is not an equilibrium, the highest yield and rate of formation of nitric acid will occur when reactants undergo more collisions per second. This would occur when pressure is high, to increase collisions between the gas molecules and water, and temperature is high, as this increases the average kinetic energy of particles, resulting in more successful collisions. *(6 marks)*

31 The river may have been contaminated on Tuesday by a rainstorm, leading to much faster flowing water. This is supported by the high turbidity (67) and total dissolved solids (436) readings on that day. The high energy in fast flowing water helps to dissolve more materials from the river banks and bottom. High energy water also helps to support suspended matter in the water, giving the high turbidity reading.

The river may have been contaminated on Thursday by run-off from a farm. The river water contains high levels of faecal coliforms (514) and a high biochemical oxygen demand (52). The faecal coliforms reproduce in an environment where there is faeces in the water, as you would find from farm animal waste, and the high BOD indicates high levels of bacteria that reproduce in a nutrient rich faecal environment. The high levels of bacteria also use up much of the oxygen dissolved in the river, resulting in low levels of dissolved oxygen (4.2). *(4 marks)*

32 (a) It is a white, neutral and water-insoluble powder that absorbs water and the indicator so that the colour change of the indicator can be seen easily against a white background. *(1 mark)*

(b) As methyl yellow turned yellow, the pH must be between 4 and 14. As methyl red turned red, the pH must be between 0 and 4.5. Therefore, the pH of the soil must be between 4 and 4.5. The plant that will grow best at this pH is Hydrangea Blue, as it shows optimal growth between pH 4 and 5, which overlaps the soil pH range. *(2 marks)*

(c) Prepare solutions of pH 1, 3, 5, 7, 9 and 11 by starting with 1 mol L^{-1} HCl and NaOH and progressively diluting them until these pHs are measured. Set up six test tubes with 5 mL of solution of the different pHs in each. Add 1 mL of the red cabbage indicator already prepared to each test tube. Record the colour the indicator turns in each pH. *(2 marks)*

Section II—Options

Question 33—Industrial Chemistry

(a) Liquid X is concentrated sulfuric acid (conc H_2SO_4). The crystals change colour because the concentrated sulfuric acid dehydrates the copper(II) sulfate pentahydrate, which is blue, producing anhydrous copper(II) sulfate, which is white. This can be represented as:

$$CuSO_4.5H_2O(s) \rightarrow CuSO_4(s) + 5H_2O(l).$$ *(3 marks)*

(b) (i) The Solvay process for the production of sodium carbonate produces CO_2 in two steps. Solid calcium carbonate is heated to produce CaO and $CO_2(g)$:

$$CaCO_3(s) \rightarrow CaO(s) + CO_2(g).$$

Also, solid sodium hydrogencarbonate is heated to produce sodium carbonate, water and CO_2:

$$2NaHCO_3(s) \rightarrow Na_2CO_3(s) + H_2O(l) + CO_2(g).$$

The $CO_2(g)$ produced is not released to the environment, and thus it is of little environmental concern. Rather, it is pumped into the Solvay tower, where it is mixed with ammoniacal brine to produce the mixture of ions crucial to the success of this process: Na^+, HCO_3^-, NH_4^+, Cl^-. *(3 marks)*

(ii) The overall reaction for the process is $2NaCl + CaCO_3 \rightarrow CaCl_2 + Na_2CO_3$.

1×10^6 g of NaCl = $1 \times 10^6/(35.45 + 22.99) = 17\,111.6$ mol.

$n(CaCl_2) = 0.5 \times 17\,111.6 = 8555.8$ mol.

Mass $(CaCl_2) = 8555.8 \times 110.98 = 949\,521$ g = 949.5 kg.

For consistency with significant figures (1 tonne), the answer is 900 kg. *(2 marks)*

(c) (i) A solution of 0.05 mol/L NaCl is very dilute, and hence the electrolysis of water preferentially takes place in this procedure.

At the anode (positive electrode) oxidation takes place:

$$2H_2O(l) \rightarrow O_2(g) + 4H^+(aq) + 4e^-.$$

The production of H^+ ions at this electrode makes the surrounding solution acidic, and hence the litmus paper turned red. The bubbles observed were due to the production of oxygen gas.

At the cathode (negative electrode) reduction takes place:

$$2H_2O(l) + 2e^- \rightarrow H_2(g) + 2OH^-(aq).$$

The production of OH^- ions makes the surrounding solution basic, and hence the litmus paper is blue. The bubbles produced are due to hydrogen gas production. *(3 marks)*

(ii) A galvanic cell based on the reaction of oxygen with hydrogen would involve:

$$\tfrac{1}{2}H_2(g) + OH^-(aq) \rightarrow H_2O(l) + e^- \qquad E^o = 0.83\ V$$

$$\tfrac{1}{2}O_2(g) + 2H^+(aq) + 2e^- \rightarrow H_2O(l) \qquad E^o = 1.23\ V$$

and overall the galvanic cell would *produce* 2.06 V of electrical energy.

In the electrolytic cell shown above this is minimum voltage *requirement* for the cell to operate. Galvanic cells produce electrical energy. Electrolytic cells require electrical energy, which is then converted to chemical energy. *(2 marks)*

(d) (i) To model an equilibrium reaction we used a plastic bottle of water suspended above a tank of water. The plastic bottle had a stoppered hole in it so that water flowed from it into the tank. Water from the tank was fed back into the top of the bottle by tubing connected to a small water pump. We turned the pump on and took the stopper out of the bottle to model the equilibrium reaction. To model the effect of increasing the concentration of a reactant, we added water to the bottle. To model the effect of a catalyst, we increased the speed of the pump. *(2 marks)*

(ii) The information collected in this investigation is highly valid, given that it is a qualitative model of an equilibrium reaction. For example, in this procedure it can be seen that the process occurs in two directions (water into the tank and water out of the tank), just like an equilibrium reaction. It is also observed that the overall level of water in the tank is constant (like an equilibrium reaction: no macroscopic change). The characteristic of constant microscopic change of an equilibrium system is also observed: water continuously flows into and out of the tank. Like an equilibrium system, this model is also a closed system: no water is removed from the system. *(3 marks)*

(e) Methods of production of industrially important chemicals include the mercury and diaphragm processes for producing NaOH, and the Solvay process for producing Na_2CO_3, all of which involve brine, saturated $NaCl(aq)$.

The production of NaOH in both the mercury and diaphragm processes involves the electrolysis of brine. In the mercury process, oxidation of chloride ions to chlorine gas occurs at the steel anode ($2Cl^- \rightarrow Cl_2(g) + 2e^-$). Reduction of Na^+ to Na occurs at a flowing mercury cathode, producing an amalgam, which when sprayed into water in a separate compartment produces $NaOH(aq)$ of very high purity: $2Na(s) + 2H_2O(l) \rightarrow 2NaOH(aq) + H_2(g)$. The chlorine gas and hydrogen gas are valuable by-products, and the mercury is recycled for further use in the electrolysis compartment. However, this process results in significant amounts of mercury leaking into the environment. A recent example of this occurred in Dryden, Ontario, where between 1962 and 1970 the Dryden Chemical Company discharged over 9000 kg of mercury into the local river system as a result of NaOH production. The social effects of mercury contamination include health impacts and direct economic impacts. Mercury poisoning damages the lungs, brain and kidneys, causing conditions such as memory loss, muscle spasms, paralysis, insanity and death. It can also damage the developing foetus. All of these illnesses have economic impacts in the cost of treatment and patient care, as well as devastating social impacts on individuals, families and entire communities. There are also terrible economic consequences related to the closure of commercial fisheries and tourist activities in contaminated areas.

In the diaphragm process brine flows into the anode compartment, where the oxidation reaction is the same, producing chlorine gas. The brine then flows into the cathode compartment, where reduction of water produces hydrogen gas and a solution of NaOH (containing NaCl):

$$2H_2O + 2e^- \rightarrow H_2 + 2OH^-.$$

In this process, the anode and cathode compartments are separated by an asbestos diaphragm. Although this process does not produce mercury contamination, it does result in exposure to asbestos. This can lead to a lung disease called asbestosis, which also has devastating social effects, and for which there is no cure. It is characterised by shortness of breath and respiratory failure, as well as various types of lung cancer. Again, there is the enormous social and economic cost of treatment and care, as well as the suffering caused to individuals and families. In 1991 the Australian government completely banned the use of any materials containing asbestos, and one Western Australian asbestos-mining town, Wittenoom, has been removed from maps and road-signs because it remains contaminated.

Another industrial process that has significant amounts of waste products is the Solvay process for Na_2CO_3 production. The overall reaction for this process is:

$$2NaCl(aq) + CaCO_3(s) \rightarrow Na_2CO_3(s) + CaCl_2(s).$$

It can be seen that $CaCl_2$ is a waste product, and another important waste product produced in a Solvay plant is the large volume of hot water, which must be disposed of. The social effects of these environmental issues depend upon the location of the plant. For coastal plants, there is little environmental or social impact, because $CaCl_2$ (or the quantity that is not sold as a road de-icing agent) can be discharged into the ocean, where it has an insignificant impact on the concentration of dissolved ions. Similarly, the heated water can be slowly discharged into the ocean, where rapid mixing with the large volume of sea water removes the possibility of thermal pollution. For plants not located near the coast, the environmental and social impacts of a Solvay plant can be mitigated by burial of the $CaCl_2$ in sealed tanks, preventing leakage into the environment, and by allowing heated cooling water to cool in artificial ponds before discharging it into the environment. The social effects of these environmental issues are insignificant when compared to the devastating effects of mercury or asbestos poisoning. *(7 marks)*

Question 34—Shipwrecks, Corrosion and Conservation

(a) The anti-corrosion rod could be made of zinc. Zinc would serve as a sacrificial anode. This is because zinc has a higher oxidation potential than iron and will corrode in preference to the steel tank.

$$Zn(s) \rightarrow Zn(aq) + 2e^- \qquad E^o = 0.76\ V$$

$$Fe(s) \rightarrow Fe^{2+}(aq) + 2e^- \qquad E^o = 0.44\ V$$

(3 marks)

(b) (i) The tarnish is due to the formation of a silver compound (e.g. silver sulfide) on the silver surface. When the cutlery is placed on the aluminium tray a galvanic cell is set up with the cutlery acting as the cathode and the more reactive aluminium acting as the anode. The sodium hydrogen carbonate solution is the electrolyte and conducts the current between the two electrodes. Silver ions (in the tarnish) are reduced to silver metal, removing the tarnish and not losing any original silver in the process. The aluminium is oxidised.

$$Al(s) \rightarrow Al^{3+}(aq) + 3e^- \qquad E^o = 1.68\ V$$

$$Ag^+(aq) + e^- \rightarrow Ag(s) \qquad E^o = 0.8\ V$$

$$Al(s) + 3Ag^+(aq) \rightarrow Al^{3+}(aq) + 3Ag(s) \qquad E^o = 2.48\ V$$

(4 marks)

(ii) The advantage of the aluminium tray method is that no original silver is lost as would happen with the polishing method, where the abrasion actually removes the tarnish and also some original metal. *(1 mark)*

(c) (i)

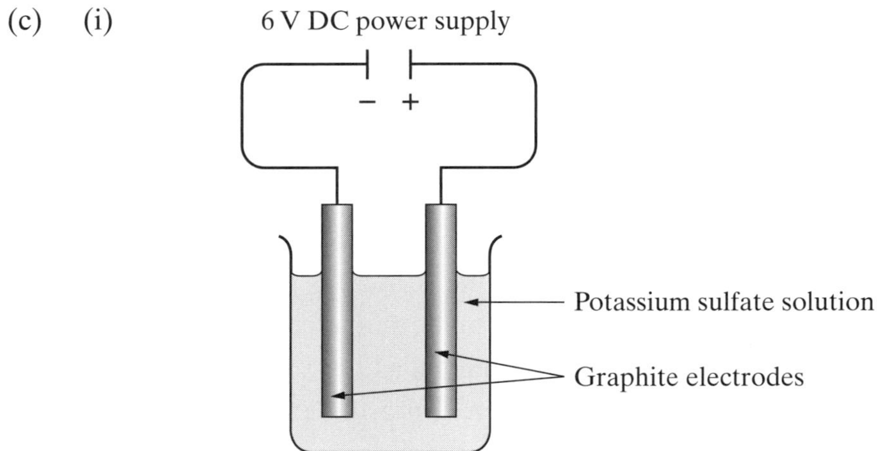

(3 marks)

(ii) Anode: $H_2O(l) \rightarrow \frac{1}{2}O_2(g) + 2H^+(aq) + 2e^-$

Cathode: $H_2O(l) + e^- \rightarrow \frac{1}{2}H_2(g) + OH^-(aq)$

The products formed at the anode are oxygen gas and hydrogen ions.

The products formed at the cathode are hydrogen gas and hydroxide ions. *(2 marks)*

(d) (i) The student should have removed all signs of corrosion from the alloys and metals by rubbing them with emery paper or steel wool. This process should be followed by rinsing the objects with distilled water. This is important as it removes a variable. All metals start with no signs of corrosion so that any corrosion that does form has happened during the investigation, allowing for a fair comparison. *(2 marks)*

(ii) On the positive side the student has addressed some of the investigation outcomes. He has chosen two metals and two alloys, has submerged them in sea water to replicate a marine environment and has addressed the need to measure rate by photographing them on a daily basis for several weeks (or by weighing them to observe mass changes due to corrosion). However, on the negative side, this is not a fair test because the alloys and metals cannot be compared when more than one variable is changed. The material of which they are made varies, but so does the shape and surface area of the metals and alloys. As the surface area is such a vital element to hold constant, the investigation is not valid. *(3 marks)*

(e) Prior to some recent discoveries of shipwrecks at great depths in the ocean, corrosion at great depth was predicted to be slow because the conditions there do not promote fast reaction rates. The temperature is cold, at just above freezing point, and this results in water molecules having low average kinetic energy, leading to a low rate of successful collisions and slow reaction rate. Dissolved oxygen is a requirement for corrosion as oxygen is the oxidant in the reaction. Oxygen enters the ocean at the surface and diffuses down; however, as it is used in respiration by living things in the ocean the concentration of dissolved oxygen at great depth is very low. This would also dramatically reduce the rate of corrosion.

However, recent discoveries of shipwrecks at great depth have allowed corrosion to be investigated and extensive corrosion has been found on vessels such as the *Titanic*, which was discovered at about 4 km in 1985. Red/brown stalactite-looking rusticles were seen hanging from the hull and railings. Chemical examination of the rusticles has found them to be composed of iron compounds and colonies of sulfate reducing bacteria. These anaerobic bacteria promote the corrosion of the iron hull of ships by reducing sulfate ions in the ocean and using the electrons from the oxidation of iron.

Oxidation: $Fe(s) \rightarrow Fe^{2+}(aq) + 2e^-$

Reduction: $SO_4^{2-}(aq) + 10H^+(aq) + 8e^- \rightarrow H_2S(aq) + 4H_2O(l)$

To obtain the electrons for this reduction the bacteria oxidise the iron at corrosion sites like shipwrecks. Although these sulfate reducing bacteria are present in low concentrations in the open ocean, they are present in very high concentrations on submerged iron objects. The hydrogen sulfide produced reacts with other metals like silver and lead, forming black metal sulfides on their surface. The hydrogen sulfide also lowers the pH of the surrounding environment because it is a weak acid.

The acidic environment around shipwrecks is created by the weak acid produced by sulfate reducing bacteria. Most wrecks are also covered by encrusting calcareous organisms.

These respire and produce carbon dioxide, which forms carbonic acid. The lower pH speeds up corrosion in two ways. Firstly, oxygen is a stronger oxidant in acidic environments:

$$O_2 + 4H^+ + 4e^- \rightarrow 2H_2O \qquad E^o = 1.23\ V$$

$$O_2 + 2H_2O + 4e^- \rightarrow 4OH^- \qquad E^o = 0.40\ V$$

This reaction has a higher reduction potential than the reduction of water and will, therefore, speed up the oxidation and reduction reactions.

Second, the hydrogen ions also act as a catalyst speeding up the reactions and as oxidants directly causing the oxidation of metals like iron:

$$Fe(s) + 2H^+(aq) \rightarrow Fe^{2+}(aq) + H_2(g)$$

Owing to these factors corrosion at depth is accelerated and deep iron objects like shipwrecks are highly corroded when we may have expected them to be much less so.

Therefore, the statement is valid, in that corrosion was predicted to be slow prior to the discovery of the *Titanic* and other recent shipwreck discoveries. However, corrosion is now known to proceed more rapidly than predicted, based on investigations into these wrecks. *(7 marks)*

Question 37—Forensic Chemistry

(a) The structure shown is a disaccharide, since it is made of two sugar molecules that have reacted together in a condensation reaction, eliminating water. The two sugar molecules that have formed this disaccharide are glucose. The molecular formula of glucose is $C_6H_{12}O_6$ and the molecular formula of the disaccharide is $C_{12}H_{22}O_{11}$ because H_2O is eliminated in the condensation reaction. It can also be said that this structure would represent a reducing sugar, since a reactive carbonyl group is shown, C=O (−CHO). This reduces species such as Cu^{2+} to Cu^+ and is oxidised to the alkanoic acid group in the process. *(3 marks)*

(b) (i) $H_2N{-}CH(R){-}COOH + H_2N{-}CH(R){-}COOH \rightarrow$

$H_2N{-}CH(R)\mathbf{{-}C(O){-}N}H{-}CH(R){-}COOH$

where the peptide bond is shown in bold. *(2 marks)*

(ii) Paper chromatography works on the principle that a mixture of solutes can be separated by applying them to a stationary phase, and moving them along it at different rates by using a suitable mobile phase. In paper chromatography the stationary phase is a simple cellulosic matrix such as filter paper. The mobile phase is a solvent or mixture of solvents, such as ethanol and water. Each component of the mixture, for example of a mixture of amino acids, has a different affinity for the stationary phase (adsorbs onto it with different strengths), and so moves with the solvent up the stationary phase, at a specific rate different from that of the other solutes. The R_f value of each amino acid in the mixture is given by the distance along the stationary phase that the solvent moves, divided by the distance that it (the amino acid) moves. This is a characteristic of a given solute for a given stationary phase and mobile phase and can therefore be used to identify the solute. For substances such as amino acids, which are colourless, an extra step is required—the use of a locating agent. These are substances that are sprayed onto the chromatogram after the separation and that react with solutes to enable them to be visualised (e.g. by visible or ultraviolet light). *(3 marks)*

(c) (i) DNA is analysed using a technique called 'DNA fingerprinting'. There are portions of DNA between genes that do not code for specific proteins and are as unique to an individual as their fingerprint. The first step is to obtain the DNA from cellular material (for example that left at a crime scene) by breaking the cell membranes and centrifuging the sample to separate the DNA.

In small samples, the DNA may be amplified (replicated to increase the number of copies) by a process called the polymerase chain reaction (PCR). In this process the sample is heated to separate the strands of DNA, which are then used as a template for constructing the complementary strand by the addition of a suitable enzyme. The process of separating the strands and replication is repeated many times.

The DNA strands are then cut by restriction enzymes to produce short base sequences (DNA fragments), which are separated by electrophoresis, based on their size. The separated base sequences are treated so that they become visible, radioactive or luminous. The visualised base sequences are then compared to standards, for example the sequences from a suspect or suspected parent, so that a positive identification can be made, or so that someone can be excluded as a suspect or parent. Because close relatives (e.g. family members, parents and children) will have more similarities in their DNA fingerprints than unrelated or more distantly related people, this fingerprinting can be used to determine relationships such as paternity. *(4 marks)*

(ii) The child is W. *(1 mark)*

(d) (i) A suitable procedure for identifying each substance would be:

Step 1. Add a small sample of each solid to water. The insoluble substance is asprin. (This step is based on the fact that all of the given substances except asprin are water soluble: sucrose because of its hydrogen-bonding ability; glycine because of H-bonding ability and because of possible zwitter-ion formation, leading to ion-dipole interactions with water; and KCl because of ion-dipole interactions with water.)

Step 2. To the solutions produced in step 1, add a few drops of Ninhydrin reagent. The solution that produces a blue or blue-violet colour is glycine. (This reagent reacts selectively with most amino acids, including glycine; there are no ammonium salts present to give a false positive result.)

Step 3. To a fresh solution of the remaining substances, add a few drops of $AgNO_3(aq)$. The solution that produces a white precipitate is KCl. $[Ag^+ + Cl^- \rightarrow AgCl(s)]$

Step 4. The remaining substance is sucrose (by elimination). *(4 marks)*

(ii) A non-destructive test is the addition of water to test water solubility. Producing a solution of the substance does not preclude further testing, and if necessary the water can be removed by evaporation. A destructive test is the addition of silver nitrate solution to produce a silver chloride precipitate. This sample cannot be re-tested because $AgCl(s)$, $Ag^+(aq)$ and $NO_3^-(aq)$ have been added to the sample. *(1 mark)*

(e) Three instrumental techniques studied in this option are mass spectroscopy, atomic emission spectroscopy and electrophoresis.

Mass spectroscopy is a technique in which a sample is vaporised and bombarded with electrons, forming a mixture of ions of various masses (because of some of the molecule(s) in the sample fragment). After the instrument discards the negative ions, the positive ions are accelerated by a magnetic and electric field towards the detector. The fragments of the original substance(s), as well as the molecular ion(s), travel towards the detector at different rates and are separated on the basis of their mass/charge ratio. Different organic molecules have a characteristic mass/charge ratio for the molecular ion and characteristic fragmentation patterns. Thus, comparison with computerised data bases of known compounds allows identification of the substance in the sample. This is of tremendous value in forensic science, because it allows the identification of minute samples of complex organic molecules, such as those that could be found in fuels, poisons and drugs, on fabrics and clothing, and even in gaseous samples. The versatility and sensitivity of this technique makes it invaluable to forensic science.

Atomic emission spectroscopy is another of the fundamental tools of a forensic scientist. It is based on the way electrons in atoms absorb and emit energy, due to the fact that the energy levels of electrons in atoms are quantised. Because of this, the emission spectrum of an element is a unique set of lines corresponding to the energies absorbed and emitted by electrons in atoms of that element. Each element has a different atomic number and electron configuration, and so the energies of electrons and the energy differences between shells are different. Thus, the spectrum is unique for each element. This makes atomic emission spectra very useful in forensic analysis, because it allows for the identification of specific elements, particularly metals, in samples being analysed. This could be used, for example, to determine the elements and their ratios in a particular mix of gunpowder, or in the casing of a bullet. This information can be related back to manufacturers' records, or databases of known products, to determine manufacturing location, time and company, for example.

DNA fingerprinting has become one of forensic science's most useful tools, and the instrumental technique that allows it is electrophoresis. In this technique, a potential difference is applied to a mixture, which has been placed in the centre of a gel or other electrolyte-soaked medium. The applied voltage causes charged particles to move towards the positive or negative electrode, depending on their charge. The rate at which they move depends on their size (mass), and thus components of the mixture are separated in two directions, towards each electrode. The technique is extremely useful to forensic scientists, as comparison with known standards allows the identification of the origins of biological samples, such as proteins, found at crime scenes. It is used for DNA fingerprinting, because

DNA fragments, produced by the action of restriction enzymes on DNA extracted from cellular material, have different masses and charges and are thus well separated by this technique. Comparison of the chromatogram of a crime-scene sample with those of suspects, for example, enables either a positive identification or the elimination of a suspect. Thus, this technique is one of the most useful to the forensic scientist. *(7 marks)*

CHAPTER 10

BOARD OF STUDIES
NEW SOUTH WALES

2012

HIGHER SCHOOL CERTIFICATE EXAMINATION

Chemistry

General Instructions

- Reading time – 5 minutes
- Working time – 3 hours
- Write using black or blue pen Black pen is preferred
- Draw diagrams using pencil
- Board-approved calculators may be used
- A data sheet and a Periodic Table are provided at the back of this paper
- Write your Centre Number and Student Number where required

Total marks – 100

Section I

75 marks

This section has two parts, Part A and Part B

Part A – 20 marks

- Attempt Questions 1–20
- Allow about 35 minutes for this part

Part B – 55 marks

- Attempt Questions 21–33
- Allow about 1 hour and 40 minutes for this part

Section II

25 marks

- Attempt ONE question from Questions 34–38
- Allow about 45 minutes for this section

Section I
75 marks

Part A – 20 marks
Attempt Questions 1–20
Allow about 35 minutes for this part

Use the multiple-choice answer sheet for Questions 1–20.

---

1 Which of the following is a measure of the clarity of water?

(A) Hardness

(B) Turbidity

(C) Total dissolved solids

(D) Biochemical oxygen demand

2 C_2H_4 → X → polymer

Which of the following compounds is represented by X in the flowchart?

(A) Cellulose

(B) Ethanol

(C) Glucose

(D) Styrene

3 What effect does a catalyst have on a reaction?

(A) It increases the rate.

(B) It increases the yield.

(C) It increases the heat of reaction.

(D) It increases the activation energy.

4 Which pieces of glassware should be used when preparing a primary standard solution?

(A) Pipette, burette and conical flask

(B) Dropper, watch glass and pipette

(C) Beaker, filter funnel and volumetric flask

(D) Measuring cylinder, stirring rod and conical flask

5 Which of the following is a balanced equation representing the fermentation of glucose?

(A) $C_6H_{12}O_6(aq) \rightarrow 2C_3H_6O_3(aq)$

(B) $C_6H_{12}O_6(aq) \rightarrow 2C_2H_5OH(aq) + 2CO_2(g)$

(C) $C_6H_{12}O_6(aq) + 6O_2(g) \rightarrow 6CO_2(g) + 6H_2O(l)$

(D) $C_6H_{12}O_6(aq) + 3O_2(g) \rightarrow C_2H_5OH(aq) + 4CO_2(g) + 3H_2O(l)$

6 Cobalt-60 is produced according to the equation:

$$^{59}_{27}Co + ^{1}_{0}n \rightarrow ^{60}_{27}Co$$

Where would a commercial quantity of cobalt-60 be produced?

(A) Cyclotron

(B) Scintillator

(C) Nuclear reactor

(D) Particle accelerator

7 Methyl orange, bromothymol blue and phenolphthalein indicators were mixed together to form a solution.

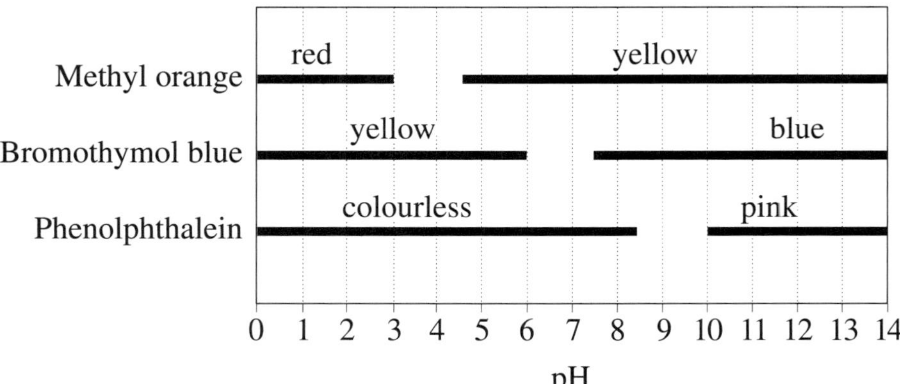

Over what pH range would the solution be yellow?

(A) 0 – 14

(B) 3 – 4.5

(C) 3 – 7.5

(D) 4.5 – 6

8 Which acid / base pair could act as a buffer?

(A) H_3O^+ / H_2O

(B) H_2O / OH^-

(C) HNO_3 / NO_3^-

(D) $H_2PO_4^- / HPO_4^{2-}$

9 Which of the following contains a coordinate covalent bond?

(A) NH_3

(B) NH_4^+

(C) H_2O

(D) OH^-

10 Samples of a solution of barium nitrate were independently tested with chloride ions, with sulfate ions and also for flame colour.

Which row of the following table would represent the results?

	Chloride	*Sulfate*	*Flame test*
(A)	No precipitate	No precipitate	Red
(B)	No precipitate	Precipitate	Green
(C)	Precipitate	Precipitate	Green
(D)	Precipitate	No precipitate	Red

11 The pH of 0.1 mol L^{-1} solutions of acetic, citric and hydrochloric acids was measured.

Which solution has the highest pH?

(A) Citric acid

(B) Acetic acid

(C) Hydrochloric acid

(D) The pH of the three solutions is the same.

12 What is the correct IUPAC name for the following compound?

```
     H   F   Cl  H
     |   |   |   |
 H — C — C — C — C — H
     |   |   |   |
     H   H   H   H
```

(A) 2-chloro-2-fluorobutane

(B) 2-fluoro-3-chlorobutane

(C) 3-fluoro-2-chlorobutane

(D) 3-chloro-2-fluorobutane

Use the information provided to answer Questions 13 and 14.

> This equation represents a common redox reaction.
>
> $$Cr_2O_7^{2-}(aq) + 14H^+(aq) + 6Fe^{2+}(aq) \rightarrow 2Cr^{3+}(aq) + 6Fe^{3+}(aq) + 7H_2O(l)$$

13 What is the oxidising agent in the reaction?

(A) H^+

(B) Cr^{3+}

(C) Fe^{2+}

(D) $Cr_2O_7^{2-}$

14 What is the value of $E^{\ominus}_{cell}$ for the reaction?

(A) 0.59 V

(B) 0.92 V

(C) 1.90 V

(D) 2.13 V

15 In which row of the following table are the listed oxides correctly classified?

	Acidic	*Basic*	*Neutral*	*Amphoteric*
(A)	CO_2	Na_2O	SO_3	Al_2O_3
(B)	Na_2O	CO_2	H_2O	Al_2O_3
(C)	CO_2	MgO	H_2O	ZnO
(D)	SO_2	K_2O	CO	CO_2

16 The graph shows the concentrations over time for the system:

$$CO(g) + 2H_2(g) \rightleftharpoons CH_3OH(g)$$

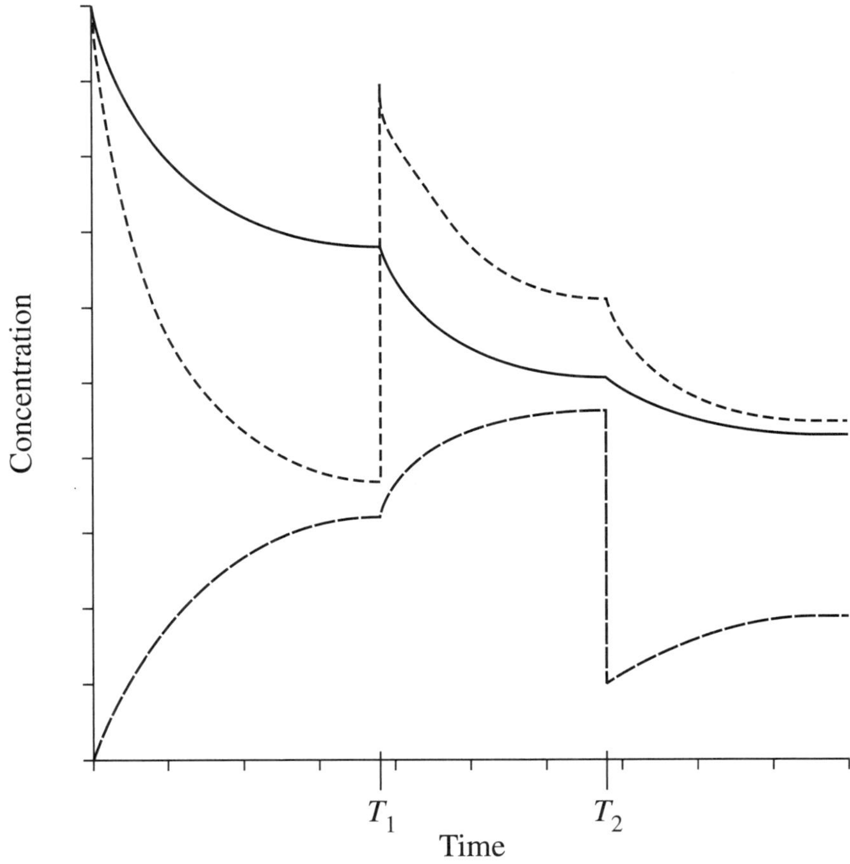

What has happened at times T_1 and T_2?

	T_1	T_2
(A)	H_2 added	CH_3OH removed
(B)	CO added	CH_3OH removed
(C)	H_2 added	CO removed
(D)	CO added	CO and H_2 removed

Question 22 (3 marks)

A student created the following models to demonstrate a chemical process.

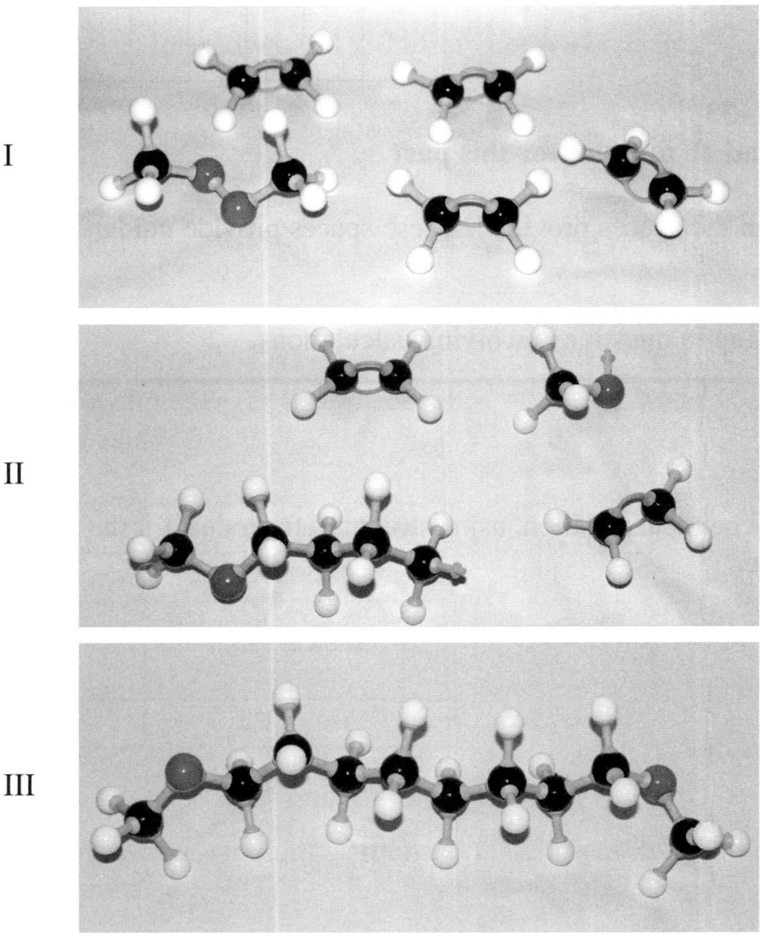

(a) What is the chemical process being modelled? **1**

....

(b) Why are models such as these useful? **2**

....

....

....

....

....

2012 HIGHER SCHOOL CERTIFICATE EXAMINATION

Chemistry

Centre Number

Section I – Part B (continued)

Student Number

Question 23 (3 marks)

Explain the impact of an increase in pressure and an increase in temperature on the solubility of carbon dioxide in water. Include a relevant equation in your answer. **3**

Question 24 (3 marks)

Explain why ammonia is such an important raw material in industry today. **3**

2012 HIGHER SCHOOL CERTIFICATE EXAMINATION

Chemistry

Centre Number

Section I – Part B (continued)

Student Number

Question 25 (3 marks)

Describe the process of monitoring waterways for eutrophication. **3**

Question 26 (8 marks)

Petroleum and sugar cane are both raw materials used for the production of ethanol.

(a) Construct separate flow diagrams for the production of ethanol from each raw material. **5**

petroleum	sugar cane
↓	↓

Question 26 continues

Question 26 (continued)

(b) Compare the environmental sustainability of producing ethanol from these two raw materials. **3**

....

....

....

....

....

....

....

....

End of Question 26

2012 HIGHER SCHOOL CERTIFICATE EXAMINATION

Chemistry

Centre Number

Section I – Part B (continued)

Student Number

Question 27 (3 marks)

Iodine-131 decays through both beta and gamma emission. Iodine-123 decays through gamma emission only.

(a) Iodine-131 is used for diagnosis and therapy whereas Iodine-123 is used only for diagnosis. **2**

	beta emission	gamma emission
Emitted particle	electron	gamma-ray
Ability to pass through biological tissue	low	high

With reference to the information and the table, justify the different uses of these two radioisotopes.

....

....

....

....

....

....

(b) Write the equation representing the decay of Iodine-131 by beta emission. **1**

....

....

Question 28 (3 marks)

A solution was made by mixing 75.00 mL of 0.120 mol L^{-1} hydrochloric acid with 25.00 mL of 0.200 mol L^{-1} sodium hydroxide. **3**

What is the pH of the solution?

....

....

....

....

....

....

....

....

2012 HIGHER SCHOOL CERTIFICATE EXAMINATION

Chemistry

Centre Number

Section I – Part B (continued)

Student Number

Question 29 (5 marks)

Draw a labelled diagram to show the layered structure of the atmosphere. In your diagram include: **5**

- the names of TWO atmospheric pollutants, positioned in the layers where the detrimental impact occurs
- the names of the sources of the two pollutants identified.

Question 30 (6 marks)

A chemist analysed aspirin tablets for quality control. The initial step of the analysis was the standardisation of a NaOH solution. Three 25.00 mL samples of a 0.1034 mol L^{-1} solution of standardised HCl were titrated with the NaOH solution. The average volume required for neutralisation was 25.75 mL.

(a) Calculate the molarity of the NaOH solution. **2**

....

....

....

....

Three flasks were prepared each containing a mixture of 25 mL of water and 10 mL of ethanol. An aspirin tablet was dissolved in each flask. The aspirin in each solution was titrated with the standardised NaOH solution according to the following equation:

$$C_9H_8O_4(aq) + NaOH(aq) \rightarrow C_9H_7O_4Na(aq) + H_2O(l)$$

The following titration results were obtained.

Tablet	*Volume* (mL)
1	16.60
2	16.50
3	16.55

(b) (i) Calculate the average mass (mg) of aspirin per tablet. **3**

....

....

....

....

....

....

....

(ii) Why was it necessary to include the ethanol in the mixture? **1**

....

Question 31 (5 marks)

The boiling points of some alkanols are given in the table.

Alkanol	*Boiling point* (°C)
Methanol	65
Ethanol	79
Propan-1-ol	97
Pentan-1-ol	138
Hexan-1-ol	157
Heptan-1-ol	176

(a) Using the data provided, construct a graph that shows the relationship between carbon chain length and boiling point. **3**

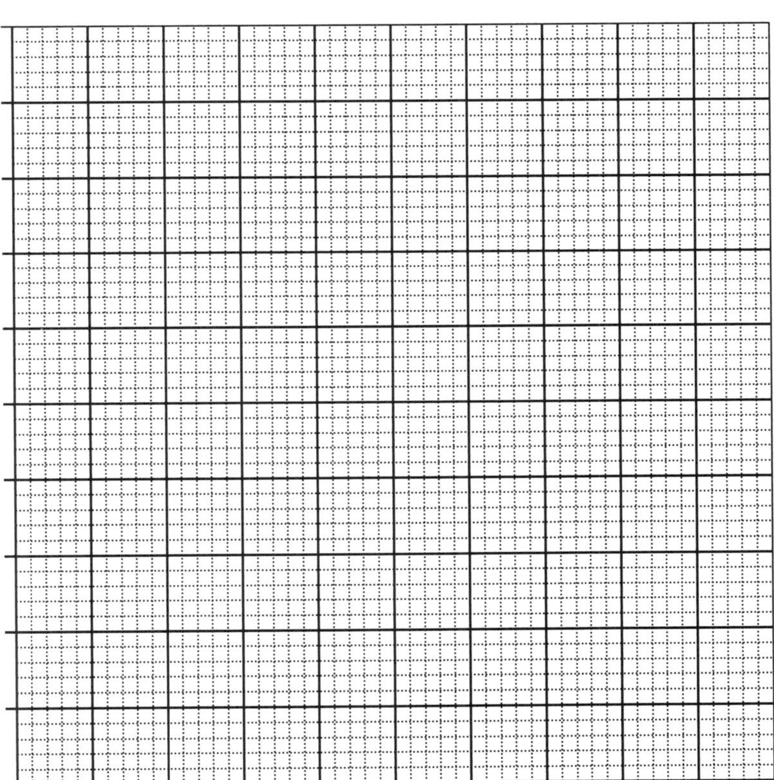

(b) Using the graph, predict the boiling point of butan-1-ol. **1**

....

(c) What is the intermolecular force responsible for the trend shown in the graph? **1**

....

2012 HIGHER SCHOOL CERTIFICATE EXAMINATION

Chemistry

Section I – Part B (continued)

Centre Number

Student Number

Question 32 (3 marks)

Question 32 (3 marks)

The mercury concentration of a certain fish species was determined by atomic absorption spectroscopy. The sample data are: **3**

Mass of fish (g)	18.6
Final sample volume (mL)	25.0
Absorbance (mean)	0.280

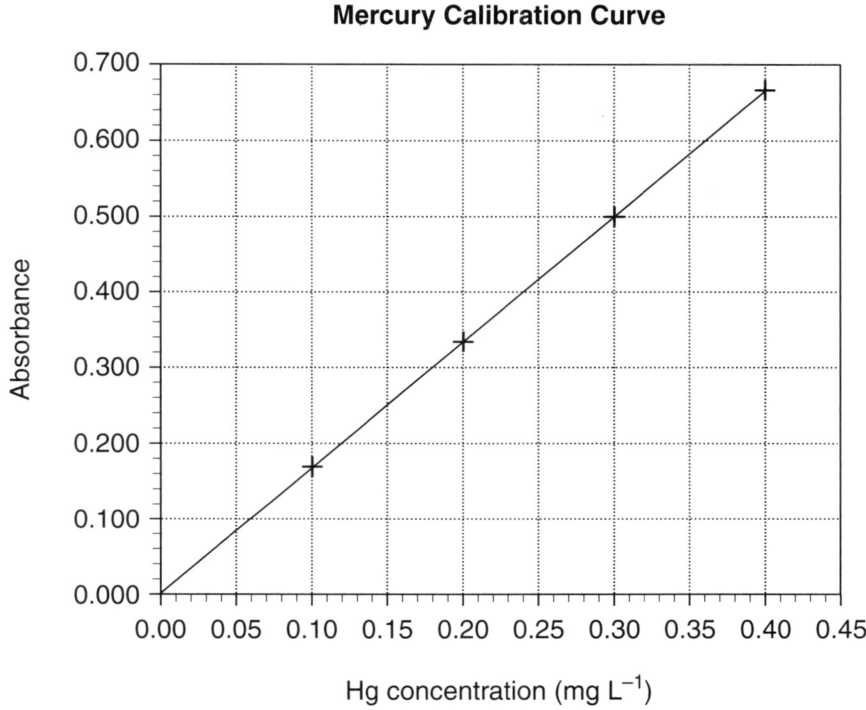

A consumer wants to avoid eating fish with a mercury concentration greater than 0.5 mg/kg of fish.

Calculate the concentration of mercury in the fish sample and state whether the consumer can eat this fish species.

....

....

....

....

....

....

....

....

Question 33 (6 marks)

Chemists can assist in reversing or minimising the environmental problems caused by technology and the human demand for products and services. **6**

With reference to this statement, assess the need for chemists to collaborate when monitoring the environmental impact of a named electrochemical cell.

2012 HIGHER SCHOOL CERTIFICATE EXAMINATION

Chemistry

Section II

25 marks
Attempt ONE question from Questions 34–38
Allow about 45 minutes for this section

Answer parts (a)–(c) of the question in Section II Answer Booklet 1.
Answer parts (d)–(e) of the question in Section II Answer Booklet 2.
Extra writing booklets are available.

Show all relevant working in questions involving calculations.

Question 34	Industrial Chemistry
Question 35	Shipwrecks, Corrosion and Conservation
Question 36	The Biochemistry of Movement *(Not included in this reproduction)*
Question 37	The Chemistry of Art *(Not included in this reproduction)*
Question 38	Forensic Chemistry

Question 34 — Industrial Chemistry (25 marks)

Answer parts (a)–(c) in Section II Answer Booklet 1.

(a) The following equipment was set up and the reaction allowed to proceed. Gases were produced at both electrodes. **3**

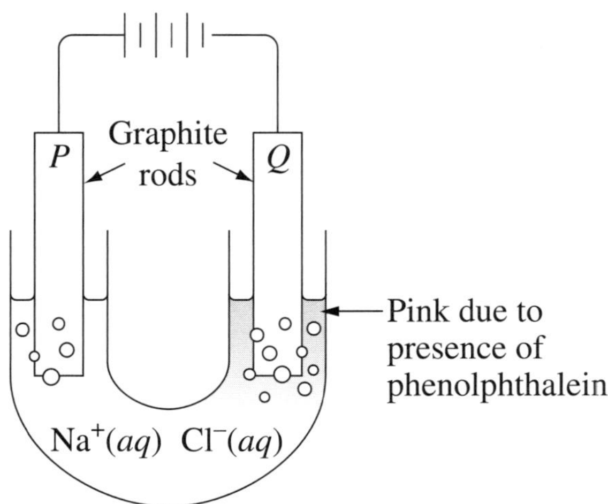

Name this process and identify the gas at each electrode.

(b) The equilibrium constant expression for a gaseous reaction is as follows:

$$K = \frac{[N_2][O_2]}{[NO]^2}$$

(i) Write the equation for this reaction. **1**

(ii) 0.400 moles of NO was placed in a 1.00 L vessel at 2000°C. The equilibrium concentration of N_2 was found to be 0.198 mol L^{-1}. **3**

Calculate the equilibrium constant for this reaction and use this value to describe the position of the equilibrium.

(iii) What could be changed that would result in a different value of K for this equilibrium? **1**

Question 34 continues

Question 34 (continued)

(c) The production of sulfuric acid is shown.

$$S(s) \longrightarrow SO_2(g) \longrightarrow SO_3(g) \longrightarrow \boxed{\text{oleum}} \longrightarrow H_2SO_4(l)$$

(i) Describe the production of oleum and its conversion to concentrated sulfuric acid. Include chemical equations in your answer. **3**

(ii) SO_3 can react with water to produce a solution of H_2SO_4. **2**

Why is it essential to convert SO_3 to oleum before the formation of H_2SO_4?

Answer parts (d)–(e) in Section II Answer Booklet 2.

(d) (i) Outline how one of the steps involved in the Solvay process can be chemically modelled in the school laboratory. Include a balanced chemical equation in your answer. **3**

(ii) Identify ONE risk factor and ONE difficulty associated with the laboratory modelling of the step. **2**

(e) Initially soap was the only product of the surfactant industry. Due to societal pressures and chemical developments, production in this industry has evolved to include a wide range of products. **7**

Account for these changes over time with reference to the structure and uses of surfactants.

End of Question 34

Question 35 — Shipwrecks, Corrosion and Conservation (25 marks)

Answer parts (a)–(c) in Section II Answer Booklet 1.

(a) A sealed container transporting a recently recovered artefact is damaged, allowing seawater to escape while appearing to leave the artefact intact. **3**

Why would the loss of seawater be of concern to the maritime archaeologists receiving the artefact?

(b) The diagram illustrates one method of protecting a steel pipe.

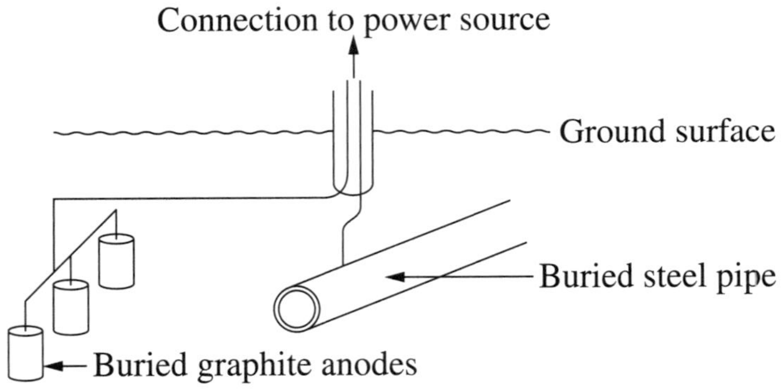

(i) Explain how this method works. **3**

(ii) Suggest an alternative way the pipe can be chemically protected. Use half-equations to support your answer. **2**

Question 35 continues

Question 35 (continued)

(c) Corrosion in some wrecks can be represented in part by the following half-equation.

$$SO_4^{2-} + 10H^+ + 8e^- \rightarrow H_2S(aq) + 4H_2O$$

(i) Under what conditions would this process occur naturally? **2**

(ii) These conditions were simulated in a laboratory by placing a piece of steel in an appropriate solution. The amount of hydrogen sulfide produced was monitored over time. **3**

Calculate the loss of iron, in grams, from the piece of steel if 0.76 g of hydrogen sulfide was produced. Include a balanced equation in your answer.

Answer parts (d)–(e) in Section II Answer Booklet 2.

(d) A student set up the following two beakers to investigate factors affecting electrolysis.

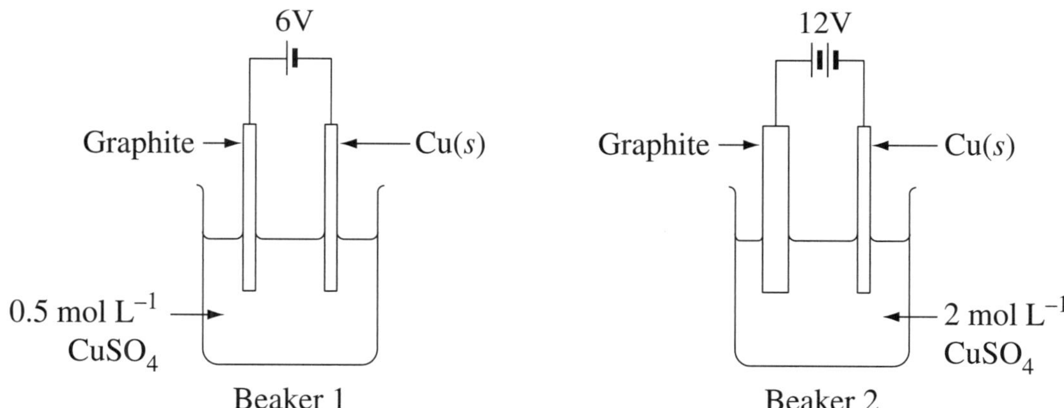

(i) Why is this investigation not valid? **1**

(ii) Draw diagrams to show how this investigation should be carried out to accurately identify TWO factors affecting electrolysis. Indicate the expected results on your diagrams. **4**

(e) Critically evaluate the continued use of steel in ship building with reference to the history of ocean-going vessels, the composition of steel and ways in which its interaction with the environment can be controlled. **7**

End of Question 35

Question 38 — Forensic Chemistry (25 marks)

Answer parts (a)–(c) in Section II Answer Booklet 1.

(a) Describe how the different characteristics of a soil sample would allow forensic chemists to trace its origin. **3**

(b) An athlete is required to give two urine samples. Sample A is tested for evidence of substances on the banned substance list. If a banned substance is detected in Sample A then Sample B is tested. If both samples are positive for a banned substance then the athlete is asked to explain the findings at a court hearing.

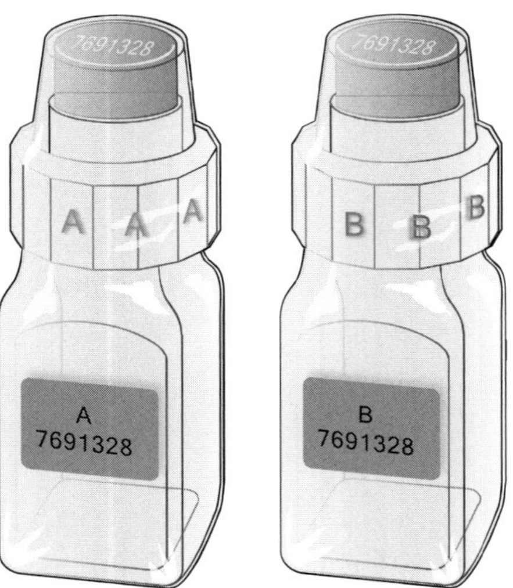

(i) Describe what precautions should be undertaken to ensure that the evidence collected for this purpose is accepted by the court. **3**

(ii) New technology is constantly required in forensic chemistry to keep up with drug cheats or improve the outcomes of a criminal investigation. **2**

Outline how recent advances in technology could have altered the outcome of a specific case.

Question 38 continues

Question 38 (continued)

(c) The diagram shows the separation of plant pigments by chromatography using hexane as a solvent.

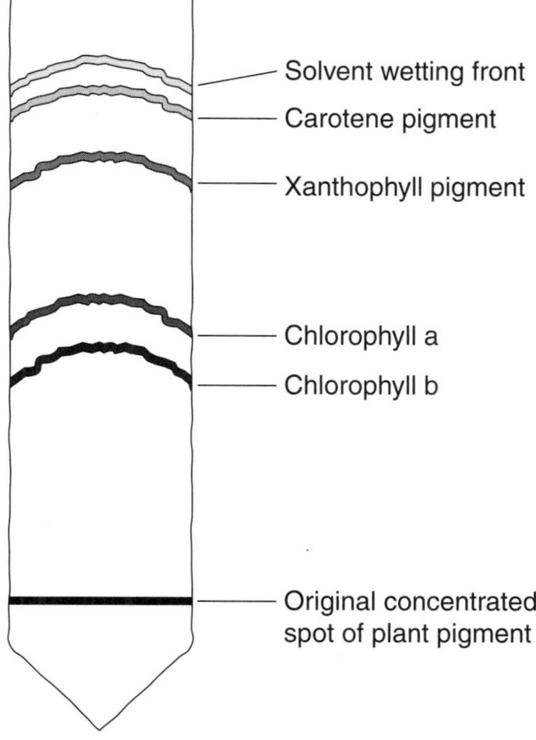

(i) State which of the pigments is the least polar and justify your choice. **2**

(ii) Explain how changing the solvent would affect the separation process. **3**

Question 38 continues

Question 38 (continued)

Answer parts (d)–(e) in Section II Answer Booklet 2.

(d) Sucrose is a carbohydrate composed of fructose and glucose.

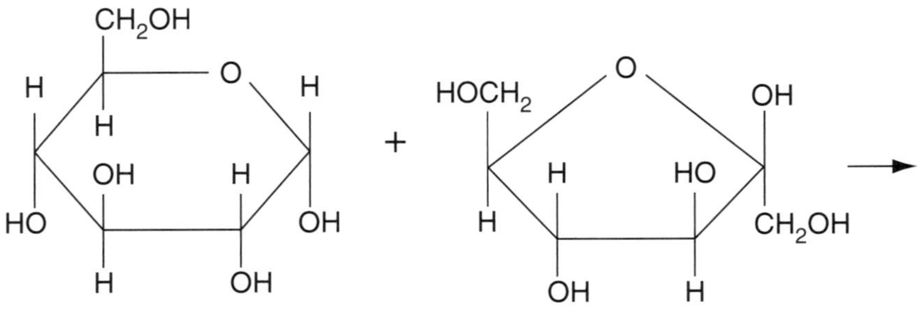

α-glucose β-fructose

(i) Complete the structural equation for the formation of sucrose. **2**

(ii) Outline a chemical test that could be performed to distinguish between sucrose and the monomers fructose and glucose. Include a chemical equation in your answer. **3**

(e) Analyse how modern techniques utilise the features of DNA and manipulate it to successfully generate an individual's DNA profile. **7**

End of Question 38

End of paper

2012 HSC Examination Paper

Sample Answers

Section I Part A

(Total 20 marks)

1 B Turbidity is due to undissolved material suspended in water.

2 D C_2H_4 is a monomer from which addition polymers are formed. It can be transformed into other monomers by substituting hydrogen atoms with other atoms or molecules to form other monomers such as styrene. Styrene is formed by substituting a hydrogen with benzene.

3 A Catalysts increase the rate of a reaction by providing an alternative pathway for a reaction that has a lower activation energy. Catalysts lower the activation energy for forward and reverse reactions in equilibrium equally, and therefore have no effect on the yield.

4 C A primary standard is one that is first weighed in a beaker and then dissolved in a small amount of deionised water. The solution is transferred to a volumetric flask through a filter funnel to avoid spillage. The volumetric flask is filled up to a graduated mark with deionised water.

5 B Fermentation involves the decomposition of glucose to ethanol and carbon dioxide in an anaerobic environment (without oxygen).

6 C Particle accelerators such as scintillators and cyclotrons accelerate positive particles into target nuclei in order to produce neutron-poor isotopes. Neutron-emitting chemicals are used to provide neutrons to make neutron-rich isotopes such as Co-60 in a nuclear reactor.

7 D The solution would be yellow between pH 4.5 and 6 for methyl orange and bromothymol blue and phenolphthalein would be colourless. Thus the mixture would be yellow.

8 D A buffer is a weak acid in equilibrium with an approximately equal concentration of its conjugate base. This is the only answer with this combination.

9 B Three hydrogen atoms share a pair of electrons each with nitrogen and with both atoms supplying an electron. These are covalent bonds. The *fourth* hydrogen shares nitrogen's lone pair of electrons but doesn't provide any electrons itself. This is a coordinate covalent bond.

10 B This answer requires students to know some solubility rules, that is, that barium chloride is soluble and barium sulfate is insoluble and that barium has a green flame colour.

11 B The highest pH is due to the lowest concentration of hydrogen ions in solution. Given that all three acids are the same concentration, the one with the lowest concentration of hydrogen ions is the weakest as it ionises least. Of the three acids given, acetic is the weakest.

12 D Branches are named in alphabetical order, therefore chloro is named before fluoro. Branches are identified by the carbon they are bonded to by number and there is a branch on carbon number 2 and number 3. *Note*: the answer (D) has been accepted by the Board of Studies for 1 mark but it is not actually the *preferred* IUPAC name. There are no correct answers in this question that show the preferred IUPAC name.

13 D The oxidation state of chromium in $Cr_2O_7^{2-}$ is +VI and its oxidation state in Cr^{3+} is +III. Therefore the chromium is being reduced and is accepting electrons and acting as an oxidising agent.

14 A $Fe^{2+} \rightarrow Fe^{3+} + e^-$ $E^0 = -0.77$ V

$\frac{1}{2} Cr_2O_7^{2-} + 7H^+ + 3e^- \rightarrow Cr^{3+} + 3\frac{1}{2} H_2O$ $E^0 = 1.36$ V

$1.36 + (-0.77) = 0.59$ V

15 C CO_2 and SO_3 are acidic oxides and Na_2O is a basic oxide.

16 A At T_2 one of the components was suddenly removed from the system causing the concentration of the two other components to drop, one at twice the rate of the other. This could only result from the removal of CH_3OH. The one that dropped at twice the rate would have to be hydrogen as it reacts in a 2:1 ratio with carbon monoxide. Therefore it is hydrogen that is added to the system at T_1.

17 C $n(C_3H_7OH)$ reacting $= \frac{1530}{2021} = 0.757$ mol

$n(H_2O)$ produced $= 0.757 \times 4 = 3.028$

$m(H_2O) = 3.028 \times 18.016 = 54.55$ g

18 C When acid is diluted, the concentration of hydrogen ions decreases and therefore the pH increases as $pH = -\log_{10}[H^+]$. When water is added, the equilibrium shifts to the right (see the equation below) to counteract the increase in water in the system. Thus the concentration of acetate ions and hydronium ions will increase as the equilibrium shifts to the right.

$CH_3COOH + H_2O(l) \rightleftharpoons CH_3COO^- + H_3O^+$

Therefore the degree of ionisation of acetic acid molecules increases.

19 A $Na_2CO_3(s) + 2CH_3COOH \rightarrow 2CH_3COONa + H_2O(l) + CO_2(g)$

$n(CH_3COOH) = 0.500 \times 0.100 = 0.0500 \text{ mol}$

$n(Na_2CO_3) \text{ required to neutralise } = \frac{0.0500}{2} = 0.02500 \text{ mol}$

$m(Na_2CO_3) = n.M(Na_2CO_3) = 0.02500 \times 105.99 = 2.65 \text{ g}$

20 A $Pb^{2+} + 2Cl^- \rightarrow PbCl_2(s)$

$n(PbCl_2) = \frac{m}{M} = \frac{0.595}{278.1} = 0.00214 \text{ mol}$

$n(Pb^{2+}) \text{ in original solution } = 0.00214 \text{ mol}$

$m(Pb^{2+}) \text{ in original solution } = 0.00214 \times 207.2 = 0.4434 \text{ g}$

$c(Pb^{2+}) \text{ in original solution } = \frac{0.4434}{0.0500} = 8.87 \text{ gL}^{-1}$

Section I Part B

21 (a)

```
    H  H  H   O              H  H                   H  H  H   O
    |  |  |  //              |  |          H2SO4    |  |  |  //                 O—H
H—C—C—C—C          + H—C—C—O—H  ⇌  H—C—C—C—C      H  H    +  |
    |  |  |  \               |  |          reflux   |  |  |  \     |  |          H
    H  H  H   O—H            H  H                   H  H  H   O—C—C—H
                                                                 |  |
                                                                 H  H
```

(2 marks)

(b) As the reactants are flammable and need to be heated, it is safe to heat them in the absence of a naked flame. The reactants can be heated in a flask placed in a hot water bath or in an electrical heating mantle to avoid exposure of flammable gases to the naked Bunsen burner flame. *(2 marks)*

22 (a) Addition polymerisation *(1 mark)*

(b) This model demonstrates an initiator molecule which breaks down to form an initiator radical. This then reacts with ethylene monomers, allowing them to link together to form a long chain polyethylene polymer. Manipulating the models assists in understanding the role played by the double bond in the monomers because when they are physically broken, it is easy to see that an electron is un-bonded and a new covalent bond can form to pair un-bonded electrons. This is better understood using a model than by having it explained verbally as most people learn better kinaesthetically. The use of models helps to show things that cannot be explained easily. Models are also useful in chemistry because atoms and molecules are too small to see and models allow us to see what the molecules most likely look like. *(2 marks)*

23 $CO_2(g) + H_2O(l) \rightleftharpoons H_2CO_3(aq)$ + heat

An increase in pressure increases the solubility of carbon dioxide because increasing the pressure increases the concentration of the gas. The equilibrium system shifts forward to counteract this increase and more gas dissolves in the process.

An increase in temperature decreases the solubility of carbon dioxide as the forward reaction is exothermic. Heating the system favours the endothermic reaction to counteract the change. The endothermic reaction is the reverse reaction and it favours the production of carbon dioxide gas.

(3 marks)

24 Ammonia is used in industry as a feedstock or raw material. It is used to make nitric acid and in the agricultural industry to make a variety of ammonium fertilisers. Nitric acid is one of the most produced chemicals in the world and is used for the production of explosives, nylon and other useful organic compounds. Nitric acid is also used as the feedstock to make ammonium fertilisers such as ammonium nitrate.

Ammonia is so important because other feedstocks can no longer be obtained from the natural environment as they once were, and yet the products remain important to society.

(3 marks)

25 Eutrophication is the enrichment of waterways with excessive amounts of nutrients such as phosphates and nitrates, leading to abundant growth of water plants (such as algae) that then block out light to deeper plants and interfere with the diffusion of oxygen into the water.

In order to monitor the level of eutrophication, levels of nitrogen and phosphorus can be measured directly using a titration for ammonia (produced by converting nitrogen compounds to ammonia) or colorimetry (phosphate compounds can produce blue-coloured solutions with molybdate ions), respectively. An N:P ratio below 10:1 is an indicator that phosphorus levels are too high.

Eutrophication can be measured indirectly by measuring the amount of dissolved oxygen in the waterway as eutrophication results in low levels of dissolved oxygen (DO).

(3 marks)

26 (a)

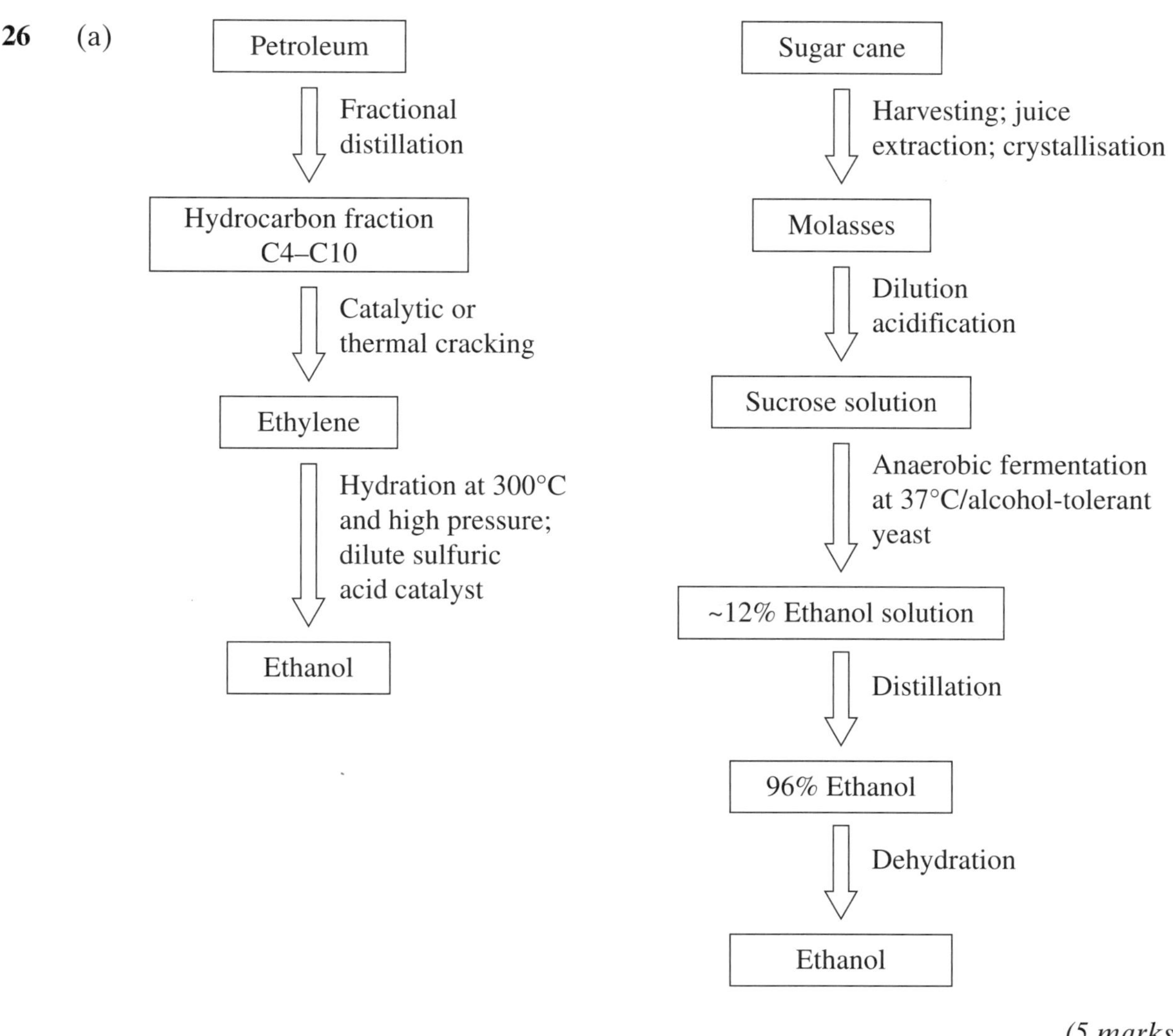

(5 marks)

(b) The production of ethanol from petroleum is not environmentally sustainable, whereas from sugarcane it is. Petroleum is a non-renewable resource and will become too expensive to extract sometime in the near future. In addition, the extraction of petroleum can lead to environmental disasters. A recent example of this was in the Gulf of Mexico, when a mine drilling resulted in an explosion and millions of tonnes of oil leaked into the environment. There are dangers with underground leaks and shipping disasters as petroleum is moved around the planet. Large amounts of water for heating and cooling are also used in the extraction of ethanol from petroleum.

Sugar cane is more sustainable as it is renewable: it can be regrown once harvested. This method releases lower net amounts of carbon dioxide into the atmosphere than the combustion of fossil fuels. Bioethanol production is not completely carbon neutral as some petroleum fuel is still used to operate crop harvesters and trucks.

(3 marks)

27 (a) Both Iodine-131 and Iodine-123 are used for diagnosis as they are both gamma emitters. Gamma radiation has high penetrating ability, a property necessary to be used in diagnosis as it needs to penetrate the human body to be detected by a gamma camera or similar instrument outside the body. Iodine-131 is used for therapy as well because it is a beta emitter. Beta radiation is a flow of electrons that penetrates a few layers of cells and can be used to destroy tissue from inside the human body, a property that Iodine-123 does not have. *(2 marks)*

(b) $^{131}_{53}I \rightarrow {}^{0}_{-1}e + {}^{131}_{54}Xe$ *(1 mark)*

28 $n(H^+) = cV = 0.120 \times 0.07500 = 0.009$ mol

$n(OH^-) = cV = 0.200 \times 0.02500 = 0.005$ mol

H^+ is in excess by $0.009 - 0.005 = 0.004$ mol

$c(H^+)$ in mixture $= \frac{n}{V} = \frac{0.004}{0.100} = 0.0400\ molL^{-1}$

$pH = -\log_{10}(0.04)$
$= 1.40$ *(3 marks)*

29

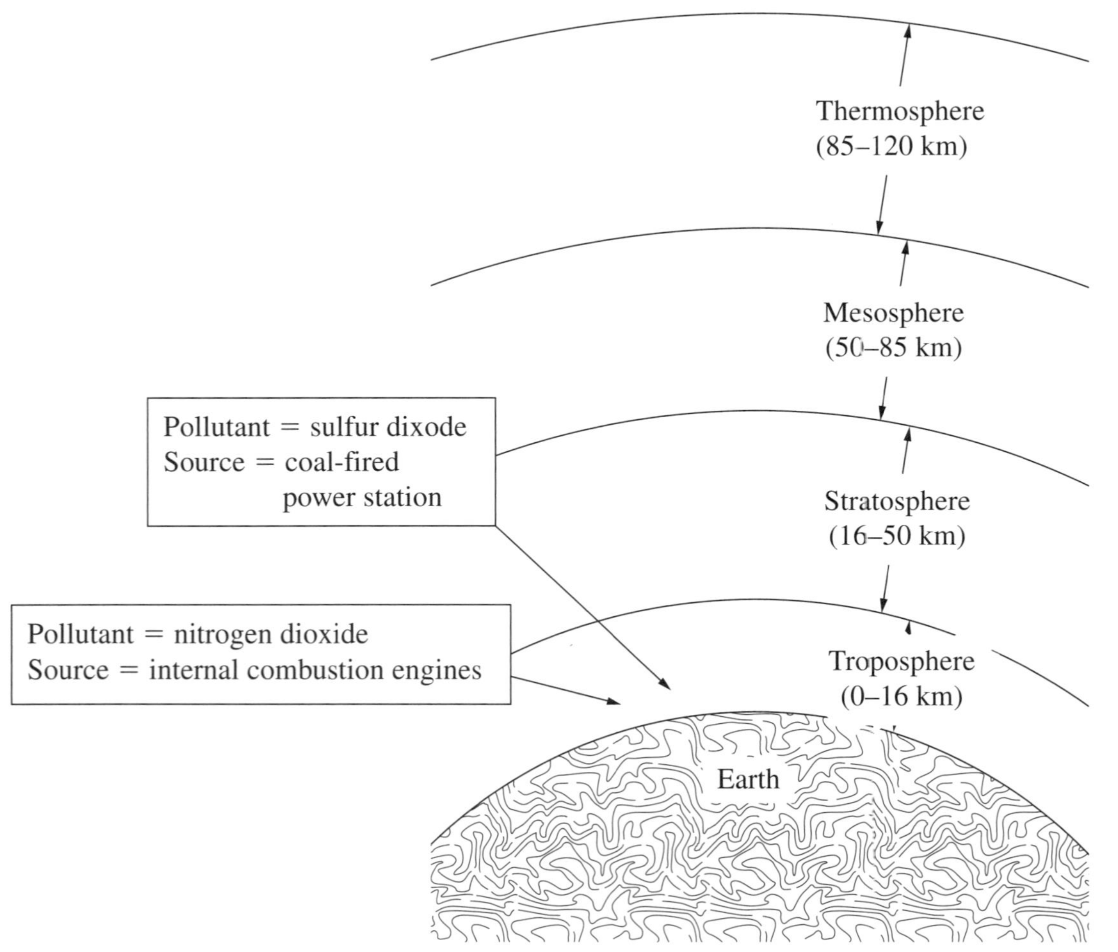

(5 marks)

30 (a) $n(HCl) = cV = 0.1034 \times 0.025 = 0.002585$ mol

$n(NaOH) = 0.002585$ mol

$$c(NaOH) = \frac{n}{V} = \frac{0.002585}{0.02575} = 0.10039\ molL^{-1}$$

$= 0.1004\ molL^{-1}$ (4 sig. fig.) *(2 marks)*

(b) (i) $n(NaOH) = cV = 0.1004 \times 0.01655 = 0.00166$ mol

$m(C_9H_8O_4) = n.M = 0.00166 \times 180.154$ g

$= 0.29932$ g

$= 299.3$ mg (4 sig. fig.) *(3 marks)*

(ii) Aspirin is an organic molecule with both non-polar and polar groups. Ethanol is added to the water solvent to assist the dissolution of the aspirin which is not very soluble in water. The non-polar tail of the ethanol will help to dissolve the non-polar groups in the aspirin and the polar OH functional group will help to dissolve the polar groups. It is important that all of the aspirin is dissolved in order to determine its mass in the tablet. *(1 mark)*

31 (a)

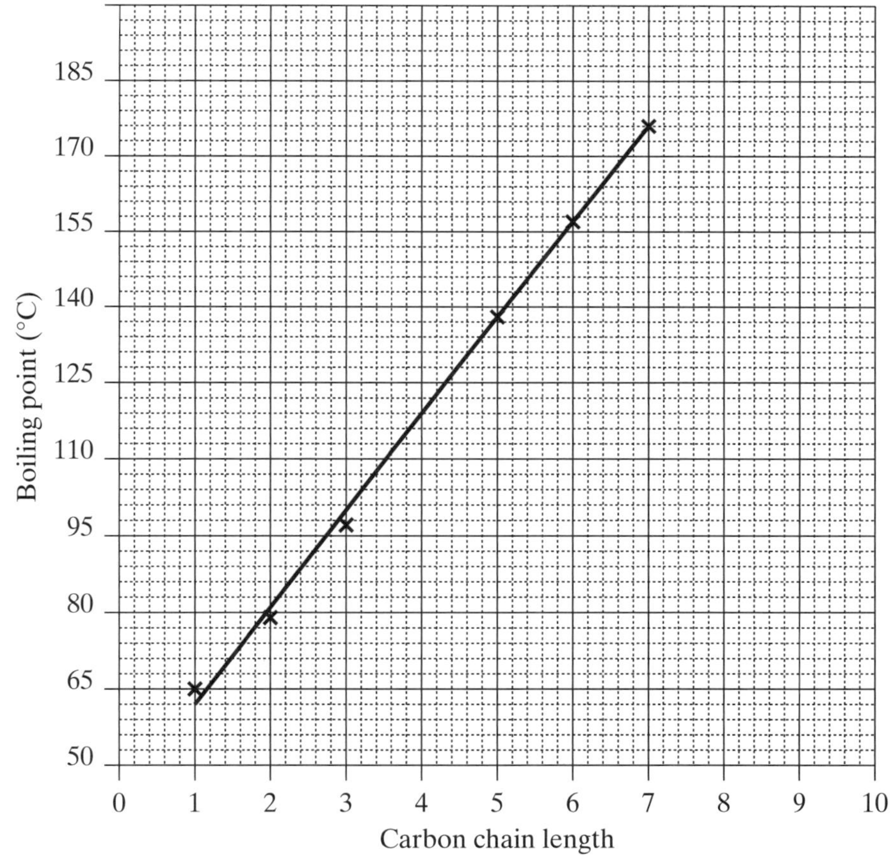

(3 marks)

(b) 118°C *(1 mark)*

(c) dispersion force *(1 mark)*

32 From the calibration curve, a mean absorbance of 0.280 corresponds to a Hg concentration of 0.17 mgL^{-1}.

Therefore, there is $0.17 \text{ mg} \times \frac{25}{1000} = 0.00425$ mg Hg in the 25-mL sample.

Therefore, there is 0.00425 mg Hg per 18.6 g fish $= \frac{0.00425}{18.6} = 0.0002285$ mg/g fish $=$ 0.23 mg/kg fish. As this concentration is below 0.5 mg/kg fish, the consumer can eat the fish. *(3 marks)*

33 The lead/acid accumulator is an electrochemical cell that can cause environmental concerns when disposed of inappropriately in landfill. This technology was developed as a portable, mobile electricity source for cars. There is a huge demand for these batteries as nearly every car in the world requires a lead/acid accumulator to function. The impact on the environment relates to the toxicity of lead and the acidity of the electrolyte used in the cell's construction. Lead is a heavy metal that accumulates in living things and increases in concentration up the food chain. In organisms towards the top of the food chain, it can lead to disruption of normal metabolic processes by affecting enzymes. This can lead to brain damage and damage to other organs and the reproductive system. Therefore, lead levels need to be monitored in our environment. Chemists collaborate in this monitoring. Lead concentration can be determined in very low concentrations by using atomic absorption spectroscopy (AAS). This device needed to be developed by chemists at the CSIRO in the 1950s. Other chemists collaborate to undertaking this monitoring by running the machinery. Soil samples need to be collected and prepared by other chemists. Analytical chemists collaborate with them to standardise the AAS so that it can determine the concentration of lead in soil samples. Other chemists collaborate once the results of sampling are obtained to report the findings and make conclusions on these findings. Research chemists are making discoveries about alternatives to car batteries that cause less impact on the environment, such as fuel cells.

ASSESSMENT

There is, therefore, a great need for chemists to collaborate to minimise the damage caused by this technology as some chemists monitor the environment while others use the results of this to develop environmentally friendly alternatives. *(6 marks)*

Section II—Options

Question 34—Industrial Chemistry

(a) The process is called electrolysis.

At the positive electrode (anode, *P*) oxygen gas will be produced from a very dilute solution of NaCl. Chlorine gas will be produced from a concentrated solution. At intermediate concentrations, a mixture of oxygen and chlorine will form.

At the negative electrode (cathode, *Q*) hydrogen gas is produced. *(3 marks)*

(b) (i) $2NO(g) \rightleftharpoons N_2(g) + O_2(g)$ *(1 mark)*

(ii) (In the workings below, note that the volume of the vessel is 1.00 L.) *(3 marks)*

	[NO] mol/L	$[N_2]$ mol/L	$[O_2]$ mol/L
Initially	0.400	0	0
At equilibrium	$0.400 - (2 \times 0.198) = 0.00400$	0.198	0.198

$$K = \frac{(0.198)^2}{(0.004)^2} = 2450.25 = 2450$$

This value indicates that the equilibrium position at 2000°C favours the right-hand side.

(iii) Change the temperature. *(1 mark)*

(c) (i) Oleum is produced when SO_3 is bubbled into concentrated sulfuric acid. The reaction that occurs is $SO_3(g) + H_2SO_4(l) \rightarrow H_2S_2O_7$. The oleum is then reacted with water to produce sulfuric acid: $H_2S_2O_7(l) + H_2O(l) \rightarrow 2H_2SO_4(l)$. *(3 marks)*

(ii) Sulfuric acid can be made directly from sulfur trioxide reacting with water: $SO_3(g) + H_2O(l) \rightarrow H_2SO_4(aq)$. However, this reaction is highly exothermic and the heat generated causes a mist of sulfuric acid/water to form, which is dangerous and difficult to handle. In order to avoid this, oleum is formed first. *(2 marks)*

(d) (i) The conversion of sodium hydrogen carbonate to sodium carbonate can be modelled in the laboratory. The reaction is $2NaHCO_3(s) \rightarrow Na_2CO_3(s) + H_2O(g) + CO_2(g)$. The reaction is set up as shown below. The sodium hydrogen carbonate is strongly heated in a side-arm test tube. The products are transferred to a test tube in an ice bath to collect the water, and then into limewater to identify the $CO_2(g)$. *(3 marks)*

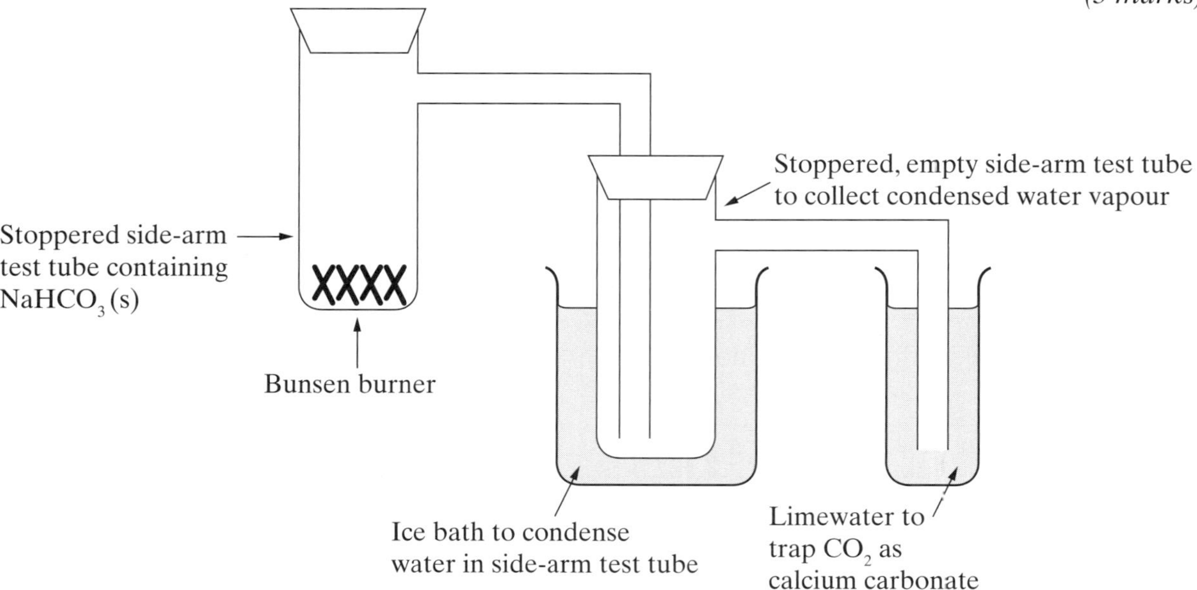

(ii) One risk factor: The limewater could be forced backwards into the heated tube of sodium hydrogen carbonate due to pressure imbalances with the atmosphere. Heating drives gases out of the tube and creates a partial vacuum. In order to avoid this problem, the limewater tube must be disconnected before the heating ceases.

One difficulty: Because $NaHCO_3$ and Na_2CO_3 are both white solids, there is no visual cue about the progress of the reaction. As a result the reaction was only carried out until some water formation was observed, or until bubbles stopped passing through the limewater. *(2 marks)*

(e) A surfactant is a substance which reduces the surface tension of water so that a homogeneous mixture can be formed between water and a non-polar substance such as oil. The mixture thus formed (water, oil, surfactant) is known as an emulsion. Soaps are examples of surfactants, and they clean because they have a non-polar, hydrophobic tail which 'dissolves' in fat/oil droplets, and a polar (or ionic), hydrophilic head which dissolves in water. Agitation of the mixture results in the formation of micelles, which are 3-D spheres composed of fat/oil droplet, surfactant and water. The fat/oil droplets are thus dispersed in the water, allowing the object to be cleaned.

Until about the 1940s, soaps were the only product of the surfactant industry, and animal fats and vegetable oils were the main source of raw materials for soap production. Soaps are produced in saponification reactions, using fats or oils as starting materials. The fats/oils are triesters of glycerol and fatty acids. They are heated with sodium hydroxide to form glycerol and the sodium salt of the fatty acids. It is the sodium salt of the fatty acid which is the soap, such as $NaCH_3(CH_2)16COO$. However, societal pressures in the early

20th century such as competing uses of the starting materials (animal fats, vegetable oils) for food, and the problems associated with using soap in hard water, spurred interest in developing synthetic detergents. (Soaps form an insoluble scum in hard water due to their precipitation with Mg^{2+} and Ca^{2+} ions, and are thus ineffective at cleaning in hard water.)

The first synthetic detergents developed were highly branched, alkylbenzene sulfonates. Because of the highly branched structure, these detergents were not easily biodegraded, and so in the 1960s it was common to see large bodies of froth and foam on beaches, in rivers and lakes. Societal pressure because of environmental concerns, and advances in catalytic chemistry, enabled the production of straight-chain anionic detergents. These are biodegradable and are still in current production and use as laundry detergents, dishwashing liquids and powders. Their structure is shown below.

The next developments came with the production of cationic detergents. These are quaternary ammonium salts, shown below. The cationic 'head' is attracted to hair and clothing fibres, and as a result they find application as hair conditioners and fabric softeners. They also have mild antiseptic properties, and as a result are also used in nappy washing powders and mouthwashes. Cationic detergents do not interact with Mg^{2+} or Ca^{2+} and so are excellent detergents for use in hard water.

Another type of surfactant in general production and use is the non-ionic detergent (shown below). This type of detergent has a polar, polyether head, and a non-polar, hydrocarbon tail. These are very low-froth detergents so are used in front-loading washing machines and dishwashers, for example. They are also an important surfactant in products such as cosmetics, adhesives, paints and pesticides.

It is clear then that the surfactant industry has broadened the scope of its products as a result of societal pressures (competing uses of animal fats/vegetable oils, issues with washing in hard water, concern over frothing of waterways) and chemical developments (discovery of catalysts to produce straight-chain anionic detergents). As well as soaps, it now also produces anionic, cationic and non-ionic detergents.

Structures of detergents:

Anionic detergent

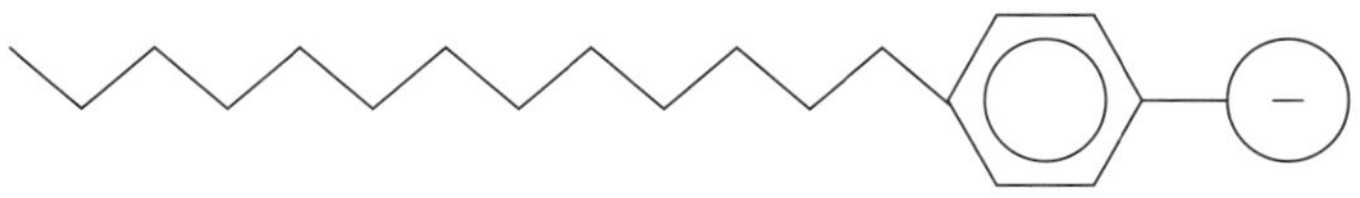

−

Cationic detergent

+

Non-ionic detergent

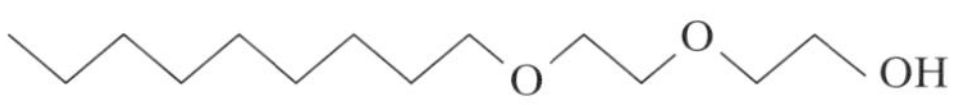

O O OH

(7 marks)

Question 35—Shipwrecks, Corrosion and Conservation

(a) The artefact would have been permeated with salt water. If wooden, all the cells and, if metal, small cracks would have salt water within. By letting the artefact lose water and dry out, the water would evaporate, leaving the salts behind. The salts will crystallise and as the crystals grow they expand and cause the cell walls to break if the artefact is wooden. If metallic, the salt crystals expand pushing the cracks open. Either way, expanding salt crystals reduce the structural integrity of artefacts, making them less strong and likely to crumble. *(3 marks)*

(b) (i) This method of protection is called cathodic protection. The steel pipe is protected from corroding by forcing it to become the cathode. This is accomplished by using a power source to force electrons into the pipe. Corrosion involves losing electrons, and therefore the pipe cannot corrode. As a cathode, water and oxygen are reduced on its surface but the steel itself remains unaffected.

Reaction at the graphite anodes: $H_2O(l) \rightarrow \frac{1}{2}O_2(g) + 2H^+ + 2e^-$

Reaction at the steel pipe cathode: $Fe^{2+} + 2e^- \rightarrow Fe(s)$ *(3 marks)*

(ii) Alternatively, the pipe could be protected by using sacrificial anodes. Metals, more reactive than iron, can be attached along and to the pipe surface. Zinc is often used as it will corrode in preference to iron as it is more reactive.

$Fe(s) \rightarrow Fe^{2+}(aq) + 2e^-$ $\quad E^\circ = 0.44$ V

$Zn(s) \rightarrow Zn^{2+}(aq) + 2e^-$ $\quad E^\circ = 0.76$ V

The steel pipe will become the cathode and will, therefore, be protected from corroding. *(2 marks)*

(c) (i) This process represents the reduction of sulfate ions found in the ocean by sulfate-reducing bacteria. The bacteria gains the electrons for this reaction by taking them from species in the ocean that are easily oxidised. These bacteria are anaerobic. Therefore, the conditions necessary for this process are a deep ocean environment where there is little oxygen present and the presence of matter that can readily be oxidised such as may occur on the ocean floor where dead organisms may settle.

The process is assisted by an acidic environment. The conditions that may create this are a surface where encrusting animals can live and produce carbon dioxide as a product of cellular respiration. The carbon dioxide reacts with water, forming carbonic acid. *(2 marks)*

(ii) $4Fe(s) + SO_4^{2-}(aq) + 10H^+(aq) \rightarrow 4Fe^{2+}(aq) + H_2S(aq) + 4H_2O(l)$

$n(H_2S) \text{ produced} = \frac{0.76}{34.086} = 0.0222965 \text{ mol}$

$n(Fe) \text{ reacting} = 0.0222965 \times 4 = 0.089186 \text{ mol}$

$m(Fe) = 0.089186 \times 55.85$

$= 4.981$ g

$= 5.0$ g (2 sig. fig.) *(3 marks)*

(d) (i) In a valid investigation, there can only be one variable that is changed. In this investigation, voltage, concentration of electrolyte and the surface area of the anode are all changed. *(1 mark)*

(ii)

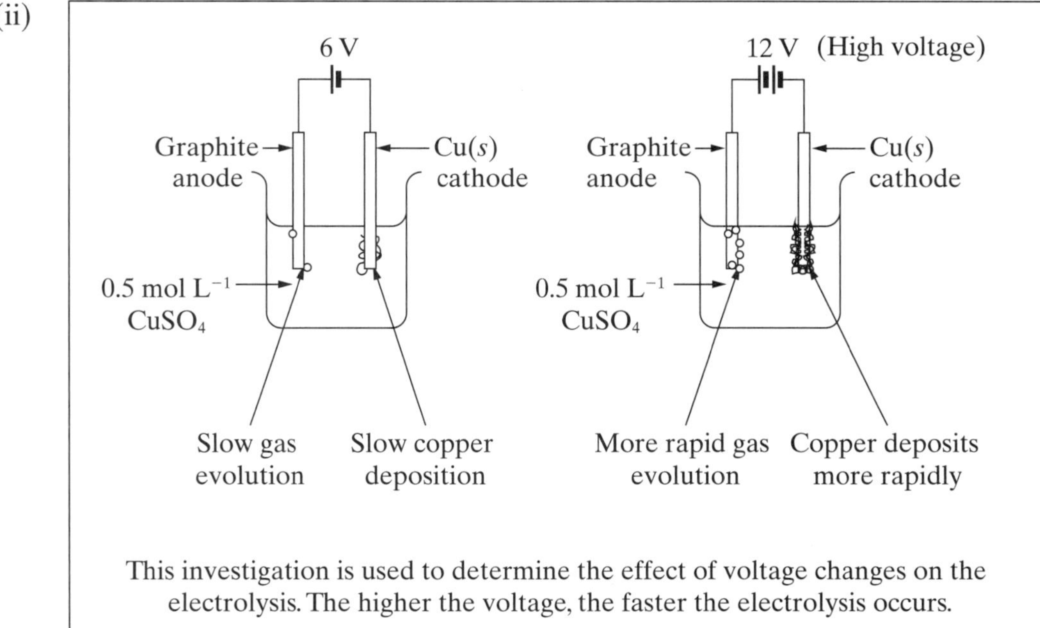

This investigation is used to determine the effect of voltage changes on the electrolysis. The higher the voltage, the faster the electrolysis occurs.

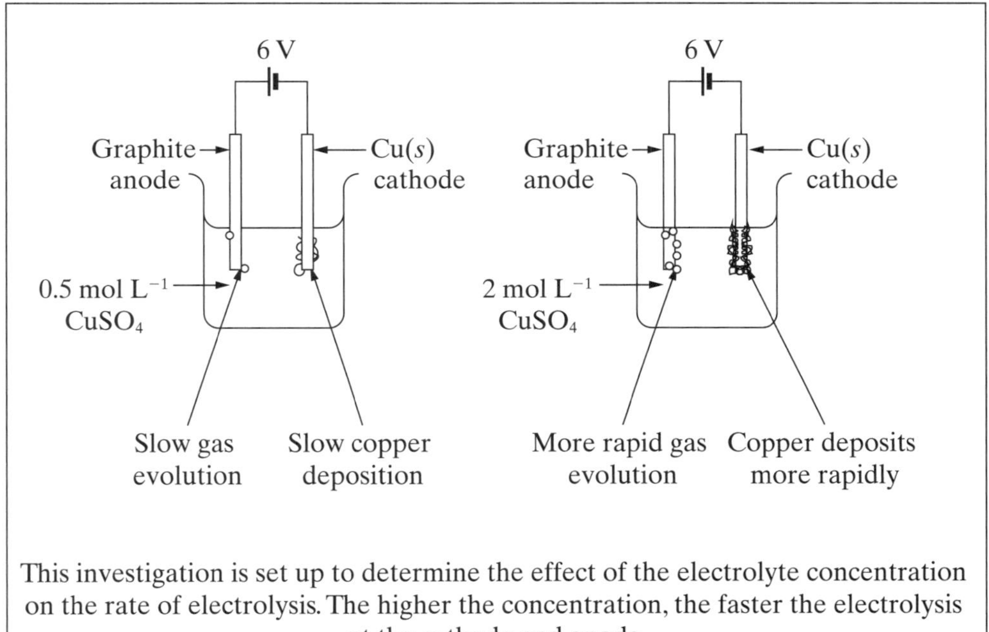

This investigation is set up to determine the effect of the electrolyte concentration on the rate of electrolysis. The higher the concentration, the faster the electrolysis at the cathode and anode.

(4 marks)

(e) Metals were first used in ship building by the Vikings on their longboats. They used wrought iron and the tough, durable alloy bronze. Some early ships used lead sheathing but this was replaced with copper soon afterwards as copper protected wood from attack by marine boring worms. Around 1500 AD, iron nails were developed to connect wooden planks together allowing for larger boats with stronger hulls. By the 1800s, iron was widely available and ships were being built of wooden planks over iron frames. The first fully iron ship was built in 1818, and by the late 1800s most ships were made completely of iron. Iron rusts readily and these early iron ships needed to be dry-docked often to have rust removed. Iron ships could be built longer, safer and with more room inside. Warships began to be constructed of the iron alloy, steel, because this was much stronger and lighter and these ships could withstand shells and cannon balls fired at them. Large modern ships continue to be constructed from steel and alloyed steel as these ships often have to transport large loads.

In the 20th century, there was a progressive improvement in steel alloys to incorporate aluminium, chromium, titanium, zinc and nickel. Aluminium is not only alloyed with iron in ships but is also used in yachts and dinghies because it is light and strong, and allows for improvements in economy and speed. Chromium is used in surface alloying. A laser is used to bind a protective chromium coat to the surface layers of steel ships, giving a great deal of corrosion protection to the steel hull. Zinc is used as sacrificial anodes on the surface of large ships as zinc will oxidise in preference to iron, thus protecting the iron from corroding.

$Zn(s) \rightarrow Zn^{2+}(aq) + 2e^-$ $\quad E° = 0.76V$

$Fe(s) \rightarrow Fe^{2+} + 2e^-$ $\quad E° = 0.44V$

Stainless steel containing 20% chromium and 10% nickel is used in modern ships as railings and in kitchen sinks and dining utensils, as it is corrosion resistant, hard and shiny. Brass consisting of 65% copper and 35% zinc is used to make ship's fittings like door knobs and light fittings, because it resists corrosion and polishes well. Bronze, consisting of 90% copper and 10% tin, is used to construct ship's bells and anchors because it is hard, resists wear and is easily cast. *(7 marks)*

Question 38—Forensic Chemistry

(a) Soil samples (e.g. from the shoes of a crime suspect) can be used by forensic chemists to link the suspect to soil at a crime site. Soil composition varies considerably and forensic scientists can use the characteristics of a soil sample to determine its origin. The size and composition of soil particles can be used in locating its origin. The mineral composition of a soil can be very helpful; for example, a high proportion of quartz in the sample can help narrow its origin to eastern Australia. Other minerals or inorganic compounds can be more specific if the soil's origin was close to a smelter or paint factory, for example. Microscopic organisms in soil are also very useful in determining its origin, such as foraminifera, whose shells are very resistant to decay. The presence of specific types of these organisms can be used to locate a soil's origin very specifically. Pollen grains are characteristic of each plant species and so the presence of specific pollen grains in soil can be related to specific areas where these plants grow. *(3 marks)*

(b) (i) In order to ensure that the evidence collected is accepted by the court, the athlete must be asked to provide the samples without notice, and must be observed continuously from that time, including while the samples are being provided. The athlete must be present while the samples are labelled, packaged and sealed in tamper-proof containers. Contamination must be avoided. They are then transported to the laboratory for testing under security. If a banned substance is detected in the first sample, an independent analysis can then be performed on the second sample. There must be thorough tracking of the samples throughout the collection, labelling, transportation, testing and storage of the samples, which is known as the 'chain of custody'. The laboratory carrying out testing must have current accreditation. *(3 marks)*

(ii) In the late 1880s in London, England, a new serial killer emerged, killing seven women. Nobody was convicted of this crime. With modern DNA analysis this outcome may have been very different. Forensic scientists could have analysed samples left on the victims for the killer's DNA, and compared this against samples from prime suspects. *(2 marks)*

(c) (i) Carotene is the least polar solvent because it travelled fastest in hexane, a non-polar solvent. *(2 marks)*

(ii) Changing the solvent would affect the separation process. The relative affinity of solutes for the solvent versus the stationary phase would change, and as a result the position of each separated component would be different at the end of the separation, changing the Rf values of each component. The reference data used to identify solutes in the sample must be specific to the solvent being used to account for these changes. Changing the solvent may also result in only partial separation of some solutes. For example, if a more polar solvent had been used, xanthophyll may have been the closest pigment to the solvent wetting front. *(3 marks)*

(d) (i) (Please note that the structural formula of fructose given in the *2012 HSC Examination Paper* is incorrect.)

α-glucose + β-fructose ⟶ water + sucrose

(2 marks)

(ii) Glucose and fructose are reducing sugars, whereas sucrose is not. This means that glucose and fructose will react with Benedict's solution (an alkaline solution of Cu^{2+}). To a test tube containing a solution of glucose, add 1 mL of Benedict's solution. Heat the mixture gently over a Bunsen burner or in a hot water bath. The formation of an orange to brick-red precipitate confirms that the sugar is a reducing sugar. The same result will be obtained with fructose, but not with sucrose. In this case, the pale blue colour of the Benedict's solution will persist.

The following equation shows the Benedict's test for glucose:

$$C_5H_{11}O_5 - CHO + 2Cu^{2+} + 5OH^- \rightarrow C_5H_{11}O_5 - COO^- + Cu_2O(s) + 3H_2O(l)$$

(3 marks)

(e) DNA is present in the nucleus of the cells of all living things. It has a polymeric structure, and is composed of two polymer strands, which together form a double helix structure. Each polymer strand is made up of nucleotides, which consist of a phosphate group, a sugar (deoxyribose) and one of four bases: adenine (A), thymine (T), guanine (G) and cytosine (C). The nucleotide units are bonded together via the phosphate group to produce the polymer chains. The polymer strands in the double helix are held together by complementary base pairing, shown in the diagram on the following page. In effect, hydrogen bonding between specific base pairs holds the two strands together. Adenine hydrogen bonds with thymine, and guanine with cytosine. This complementary base pairing is crucially important in the replication of DNA. The specific sequence of bases along the chain is what carries the genetic code. DNA has a number of features which make its forensic analysis possible.

The most important feature of DNA related to its use in forensic analysis is that it is unique to each individual (except for identical twins). There are sections of the base sequence which are not part of a gene, and these can be used to create a DNA profile of an individual, because they are as unique to the individual as their fingerprint. Comparison of this DNA profile with samples found at a crime scene enable (in conjunction with other evidence) suspects to be identified or eliminated in an investigation.

Another important feature of DNA is that it is found in every living cell. This means that a forensic scientist can compare DNA in a blood sample found at a crime scene with the DNA in saliva or skin cells left on the edge of a coffee cup.

DNA is a durable, robust molecule allowing evidence to be preserved. It can remain intact for very long periods of time, allowing a DNA profile to be generated from a sample that is very old.

A technique called the polymerase chain reaction (PCR) allows extremely small samples to be analysed. When a biological sample is found at a crime scene, the cells are first broken up and the DNA isolated. The sample is gently heated to separate the two strands of the double helix. Free nucleotides are also present in the mixture. An enzyme is added which assembles the complementary strand to each of the now separated strands. In this way, the number of copies of the DNA originally present is doubled. The feature of DNA which allows this to work is the complementary base pairing between nucleotides. The PCR can be repeated until the required amount of material is present.

The next stage of developing a DNA profile is cutting the long polymer strands into smaller pieces. This is possible because restriction enzymes are able to cut the DNA strand at specific places depending on the particular base sequence. The short segments of DNA thus produced are used to build up the person's DNA profile. Because DNA fragments are charged, they can be separated by gel electrophoresis, which separates them according to size. Base sequences also bind to specific marker molecules, so these are used after electrophoresis to allow the visualisation of the DNA profile after separation. In this way the features of DNA allow it to be manipulated to generate an individual's DNA profile.

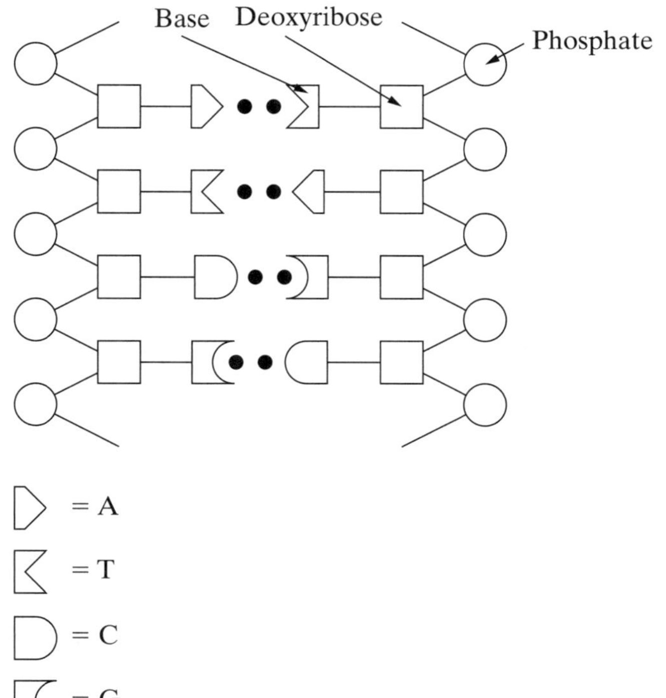

(7 marks)

CHAPTER 11

BOARD OF STUDIES
NEW SOUTH WALES

2013
HIGHER SCHOOL CERTIFICATE
EXAMINATION

Chemistry

General Instructions

- Reading time – 5 minutes
- Working time – 3 hours
- Write using black or blue pen
  Black pen is preferred
- Draw diagrams using pencil
- Board-approved calculators may be used
- A data sheet and a Periodic Table are provided at the back of this paper
- Write your Centre Number and Student Number where required

Total marks – 100

Section I

75 marks

This section has two parts, Part A and Part B

Part A – 20 marks

- Attempt Questions 1–20
- Allow about 35 minutes for this part

Part B – 55 marks

- Attempt Questions 21–31
- Allow about 1 hour and 40 minutes for this part

Section II

25 marks

- Attempt ONE question from Questions 32–36
- Allow about 45 minutes for this section

Section I
75 marks

Part A – 20 marks
Attempt Questions 1–20
Allow about 35 minutes for this part

Use the multiple-choice answer sheet for Questions 1–20.

---

1 Which pair of reactants is used to industrially synthesise ammonia?

(A) H_2 and C

(B) H_2 and N_2

(C) H_2O and N_2

(D) H_2O and NO_2

2 What is the purpose of the flame in atomic absorption spectroscopy (AAS)?

(A) To ionise the sample

(B) To produce a spectrum

(C) To atomise the substance

(D) To provide the absorption wavelength

3 What is the main environmental problem associated with chlorofluorocarbon compounds?

(A) Acid rain

(B) Eutrophication

(C) Global warming

(D) Ozone depletion

4 Butan-1-ol burns in oxygen according to the following equation.

$$C_4H_9OH(l) + 6O_2(g) \rightarrow 4CO_2(g) + 5H_2O(l)$$

How many moles of carbon dioxide would form if two moles of butan-1-ol were burnt in excess oxygen?

(A) 2

(B) 4

(C) 8

(D) 10

5 When placed in the Periodic Table, the recently discovered element 116 would be found in the same group as

(A) element 16.

(B) element 43.

(C) element 87.

(D) element 102.

6 A representation of the Periodic Table is shown. The positions of six different elements, *P*, *Q*, *R*, *S*, *T* and *U* are given.

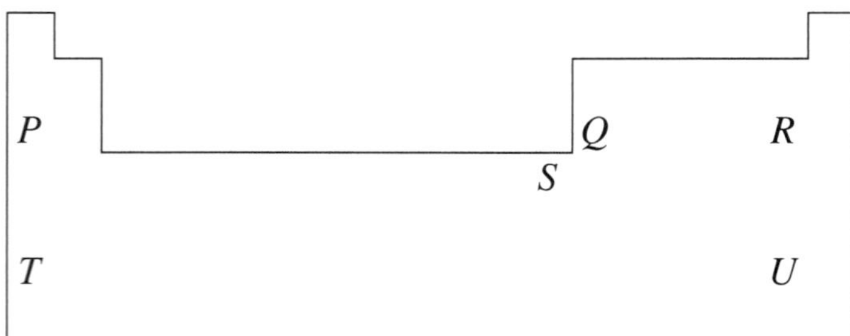

Which row of the following table shows the correct acid and base reactivities of the oxides of these elements?

	Oxide reacts with acid only	*Oxide reacts with base only*	*Oxide reacts with both acid and base*
(A)	*P*, *T*	*R*, *U*	*Q*, *S*
(B)	*P*, *R*	*T*, *U*	*Q*, *S*
(C)	*Q*, *R*	*P*, *S*	*T*, *U*
(D)	*Q*, *S*	*R*, *U*	*P*, *T*

7 Which of the following is the correct equation for an INCOMPLETE combustion of octane?

(A) $C_8H_{18}(l) + 8O_2(g) \rightarrow 8CO_2(g) + 9H_2(g)$

(B) $C_8H_{18}(l) + \frac{25}{2}O_2(g) \rightarrow 8CO_2(g) + 9H_2O(l)$

(C) $C_8H_{18}(l) + 6O_2(g) \rightarrow 3CO(g) + 5C(s) + 9H_2O(l)$

(D) $C_8H_{18}(l) + 13O_2(g) \rightarrow 4CO(g) + 4C(s) + 9H_2O(l)$

8 Which of the following structural formulae shows citric acid, also known as 2-hydroxypropane-1,2,3-tricarboxylic acid?

(A)

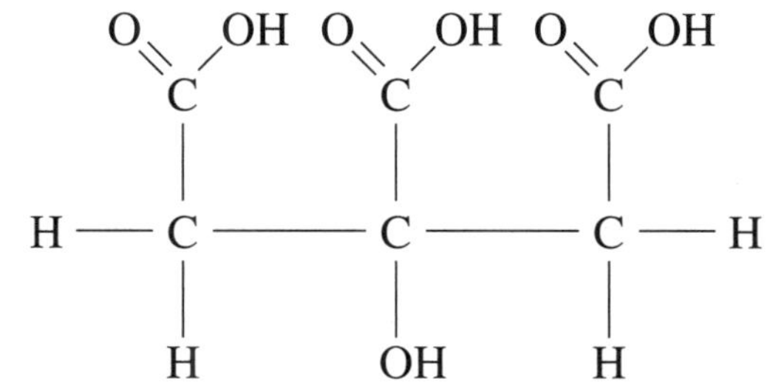

```
     O   OH  O   OH  O   OH
      \\ /    \\ /    \\ /
        C       C       C
        |       |       |
   H —  C ———— C ———— C — H
        |       |       |
        H       OH      H
```

(B)

```
     O   OH  O   OH  O   OH
      \\ /    \\ /    \\ /
        C       C       C
        |       |       |
   H —  C ———— C ———— C — OH
        |       |       |
        OH      OH      H
```

(C)

```
     O   OH  O   OH  O   OH
      \\ /    \\ /    \\ /
        C       C       C
        |       |       |
  HO —  C ———— C ———— C — H
        |       |       |
        H       H       H
```

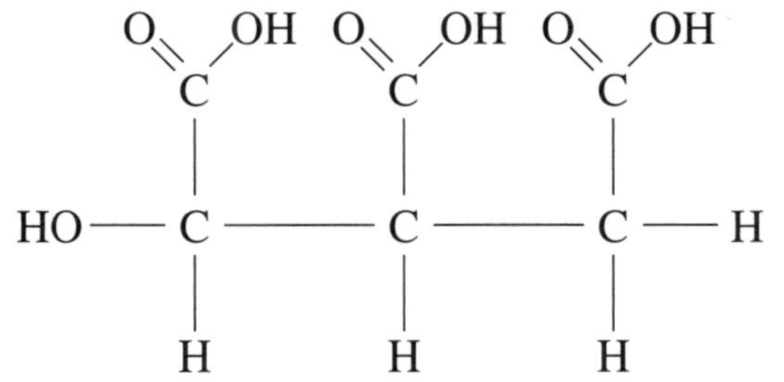

(D)

```
     O   OH  O   OH  O   OH
      \\ /    \\ /    \\ /
        C       C       C
        |       |       |
  HO —  C ———— C ———— C — OH
        |       |       |
        H       H       H
```

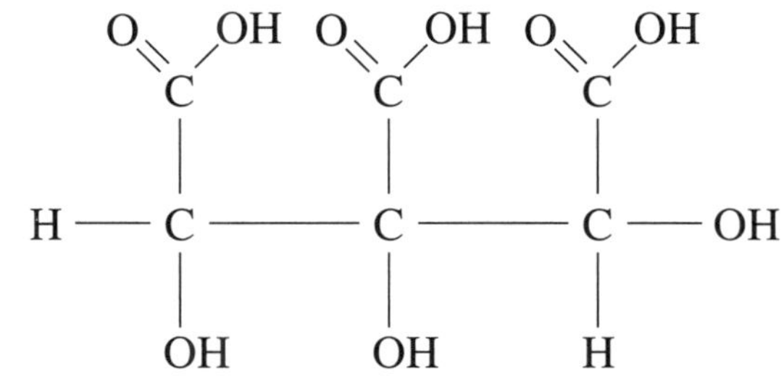

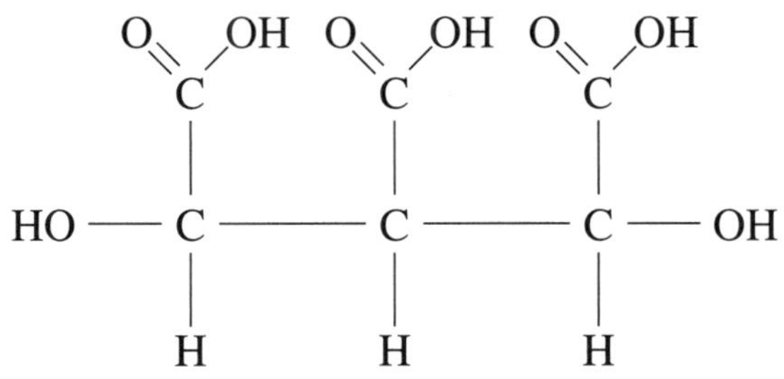

9 A portion of a resin made from acrylic acid ($CH_2{=}CHCOOH$) is shown.

$$\left[-CH_2-\underset{\underset{\displaystyle O \!\!=\!\! C-OH}{|}}{CH}-CH_2-\underset{\underset{\displaystyle O \!\!=\!\! C-OH}{|}}{CH}- \right]_n$$

Which type of reaction results in the formation of this polymer?

(A) Addition

(B) Condensation

(C) Dehydration

(D) Esterification

10 The following equilibrium is set up in a sealed reaction vessel.

$$N_2O_4(g) \rightleftharpoons 2NO_2(g) \qquad \Delta H = +54.8 \text{ kJ mol}^{-1}$$

Which of the following would INCREASE the yield of nitrogen dioxide?

(A) Adding a catalyst to the reaction vessel

(B) Decreasing the volume of the reaction vessel

(C) Raising the temperature of the reaction vessel

(D) Increasing the pressure by adding argon to the reaction vessel

11 The table shows the heat of combustion for four compounds.

Compound	*Heat of combustion* (kJ mol^{-1})
CO	233
CH_4	890
C_2H_2	1300
C_2H_6	1560

Which of these compounds would produce the greatest amount of energy if 1.00 g of each is burnt?

(A) CO

(B) CH_4

(C) C_2H_2

(D) C_2H_6

12 An experiment was set up as shown.

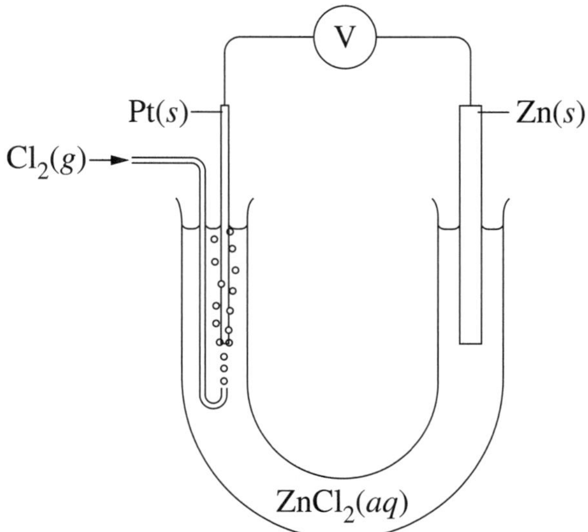

Which of the following statements is correct?

(A) The chlorine gas is the anode.

(B) The zinc electrode is the anode.

(C) The platinum electrode is the anode.

(D) There is no anode because there is no salt bridge.

13

What is the structure of the polymer most likely to have been used in the manufacture of the cup shown?

(A)

$$-\!\left[CH_2-CH_2-CH_2-CH_2-CH_2-CH_2 \right]_n\!-$$

(B)

$$-\!\left[CH(C_6H_5)-CH_2-CH(C_6H_5)-CH_2-CH(C_6H_5)-CH_2 \right]_n\!-$$

(C)

$$-\!\left[CH_2-CHCl-CH_2-CHCl-CH_2-CHCl \right]_n\!-$$

(D)

CH$_2$OH   H   OH   CH$_2$OH
C — O   C — C   C — O
H   H   O   OH   H   H   H   H   O
C   C   C   C   C   C
O   OH   H   H   H   H   O   OH   H   H
C — C   C — O   C — C
H   OH   CH$_2$OH   H   OH
n

14 Sodium reacts with water to give hydrogen gas and sodium hydroxide solution.

What volume of gas would be produced from the reaction of 22.99 g of sodium at 25°C and 100 kPa?

(A) 11.36 L

(B) 12.40 L

(C) 22.71 L

(D) 24.79 L

15 A pH 3.0 solution of HCl(*aq*) is diluted by adding water to produce a pH 5.0 solution.

Which row in the following table correctly identifies an appropriate volume of the original solution and the volume of water added for this dilution?

	Volume of original solution (mL)	*Volume of water added* (mL)
(A)	100	900
(B)	100	1000
(C)	10	990
(D)	1	1000

16 Which two species will react to form a product containing a coordinate covalent bond?

(A) Ca(*s*) and $2H^+$(*aq*)

(B) H_2O(*l*) and H^+(*aq*)

(C) Ag^+(*aq*) and Cl^-(*aq*)

(D) NH_4^+(*aq*) and OH^-(*aq*)

17 A 25.0 mL sample of a 0.100 mol L^{-1} hydrochloric acid solution completely reacted with 23.4 mL of sodium hydroxide solution.

What volume of the same sodium hydroxide solution would be required to completely react with 25.0 mL of a 0.100 mol L^{-1} acetic acid solution?

(A) Less than 23.4 mL

(B) 23.4 mL

(C) More than 23.4 mL

(D) Unable to calculate unless the concentration of the sodium hydroxide solution is also known

18 Consider the following series of reactions.

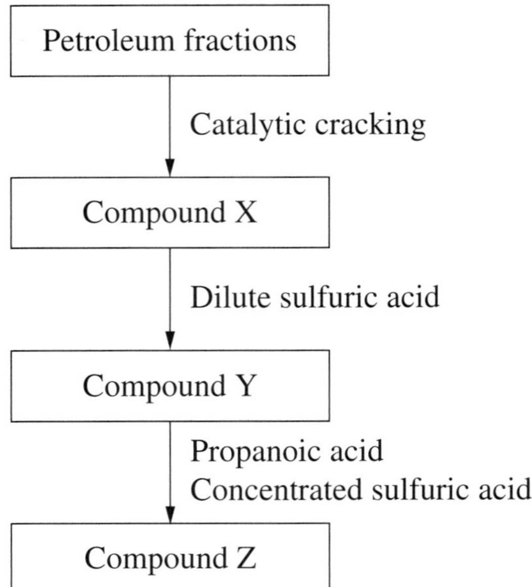

Which row in the table correctly identifies Compounds X, Y and Z?

	Compound X	*Compound Y*	*Compound Z*
(A)	Propene	Propan-1-ol	Ethyl propanoate
(B)	Propene	Ethanol	Propyl ethanoate
(C)	Ethanol	Ethylene	Propyl ethanoate
(D)	Ethylene	Ethanol	Ethyl propanoate

19 A solution was obtained by boiling flowers in water. After various substances were added to separate samples of the solution, the colour of each was noted.

Substance added	*Colour observed*
0.1 mol L^{-1} $HCl(aq)$	Bright pink
0.01 mol L^{-1} $HCl(aq)$	Bright pink
0.001 mol L^{-1} $HCl(aq)$	Pale yellow
Distilled water	Bright yellow
0.001 mol L^{-1} $NaOH(aq)$	Bright yellow
0.01 mol L^{-1} $NaOH(aq)$	Bright yellow

For which of the following titrations would it be appropriate to use this solution as an indicator?

(A) $HCl(aq) + NH_3(aq)$

(B) $HCl(aq) + NaOH(aq)$

(C) $CH_3COOH(aq) + NH_3(aq)$

(D) $CH_3COOH(aq) + NaOH(aq)$

20 The structures of ozone and molecular oxygen are shown.

$$O{=}O{-}O \qquad O{=}O$$

Ozone   Oxygen

Ozone is more easily decomposed than molecular oxygen because

(A) it is polar.

(B) it is a bent molecule.

(C) it has a greater molecular mass.

(D) it has a lower average bond energy.

2013 HIGHER SCHOOL CERTIFICATE EXAMINATION

Chemistry

Centre Number

Section I (continued)

Student Number

Part B – 55 marks
Attempt Questions 21–31
Allow about 1 hour and 40 minutes for this part

Answer the questions in the spaces provided. These spaces provide guidance for the expected length of response.

Show all relevant working in questions involving calculations.

Question 21 (6 marks)

Question 21 continues

Question 21 (6 marks)

An ester and an acid are both listed as additives on a food label.

(a) Why is acid used as a food preservative? **2**

....

....

....

....

(b) Explain, on a labelled diagram, why reflux is used to produce an ester. **4**

Question 22 (5 marks)

A solution contains three cations, Ba^{2+}, Cu^{2+} and Pb^{2+}. The flow chart indicates the plan used to confirm the identity of these cations.

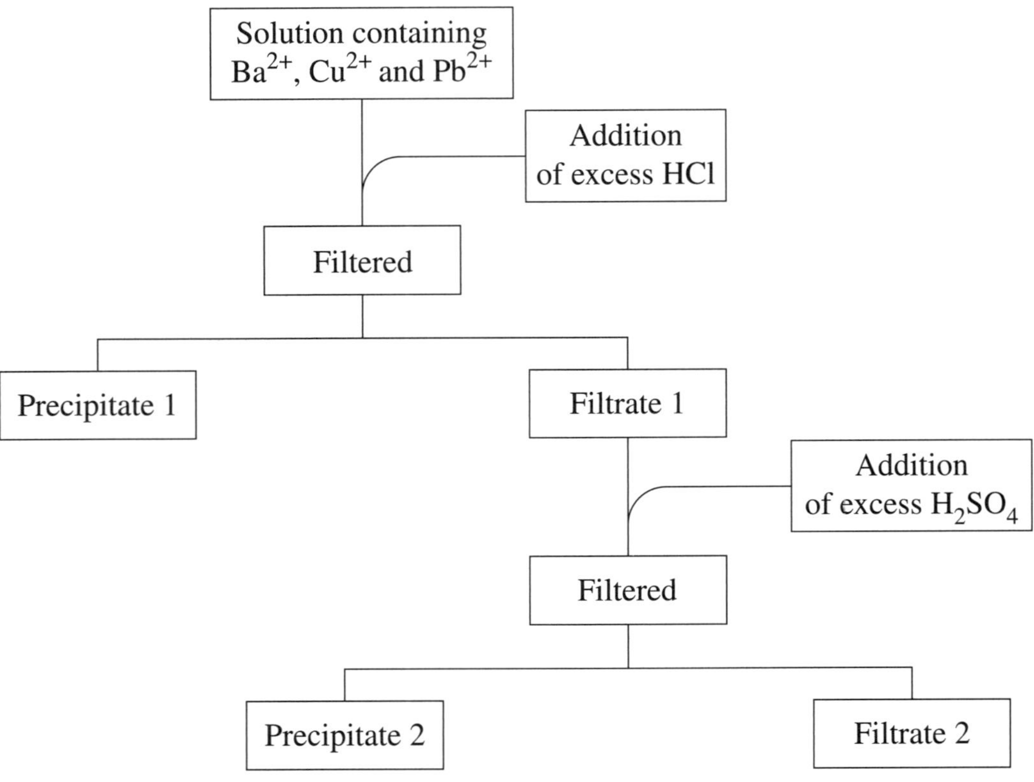

(a) Name Precipitate 2. **1**

....

(b) Write a balanced net ionic equation for the formation of Precipitate 1. **2**

....

....

(c) Suggest a test and the expected result that would confirm the identity of the metal cation remaining in Filtrate 2. **2**

....

....

....

....

2013 HIGHER SCHOOL CERTIFICATE EXAMINATION

Chemistry

Section I – Part B (continued)

Centre Number

Student Number

Question 23 (7 marks)

A 20.72 g sample of solid lead was placed into 0.100 L of 1.00 mol L^{-1} silver nitrate solution.

(a) Complete the table. Show relevant calculations in the space below the table. **5**

Chemical species	$Pb^{2+}(aq)$	$Pb(s)$	$Ag^{+}(aq)$	$Ag(s)$	$NO_3^{-}(aq)$
Moles in final mixture					
Balanced chemical equation					

....

....

....

....

....

....

....

....

....

(b) With reference to only ONE species in the product mixture, explain why care must be taken in disposing of the final mixture. **2**

....

....

....

....

Question 24 (4 marks)

Consider this chemical system which is at equilibrium.

$$X(g) + Y(g) \rightleftharpoons Z(g) + \text{heat}$$

(a) Explain the effect of decreasing the volume of the reaction vessel. **2**

....

....

....

....

(b) Explain the effect of adding a catalyst to this equilibrium mixture. **2**

....

....

....

....

Question 25 (4 marks)

An indicator is placed in water. The resulting solution contains the green ion, Ind^-, and the red molecule, $HInd$. **4**

Explain why this solution can be used as an indicator. In your response, include a suitable chemical equation that uses Ind^- and $HInd$.

....

....

....

....

....

....

....

....

2013 HIGHER SCHOOL CERTIFICATE EXAMINATION

Chemistry

Centre Number

Section I – Part B (continued)

Student Number

Question 26 (4 marks)

Explain how microscopic membrane filters purify contaminated waters, in terms of their design and composition. **4**

....

....

....

....

....

....

....

....

....

....

....

....

Question 27 (4 marks)

A 0.259 g sample of ethanol is burnt to raise the temperature of 120 g of an oily liquid, as shown in the graph. There is no loss of heat to the surroundings. **4**

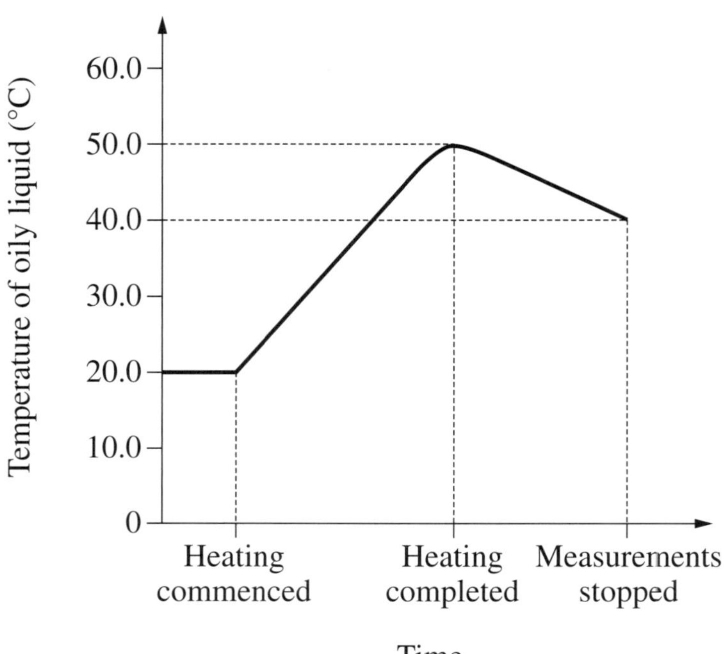

Using the information shown on the graph, calculate the specific heat capacity of the oily liquid. The heat of combustion of ethanol is 1367 kJ mol^{-1}.

....

....

....

....

....

....

....

....

....

....

....

....

....

....

2013 HIGHER SCHOOL CERTIFICATE EXAMINATION

Chemistry

Section I – Part B (continued)

Centre Number

Student Number

Question 28 (5 marks)

A student attempted to determine the concentration of a hydrochloric acid solution. The following steps were performed.

Step 1. A conical flask was rinsed with water.

Step 2. A 25.0 mL pipette was rinsed with water.

Step 3. The student filled the pipette with a standard sodium carbonate solution to the level shown in the diagram.

Gradation mark

Sodium carbonate solution

Step 4. The standard sodium carbonate solution in the pipette was transferred to the conical flask. The student ensured that all of the sodium carbonate solution was transferred to the conical flask by blowing through the pipette. Three drops of an appropriate indicator were added to the conical flask.

Step 5. A burette was rinsed with the hydrochloric acid solution and then filled with the acid. The student then carried out a titration to determine the concentration of the hydrochloric acid solution.

In steps 2, 3 and 4 above the student did not follow acceptable procedures.

(a) Identify the mistake the student made in step 4 and propose a change that would improve the validity of the result. **2**

....

....

....

....

Question 28 (continued)

(b) Explain the effect of the mistakes made in steps 2 and 3 on the calculation of the concentration of the hydrochloric acid solution. **3**

....

....

....

....

....

....

....

....

....

End of Question 28

Question 29 (2 marks)

Consider this chemical equation. **2**

$$2Cl^-(aq) + Br_2(l) \rightarrow 2Br^-(aq) + Cl_2(g)$$

Will the reaction occur spontaneously? Justify your response.

....

....

....

....

....

2013 HIGHER SCHOOL CERTIFICATE EXAMINATION

Chemistry

Centre Number

Section I – Part B (continued)

Student Number

Question 30 (6 marks)

The images below represent the concentration of ozone in the atmosphere above Antarctica near the South Pole for the years 1979, 1994 and 2012, measured at the same time each year. **6**

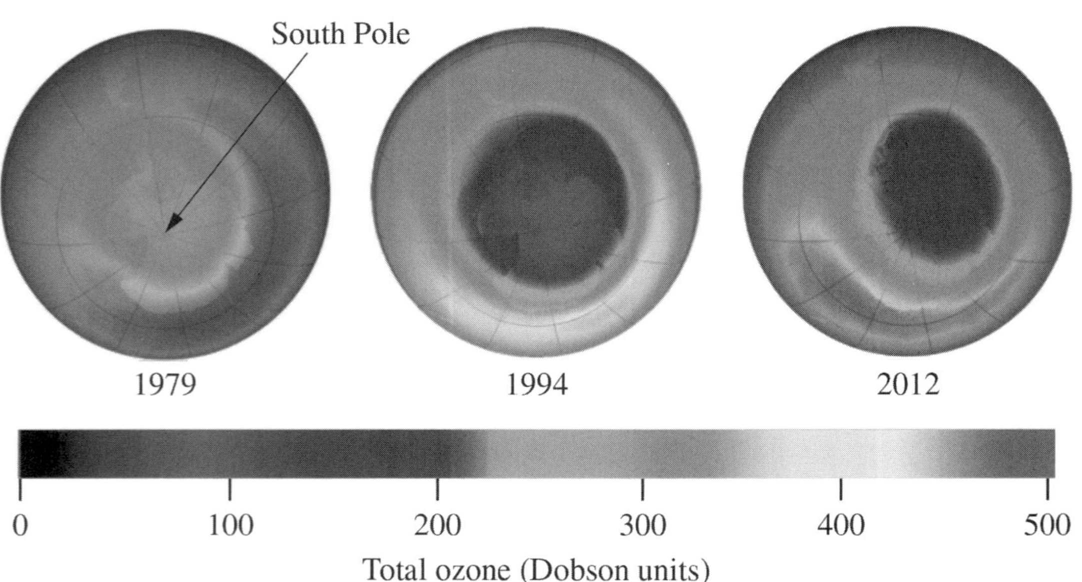

Account for the changes in ozone concentration above Antarctica between 1979 and 2012. Your response should include relevant equations.

....

....

....

....

....

....

....

....

....

Question 31 (8 marks)

(a) Construct separate flow diagrams to show the steps used in the production of polyethylene and those used in the production of a recently developed biopolymer. **5**

Polyethylene	Name of biopolymer: ....

(b) Justify the use of a recently developed biopolymer over a polymer obtained from fossil fuel. **3**

....

....

....

....

....

....

2013 HIGHER SCHOOL CERTIFICATE EXAMINATION

Chemistry

Section II

25 marks
Attempt ONE question from Questions 32–36
Allow about 45 minutes for this section

Answer parts (a)–(c) of the question in Section II Answer Booklet 1.
Answer parts (d)–(e) of the question in Section II Answer Booklet 2.
Extra writing booklets are available.

Show all relevant working in questions involving calculations.

Question 32	Industrial Chemistry
Question 33	Shipwrecks, Corrosion and Conservation
Question 34	The Biochemistry of Movement *(Not included in this reproduction)*
Question 35	The Chemistry of Art *(Not included in this reproduction)*
Question 36	Forensic Chemistry

Question 32 — Industrial Chemistry (25 marks)

Answer parts (a)–(c) in Section II Answer Booklet 1.

(a) The diagram shows a sequence of steps in the removal of grease from a surface. **3**

Explain the process shown in these steps.

Grease
Surface
Step 1 ⇒ Step 2 ⇒ Step 3

(b) Hydrogen iodide is a colourless gas that will decompose into colourless hydrogen gas and purple iodine gas according to the following endothermic reaction.

$$2HI(g) \rightleftharpoons H_2(g) + I_2(g)$$

(i) A 1.0 L glass container was filled with 0.60 moles of hydrogen iodide gas. When equilibrium was established, there were 0.25 moles of iodine gas present in the container. **3**

Calculate the equilibrium constant for this reaction.

(ii) The container was then cooled. **2**

Explain the change in the appearance of its contents.

Question 32 continues

Question 32 (continued)

(c) The diagram shows a region where construction of a Solvay plant is being considered.

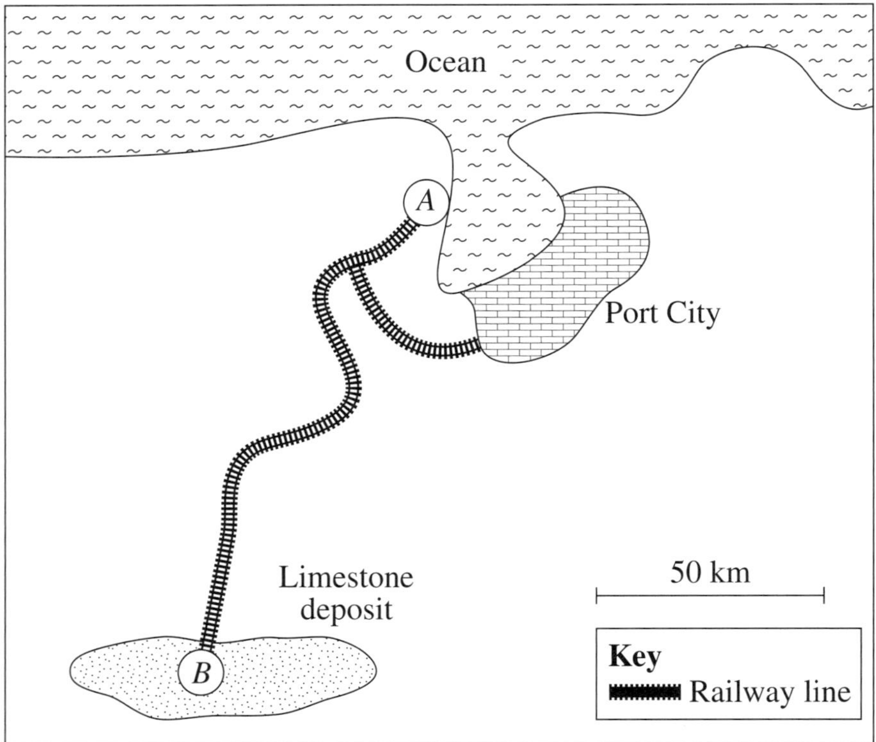

(i) Why is limestone important in the Solvay process? Include a relevant chemical equation in your response. **2**

(ii) Justify which of sites *A* or *B* would be the preferred location for a Solvay plant. **3**

Question 32 continues

Question 32 (continued)

Answer parts (d)–(e) in Section II Answer Booklet 2.

(d) A first-hand investigation is conducted on the electrolysis of an aqueous solution of sodium chloride.

(i) Justify the use of a safety precaution, other than wearing safety glasses, when carrying out this investigation. **2**

(ii) Describe how the THREE products of this electrolysis could be identified. **3**

(e) There is often a compromise between maximising yield and minimising the environmental impact of industrial processes. **7**

Justify this statement with reference to the production of sulfuric acid.

End of Question 32

Question 33 — Shipwrecks, Corrosion and Conservation (25 marks)

Answer parts (a)–(c) in Section II Answer Booklet 1.

(a) The treatment of a marine artefact is shown. 

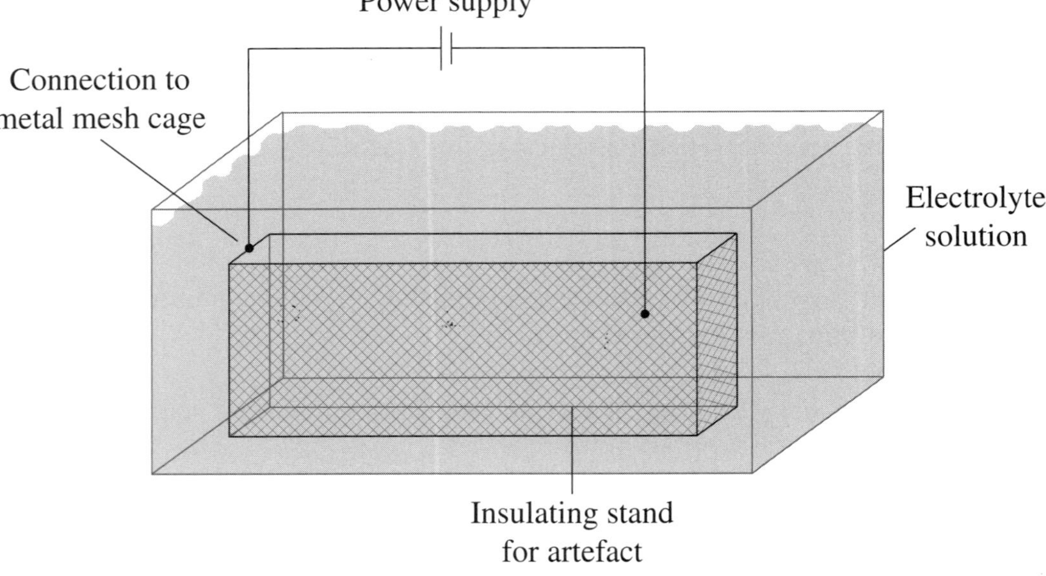

Explain why this process is used.

(b) (i) Write a balanced chemical equation for the corrosion of tin in shallow water. **2**

(ii) Summarise the role of electron transfer in corrosion reactions. Include relevant equations in your answer. **3**

Question 33 continues

Question 33 (continued)

(c) The solubility of oxygen in sea water at various temperatures is shown in the graph.

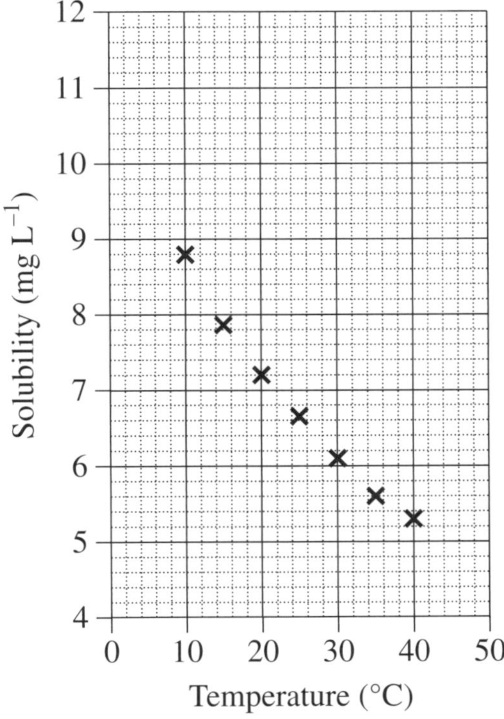

(i) Predict the solubility of oxygen at 4°C. **2**

(ii) Analyse the factors that affect the concentration of dissolved oxygen at increasing depths in the ocean. **3**

Answer parts (d)–(e) in Section II Answer Booklet 2.

(d) (i) Design a suitable first-hand investigation that could be carried out in a school laboratory to compare the effectiveness of different ways of preventing the corrosion of iron in a marine environment. Include relevant diagrams in your answer. **3**

(ii) Use the expected results of the investigation to justify the ongoing use of iron in the construction of ocean-going vessels. **2**

(e) From the scientists studied in this option, assess whose work has most changed the nature of scientific thinking about redox reactions. **7**

End of Question 33

Question 36 — Forensic Chemistry (25 marks)

Answer parts (a)–(c) in Section II Answer Booklet 1.

(a) A sample was collected from inside a reaction vessel at a suspected illegal drug laboratory. The sample was analysed by high performance liquid chromatography (HPLC). The chromatogram obtained from the collected sample is provided below. **3**

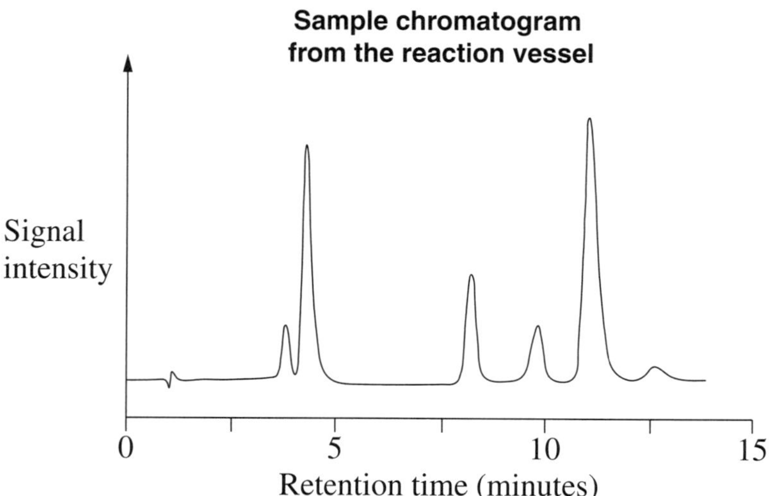

The chromatogram of a standard mixture containing four compounds commonly produced in illegal drug laboratories is also provided.

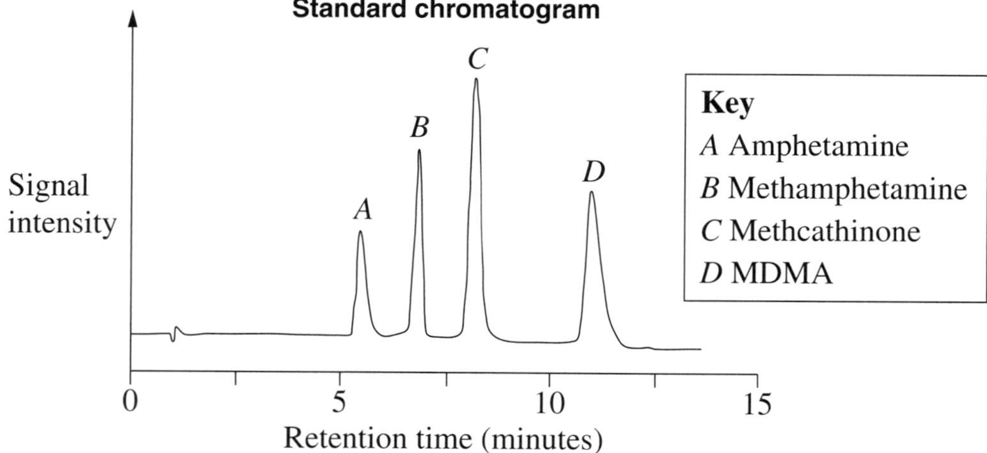

By referring to the chromatograms, explain how this information could be used as evidence to support a guilty verdict in a court case.

Question 36 continues

Question 36 (continued)

(b) (i) Information about three polysaccharides (cellulose, glycogen and starch) is provided in the table below. **2**

X	Y	Z
Plant polysaccharide	Animal polysaccharide	Plant polysaccharide
Contains branched and linear chains	Contains highly branched chains	Contains linear chains

Identify X, Y and Z.

(ii) Select ONE of cellulose, glycogen or starch and explain its primary function in terms of its composition and structure. **3**

Question 36 continues

Question 36 (continued)

(c) A pesticide manufacturer is suspected of releasing waste water contaminated with heavy metal ions into a local river. Atomic emission spectroscopy is used to identify the possible pollutants.

(i) Use the following emission spectra to identify a metal pollutant in the river. Justify your answer. **2**

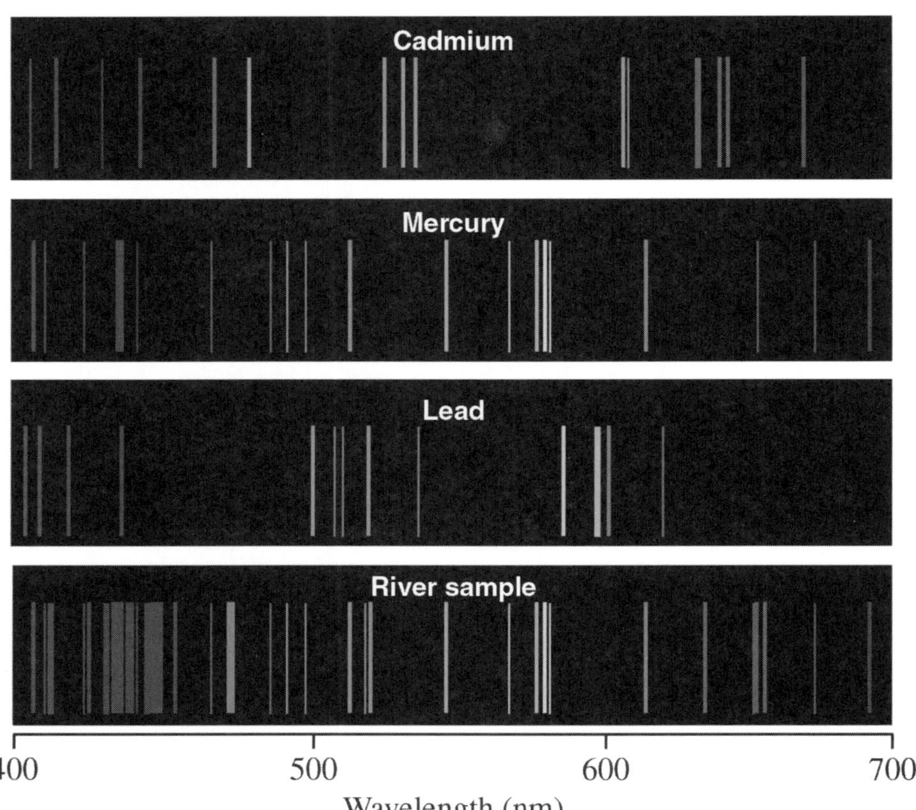

(ii) Outline a procedure that could be used in a school laboratory to SAFELY produce and analyse the emission spectrum of an element. **3**

Question 36 continues

Question 36 (continued)

Answer parts (d)–(e) in Section II Answer Booklet 2.

(d) Alkanes, alkenes, alkanols and alkanoic acids are four different classes of organic compounds.

(i) Describe a simple test that would confirm that a compound is organic. **2**

(ii) Describe a sequence of tests that could be used to distinguish between any THREE of the classes of organic compounds named above. **3**

(e) A chemist working in a forensic laboratory is asked to analyse a residue of an explosive. They selected chromatography as the preferred analytical technique over electrophoresis. Justify this decision. **7**

End of Question 36

End of paper

2013 HSC Examination Paper

Sample Answers

Section I Part A
(Total 20 marks)

1 B

2 C

3 D

4 C One mole of butan-1-ol reacts to form four moles of CO_2 according to the equation.

5 A

6 A Metal oxides form bases and react with acids. Non-metal oxides form acids and react with bases. Those on the border between metals and non-metals are amphoteric and react with both.

7 C The products of incomplete combustion are water, carbon monoxide and soot. Option D is not balanced.

8 A

9 A Addition reactions result from the breaking of double bonds, allowing new covalent bonds to form between monomers without the loss of small molecules.

10 C The forward reaction is endothermic. Raising the temperature will cause the forward reaction to be favoured as the added heat will be absorbed and NO_2 produced.

11 B $890 \div 16$ (molar mass of CH_4) $= 55.6$ kJ g^{-1} is higher than 8.3, 50.0 and 52.0 kJ g^{-1} respectively.

12 B Oxidation will occur on the zinc. An anode is defined as the site of oxidation.

13 D This represents cellulose, the major component of plant cell walls. The other three polymers are all formed from petrochemicals so are not renewable.

14 B $2Na + 2H_2O \rightarrow H_2 + 2NaOH$

$$n(\text{Na}) = \frac{22.99}{22.99} = 1.00 \text{ mol}$$

$$n(\text{H}_2) = \frac{1.00}{2} = 0.500 \text{ mol (from balanced equation)}$$

$$V(\text{H}_2) = 0.500 \times 24.79 = 12.40 \text{ L}$$

15 C The concentration of HCl at pH 3.0 = 1×10^{-3} mol L^{-1}

The concentration of HCl at pH 5.0 = 1×10^{-5} mol L^{-1}

$$V_2 = C_1 \times \frac{V_1}{C_2}$$
$$= 1 \times 10^{-3} \times \frac{10 \text{ mL}}{1 \times 10^{-5}}$$
$$= 1000 \text{ mL}$$

If the final volume is 1000 mL and the initial volume was 10 mL, 990 mL needs to be added.

16 B $H_2O + H^+ \rightarrow H_3O^+$

The H^+ ion shares a pair of electrons with the oxygen atom in water and both electrons are supplied by the oxygen.

17 B The same volume will be required to neutralise a weak acid as a strong acid because the weak acid is in equilibrium with its conjugate ion and hydrogen ion. As the hydrogen ion is neutralised by the sodium hydroxide, the equilibrium will keep shifting forward until all of the original acid molecules have ionised.

18 D Propanoic acid and concentrated sulfuric acid added to compound *Y* form *Z*, a propanoate ester. Because the ester is ethyl propanoate, the alkanol reacting must be ethanol. Ethanol is produced by adding water to ethylene using a dilute sulfuric acid catalyst.

19 A The indicator needs to change colour at or very near the equivalence point of the titration. For a strong acid–weak base titration the indicator needs to change colour between pH 3 and 7. This indicator changes colour from pink to yellow close to the equivalence point.

20 D The other three answers offer reasons for differences in physical properties. Decomposition is a chemical property and is influenced by intra-molecular bonds.

Section I Part B

21 (a) Acids, such as acetic acid, are used as food preservatives because they lower the pH to a point that bacteria cannot survive. *(2 marks)*

(b)

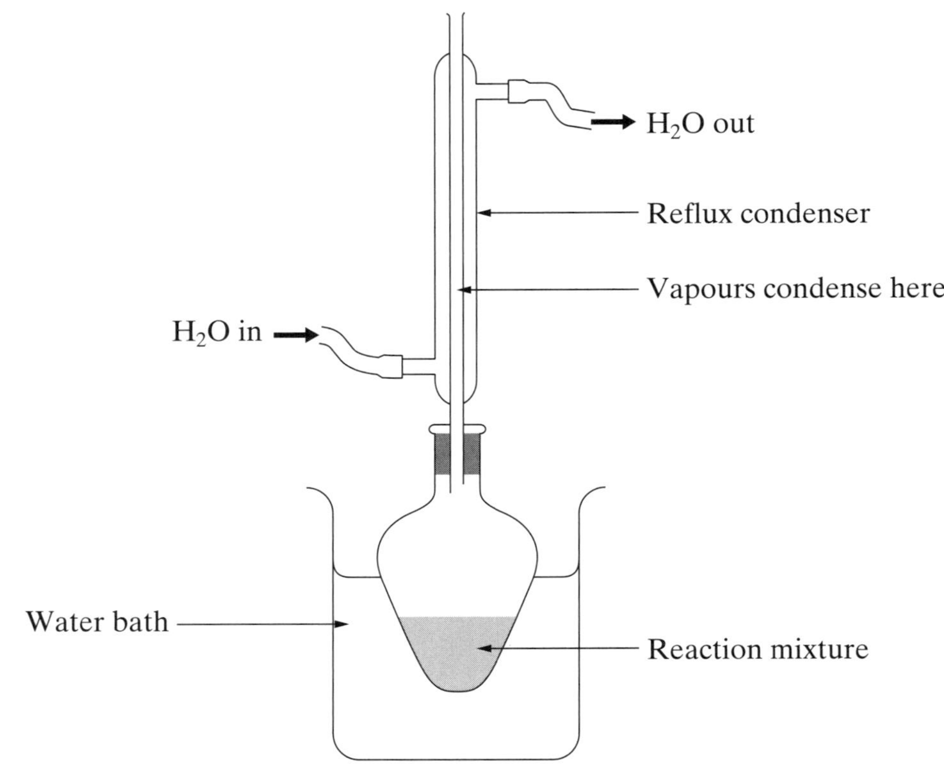

Reflux is the process whereby a reaction mixture (an alkanol and an alkanoic acid) is heated and the volatile components evaporate. Cold water running through an upright reflux condenser surrounds the vapours, which condense and return to the reaction mixture where they continue to be heated. Reflux is used in esterification because the reactants are volatile and because they need to be heated for an extended period of time due to the low reaction rate. If refluxing was not used the volatile reactants and products would escape before equilibrium had been achieved. *(4 marks)*

22 (a) Barium sulfate *(1 mark)*

(b) $Pb^{2+}(aq) + 2Cl^{-}(aq) \rightarrow PbCl_2(s)$ *(2 marks)*

(c) The Cu^{2+} ion can be identified by adding excess sodium hydroxide solution. The hydroxide will precipitate the copper as a pale blue solid (copper(II) hydroxide) which darkens on standing. *(2 marks)*

23 (a)

Chemical species	$Pb^{2+}(aq)$	$Pb(s)$	$Ag^{+}(aq)$	$Ag(s)$	$NO_3^{-}(aq)$
Moles in final mixture	0.0500	0.0500	0.00	0.100	0.100
Balanced chemical equation	$Pb(s) + 2AgNO_3(aq) \rightarrow Pb(NO_3)_2(aq) + 2Ag(s)$				

$$n(Pb) = \frac{20.72}{207.2} = 0.100 \text{ mol}$$

$$n(AgNO_3) = cV = 1.00 \times 0.100 = 0.100 \text{ mol}$$

As one mole of Pb reacts with two moles of $AgNO_3$, $AgNO_3$ is the limiting reagent and all of the 0.100 moles will be used. Therefore, in the final mixture $n(Ag) = 0.100$ mol.

$n(Pb)$ used $= \frac{0.100}{2} = 0.0500$ mol. 0.0500 mol remain unused.

$n(Pb^{2+})$ produced $= 0.0500$ mol

$n(NO_3^{-}) = 0.100$ mol *(5 marks)*

(b) Care must be taken when disposing of lead ions because lead accumulates in the bodies of organisms and is magnified up the food chain. It can reach toxic levels in organisms at the top of the food chain and can cause brain damage and affect reproductive systems. Lead ions should not be disposed of down the sink. Instead, the mixture should be evaporated and solids stored in plastic. *(2 marks)*

24 (a) Decreasing the volume of the vessel causes an increase in the total concentration of gas species present. The system will shift forward to partially counteract this change, reducing the overall concentration of species. This will reduce the concentration of species X and Y and increase the concentration of Z. The forward shift will also lead to an increase in the heat released because the forward reaction is exothermic. *(2 marks)*

(b) Adding a catalyst will increase the rate of the forward and reverse reactions equally. This will cause the species to react faster and reach equilibrium quicker but will not increase the yield of product. *(2 marks)*

25 $Ind^{-} + H_3O^{+} \rightleftharpoons HInd + H_2O$

This solution can be used as an indicator because it is an equilibrium system with the different coloured species on either side of the equilibrium. The equilibrium is sensitive to pH because hydronium ions and water form a part of the reaction. When acid is added to the system the equilibrium shifts forward to partially counteract the added H_3O^{+}, and in the process uses the green Ind^{-} ion to form the red $HInd$. When base is added to the system it reacts with the H_3O^{+}, reducing its concentration. The equilibrium shifts in reverse to counteract this change and in the process converts red $HInd$ to green Ind^{-}. *(4 marks)*

26 Microscopic membrane filters are composed of polymers such as polysulfone or cellulose acetate. The membranes have tiny holes (pores) that allow water molecules to pass through but retain particles present in the water. Different pore sizes can be used to filter different types of material. For example, a filter with pore sizes of 100 to 1000 nm will remove silt and microbes but not dissolved ions. A filter with very small pores (0.1 to 1 nm) will remove all dissolved material including ions. The filters are designed in such a way that water flows over their surface and not at right angles to it, so filtered material does not clog up the filter. *(4 marks)*

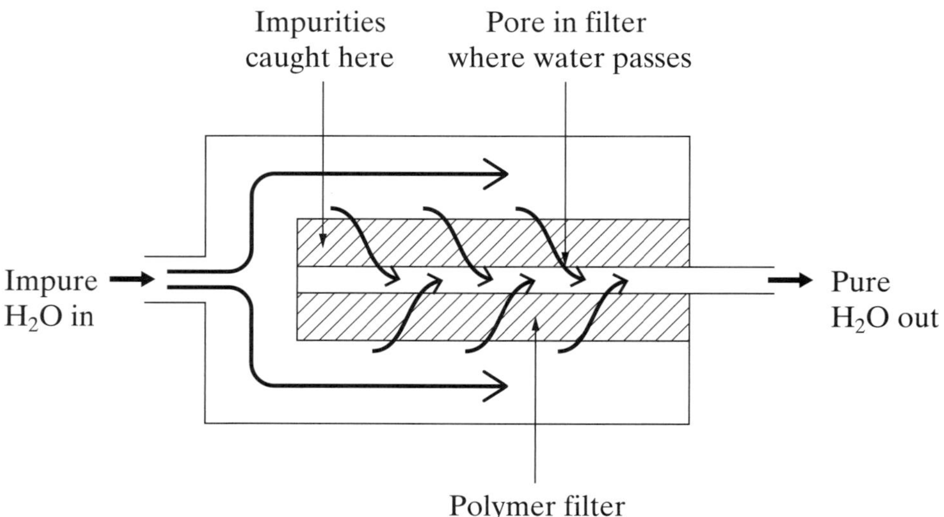

27 $\Delta_c H(\text{ethanol}) = -1367 \text{ kJ mol}^{-1} = -1\,367\,000 \text{ J mol}^{-1}$

$M(\text{ethanol}) = 46.068 \text{ g mol}^{-1}$

Mass ethanol burnt = 0.259 g

$$\Delta H \text{ for this reaction} = -1\,367\,000 \times \frac{0.259}{46.068} = -7685.44 \text{ J}$$

$$\Delta H = -m \times C \times \Delta T$$

$$C(\text{oily liquid}) = -\frac{\Delta H}{m} \times \Delta T$$

$$= -\left[\frac{-7685.44}{120 \times 30}\right]$$

$$= 2.1348$$

$$= 2.13 \times 10^3 \text{ J kg}^{-1} \text{ K}^{-1} \text{ (to 3 significant figures)}$$

(4 marks)

28 (a) The mistake made in step 4 was that the student blew into the pipette to remove all of the sodium carbonate. The pipette is designed to transfer 25.0 mL of solution by letting it run out of the pipette and touching the end of the pipette to the side of the conical flask. By blowing into the pipette to remove all of the solution the student would have transferred more than 25.0 mL of standard. *(2 marks)*

(b) By rinsing the pipette with water and not doing a final rinse with the standard, the student has diluted the standard. This means that the number of moles of standard transferred to the flask is too low. By filling the standard to below the graduated mark on the pipette the student has further reduced the amount of standard in the pipette, lowering the number of moles of standard introduced to the flask. The volume of HCl titrated to neutralise the standard will therefore be lower than it should be, and the calculated concentration of the HCl will be greater than it should be (in $c = \frac{n}{V}$, n is an accurate number of moles but the volume is smaller than it should be). *(3 marks)*

29 The reaction will not occur spontaneously as the total E^0 for the reaction is negative.

$Cl^-(aq) \rightarrow \frac{1}{2}Cl_2(g) + e^-$ $\qquad E^0 = -1.36$ V

$\frac{1}{2}Br_2(l) + e^- \rightarrow Br^-(aq)$ $\qquad E^0 = 1.08$ V

$$\text{Total } E^0 = 1.08 - 1.36$$
$$= -0.28 \text{ V}$$

(2 marks)

30 From the total ozone key provided in the question, the concentrations of ozone in the three years in question are in the following approximate ranges.

1979: 225 to 250 Dobson units

1994: 70 to 100 Dobson units

2012: 150 to 200 Dobson units

Between 1979 and 1994 the concentration of stratospheric ozone above Antarctica dropped significantly. This was due to the decomposition of ozone by chlorofluorocarbons (CFCs), which have been produced by man since the 1930s for use as refrigerants and propellants.

CFCs such as trichlorofluoromethane (CCl_3F) undergo photodissociation to form highly reactive chlorine radicals. The chlorine radicals decompose ozone to oxygen and a chlorine oxide radical. The chlorine oxide radical reacts with surrounding oxygen radicals to form oxygen gas molecules and more chlorine radicals, which in turn decompose more ozone.

$CCl_3F(g) + UV \rightarrow CCl_2F\bullet(g) + Cl\bullet(g)$

$Cl\bullet(g) + O_3(g) \rightarrow O_2(g) + OCl\bullet(g)$

$OCl\bullet(g) + O\bullet(g) \rightarrow O_2(g) + Cl\bullet(g)$

A single CFC molecule can go on and destroy thousands of ozone molecules in this way. This is what has occurred to reduce the ozone levels to the low of 70 to 100 Dobson units recorded in 1994.

The Montreal Protocol is an international treaty convened in 1987 to address the reduction in ozone. Nearly 50 countries originally signed the treaty to phase out the use of CFCs and halons (brominated fluorocarbons). Now 150 countries have signed. CFCs were gradually replaced with less destructive hydrochlorofluorocarbons (HCFCs). HCFCs are largely broken down in the troposphere, so few of these molecules reach the stratosphere to cause ozone destruction. Following subsequent meetings in London (1990) and Copenhagen (1992), it was agreed to replace HCFCs with hydrofluorocarbons (HFCs). HFCs contain no chlorine or bromine atoms, so they have zero ozone depletion potential. The recorded increase in ozone concentration to approximately 150 to 180 Dobson units in 2012 is a result of these measures. *(6 marks)*

31 (a)

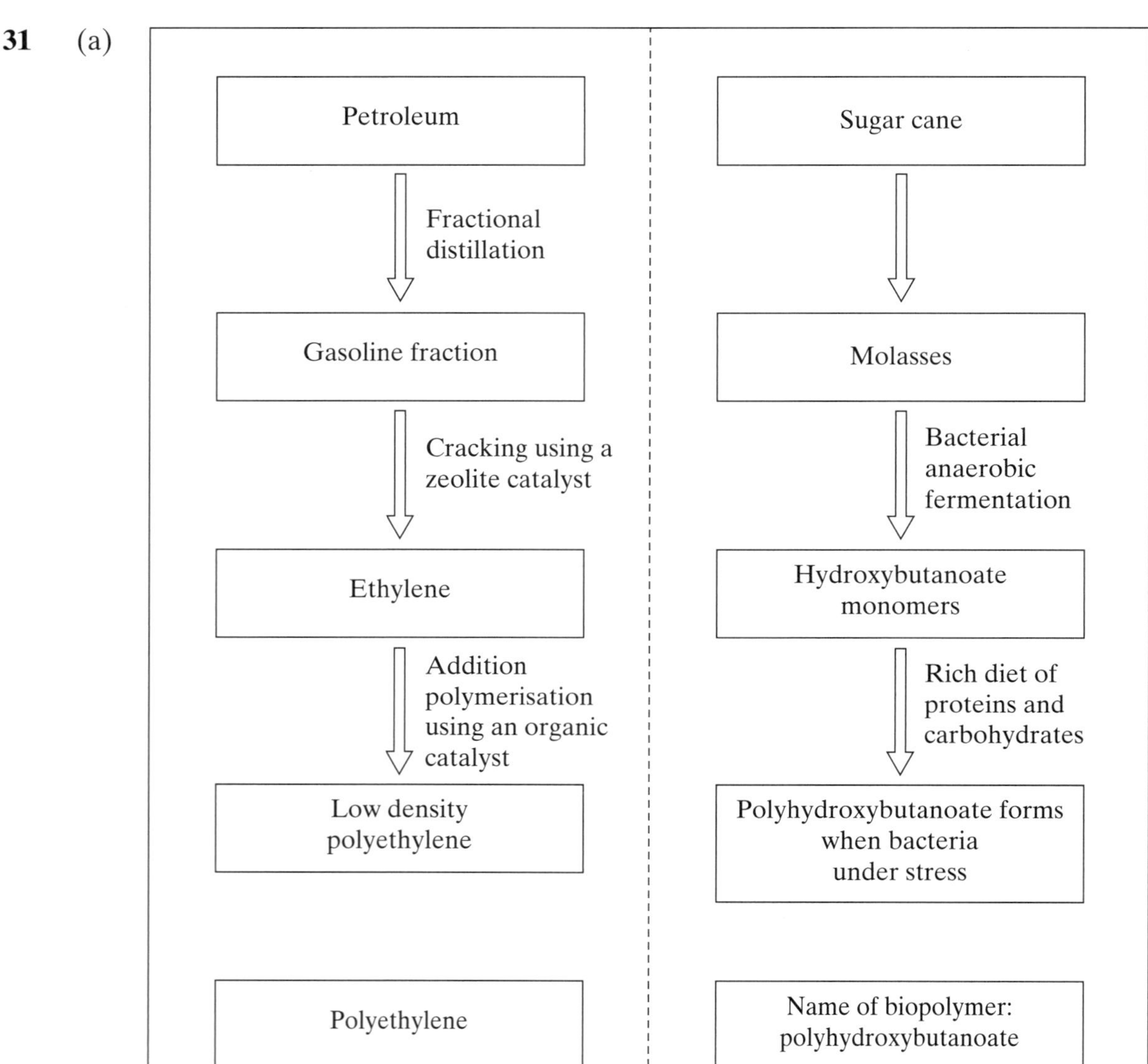

(5 marks)

(b) The recently developed biopolymer polyhydroxybutanoate (PHB) has a number of advantages over the similar petroleum-derived polymer, polypropylene. Unlike polypropylene, PHB is biodegradable and will not add to landfill. PHB is also renewable as it can be produced by bacterial fermentation of biomass, which can continually be regrown. Plastics made from petroleum, such as polypropylene, are non-renewable and their production will eventually cease as supplies of petroleum run out. *(3 marks)*

Section II—Options

Question 32—Industrial Chemistry

(a) The diagram shows the cleaning action of soap (an anionic detergent) via the formation of a micelle. The hydrophobic, non-polar tail (e.g. C_{17}) of the soap particle dissolves in the grease droplet, while the hydrophilic, ionic head (COO^-) is surrounded by water molecules due to dipole–dipole interactions. When the mixture is agitated, a 3-D micelle is formed, removing the grease from the surface so it can be washed away. The surface of the micelles is negatively charged so micelles repel each other, and do not clump together. This is how soap emulsifies grease and water. *(3 marks)*

(b) (i)

	$2HI(g)$	$\rightleftharpoons$	$H_2(g)$	+	$I_2(g)$
Initial concentration (mol/L)	0.60		0		0
Used/produced	0.50		0.25		0.25
Equilibrium concentration (mol/L)	0.10		0.25		0.25

$$K = \frac{[H_2][I_2]}{[HI]^2} = \frac{0.25 \times 0.25}{0.1^2} = 6.25 \text{ (units not required)}$$ *(3 marks)*

(ii) Given that the forward reaction is endothermic, decreasing the temperature will shift the equilibrium position to the left-hand side, favouring the exothermic direction. This will decrease the concentration of H_2 and I_2, and increase the concentration of HI. The purple colour of the equilibrium mixture (due to I_2 gas) will fade. *(2 marks)*

(c) (i) Limestone, or $CaCO_3(s)$, is important in the Solvay process because it is a raw material. It appears in the overall equation as: $CaCO_3(s) + 2NaCl(aq) \rightarrow CaCl_2(s) + Na_2CO_3(s)$. Its role is twofold. It provides a source of $CO_2(g)$ upon thermal decomposition: $CaCO_3(s) \rightarrow CaO(s) + CO_2(g)$. It also provides a source of $Ca(OH)_2(aq)$ for use in ammonia recovery: $CaO(s) + H_2O(l) \rightarrow Ca(OH)_2(aq)$ and then $2NH_4Cl(aq) + Ca(OH)_2(aq) \rightarrow CaCl_2(aq) + 2H_2O(l) + 2NH_3(g)$. *(2 marks)*

(ii) Site B could serve as a location for a Solvay plant because it is close to a source of limestone, which is a raw material in the process, and it provides potential for disposal of calcium chloride, which can be buried as a solid in specially lined pits where the limestone is removed. However, site B is approximately 150 km from the coast, where brine is obtained and from where the product can be shipped to markets. It is also about 150 km from Port City, which has markets for sodium carbonate. On the other hand, site A is close to both the ocean and Port City. Importantly, site A's proximity to the ocean is also perfect for disposal of calcium chloride, since adding Ca^{2+} and Cl^- ions to the ocean does not significantly increase their concentration in seawater. For these reasons site A is a better site for a Solvay plant. *(3 marks)*

(d) (i) Electrolysis of aqueous NaCl produces $Cl_2(g)$ which is highly toxic by inhalation. In order to avoid inhaling this gas, the experiment must be conducted in a fume cupboard. *(2 marks)*

(ii) The three products of the electrolysis of NaCl(*aq*) are $Cl_2(g)$, $H_2(g)$ and NaOH(*aq*). The experiment is conducted using Hoffmann's voltameter, which is filled with NaCl(*aq*) to which universal indicator has been added. The universal indicator is initially green because the solution is neutral.

After electrolysis, the gas that collects at the anode is tested by holding damp blue litmus paper above the gas tap as it is opened very slowly. The blue colour disappears as the paper is bleached, confirming that the gas produced at the anode is chlorine: $2Cl^-(aq) \rightarrow Cl_2(g) + 2e^-$.

The gas that collects at the cathode is collected in an inverted test tube. A lit match is inserted into the test tube and an audible pop is heard, confirming the production of H_2 at the cathode: $2H_2O(l) + 2e^- \rightarrow H_2(g) + 2OH^-(aq)$.

The universal indicator changes colour from green to purple around the cathode, confirming the production of OH^- ions. *(3 marks)*

(e) The production of sulfuric acid is a multi-step process, starting with the mining of sulfur. In most steps there are environmental considerations; however, it is possible to both maximise yield and minimise environmental impacts throughout the process.

Mining: Sulfur is mined using the Frasch process. Super-heated steam and high-pressure air is forced into an underground sulfur deposit, melting the sulfur and forcing it to the surface where it solidifies. While this process produces high-purity sulfur, it can also cause land subsidence due to the removal of the sulfur. Mining also removes natural habitat, and the burning of fossil fuels to provide the energy necessary for mining produces greenhouse gas (CO_2) emissions. In order to minimise the environmental impacts, it is necessary to reduce the yield of sulfur from the site. The environmental impacts of sulfur mining can also be minimised by using $SO_2(g)$ from the smelting of sulfide ores as a high-quality feedstock for sulfuric acid production. For example, in the processing of ZnS: $2ZnS(s) + 3O_2(g) \rightarrow 2ZnO(s) + 2SO_2(g)$.

Contact process step 1: The production of $SO_2(g)$ from sulfur involves spraying molten sulfur into a furnace where it burns in hot air: $S(l) + O_2(g) \rightarrow SO_2(g)$. This produces a yield of SO_2 of 100%, and is exothermic. The energy released is used to generate steam to produce electricity. This minimises the environmental impacts by reducing reliance on external power supplies and, in turn, reducing CO_2 emissions that would otherwise contribute to the enhanced greenhouse effect.

Contact process step 2: The next step in the production of sulfuric acid is the conversion of SO_2 to SO_3 in an exothermic reaction: $2SO_2(g) + O_2(g) \rightleftharpoons 2SO_3(g)$. The reaction conditions used are:

- temperature of approximately 400 to 550 °C
- pressure of approximately 100 to 200 kPa
- catalyst of V_2O_5 in three or four beds of successively lower temperature.

The forward reaction is exothermic, so a moderate temperature is chosen to maximise both rate and yield. Lower temperatures favour a higher yield (because the forward reaction is exothermic) but decrease the rate. Although Le Chatelier's principle predicts that high pressure favours the forward reaction, using a high pressure is more dangerous and costly. As high yields can be obtained at about 100 to 200 kPa, this expense is not necessary. The catalyst increases the reaction rate, which in turn allows the lower temperature to be used, and because the catalyst beds are successively cooler, a very high yield of SO_3 is obtained. The oxidation of SO_2 to SO_3 is also exothermic so the heat released is used to generate electricity on site. This minimises the environmental impacts associated with burning fossil fuels to generate electricity. The use of low pressure and a moderate temperature also reduces the energy requirements of the plant, minimising the environmental impacts by reducing CO_2 emissions. Another important factor which both minimises environmental impacts and increases the yield of SO_3 is the recycling of unreacted gases back into the converter. This increases the yield of SO_3 to about 99.5% and also ensures that SO_2 emissions, which lead to increased respiratory illness and acid rain, are within the acceptable limits required by environmental protection authorities.

Contact process steps 3 and 4: The $SO_3(g)$ is dissolved in concentrated H_2SO_4 to produce oleum: $SO_3(g) + H_2SO_4(l) \rightarrow H_2S_2O_7(l)$. The oleum is then diluted to produce sulfuric acid: $H_2S_2O_7(l) + H_2O(l) \rightarrow 2H_2SO_4(l)$. The yield in these reactions is 100% and they have no adverse environmental impacts, so no compromise is necessary. *(7 marks)*

Question 33—Shipwrecks, Corrosion and Conservation

(a) This process, called electrolytic reduction cleaning, is used to restore metal artefacts. By driving electrons from the power supply into the artefact, the artefact becomes cathodic. Metal ions on the surface of the artefact are then reduced back to the metal or an oxide of lower oxidation state. This restores some of the corroded material and prevents the artefact from corroding further. For example, if this artefact was iron:

$Fe^{2+}(aq) + 2e^- \rightarrow Fe(s)$

Another reason for using reduction cleaning is that small bubbles of hydrogen (created when water surrounding the artefact is reduced) help to clean the artefact of loosely attached materials. *(3 marks)*

(b) (i) Shallow water contains dissolved oxygen which enters from the surface. Tin oxidises very slowly in aerobic sea water, forming a mixture of hydrated tin(II) and tin(IV) oxides on the surface of the tin.

$Sn(s) + \frac{1}{2}O_2(aq) + H_2O(l) \rightarrow Sn^{2+}(aq) + 2OH^-(aq) \rightarrow SnO.H_2O(s)$

Hydrated tin(IV) oxide can also form because the oxygen oxidises the tin(II) ions.

$SnO.H_2O(s) + \frac{1}{2}O_2(aq) \rightarrow SnO_2.H_2O(s)$ *(2 marks)*

(ii) Corrosion is the oxidation of metals to their ions, reducing the integrity of the original metal structure.

Oxidation involves losing electrons to an oxidant. Corrosion cannot occur without an exchange of electrons from reductant to oxidant. For example, when iron corrodes iron atoms give up electrons to the oxidant, usually oxygen and water.

$Fe(s) \rightarrow Fe^{2+}(aq) + 2e^-$

$\frac{1}{2}O_2(aq) + H_2O(l) + 2e^- \rightarrow 2OH^-(aq)$

The corrosion product, iron(II) hydroxide, could only result from the electron exchange between the reductant (iron) and oxidant (O_2).

$Fe(s) + \frac{1}{2}O_2(aq) + H_2O(l) \rightarrow Fe(OH)_2(s)$ *(3 marks)*

(c) (i) From the extrapolated graph, solubility of oxygen at 4°C is approximately 10.2 mg L^{-1}. *(2 marks)*

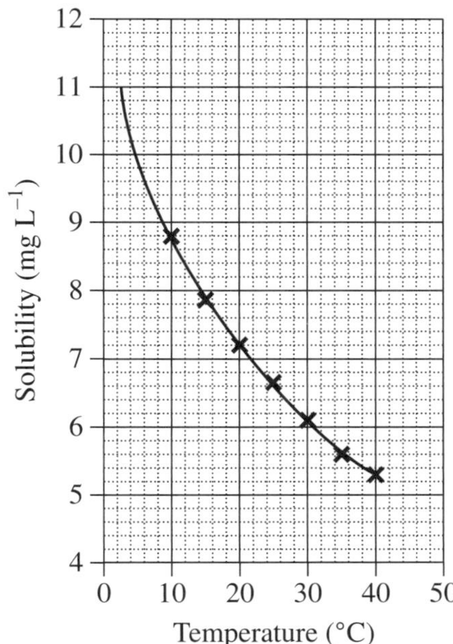

(ii) Temperature and pressure are the two main factors that affect the solubility of gases. At lower temperatures and higher pressures, gases are more soluble. As depth increases in the ocean, temperature drops and pressure increases, so you would expect the concentration of dissolved oxygen to increase. However, the concentration of dissolved oxygen drops as depth increases. The surface waters of the ocean are rich in dissolved oxygen, which enters the ocean from the atmosphere at the surface, and is also produced by the photosynthetic activity of producers in surface waters. Much of the oxygen in the ocean is used by living things, so very little makes it into deep waters. Dissolved oxygen in deep waters has either diffused there from surface waters or been carried in by deep ocean currents that carry oxygenated cold water along the ocean floor. These currents result in a slightly higher concentration of dissolved oxygen at the ocean floor. *(3 marks)*

(d) (i) 1. Set up the apparatus as in the diagram below, where the salt water is sea water and is the same concentration and temperature for all test tubes. The pieces of iron are identical.

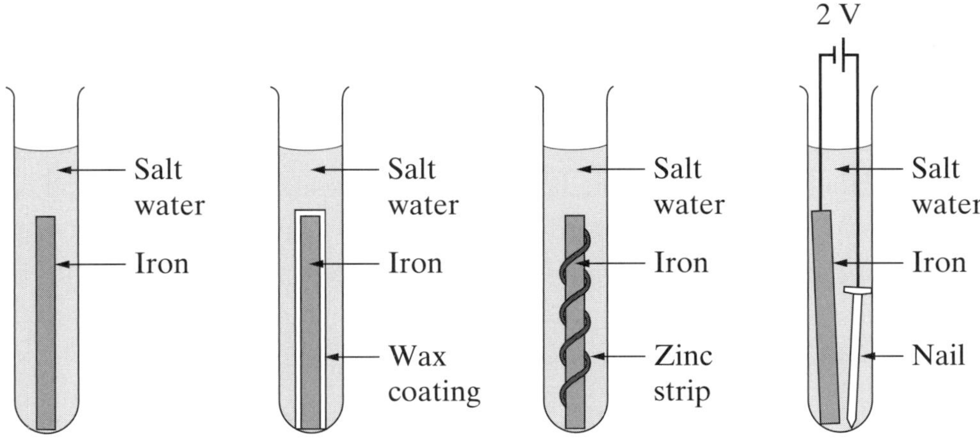

2. Record the level of corrosion on all pieces of iron every day for 10 days. Use a scale of 1 to 5 to compare corrosion, where 1 is minimal signs of rusty precipitate and 5 is extensive corrosion.

3. Repeat this investigation four times and average the result. *(3 marks)*

(ii) The expected results of this investigation are that after 10 days, the control piece of iron will show extensive evidence of corrosion (5 on the 1 to 5 scale), the iron covered in wax will show minimal levels of corrosion (2 on the scale) and the other two will show minimal, if any, signs of corrosion (1 on the scale).

Iron is an abundant, easily workable metal with very good tensile strength. The ongoing use of iron in the construction of ocean-going vessels is justified because corrosion of iron can be easily prevented by attaching zinc to the metal. The zinc will oxidise in preference to iron as it is more reactive (i.e. a stronger reductant), preventing the iron from corroding. Corrosion can also be prevented by applying voltage to the metal to be protected, which forces the iron to become cathodic. *(2 marks)*

(e) Galvani, Volta, Davy and Faraday are four scientists whose work in electrochemistry has changed scientific thinking about redox reactions.

Galvani was the first to use static electricity to induce a reaction in nerves when he set up his galvanic cell in a dissected frog leg. Although he incorrectly concluded that the current was produced by the animal (which he called *animal electricity*) and not the differing potentials of metals involved, he paved the way for later work on electrolytes and the relationship between electrolytes and nerve impulses.

Volta was the first to build a functioning battery outside of animal tissue and correctly explain how it worked. He experimented with a range of metals and showed which combinations produced the highest voltages. He built the first battery by stacking layers of zinc and tin or silver and copper, separated by cardboard soaked in brine. His battery provided future scientists with a reliable source of electricity.

Davy built on Volta's findings in electrolysis and used his battery to decompose water and then to isolate previously undiscovered elements (such as sodium, potassium and barium) via the electrolysis of their molten salts.

However, Faraday's work has most changed the nature of thinking about redox reactions because he was able to quantify them. In his first law of electrolysis, Faraday established a relationship between voltage, time and current. The amount of product in an electrolytic reaction could now be predicted as long as voltage and the time the current flowed were known. He also introduced terminology used in electrochemistry, such as *electrolyte*, *anode*, *cathode*, *anion* and *cation*. Faraday laid the groundwork for the future study of electrolysis and redox reactions by showing that a certain amount of electricity was used to produce a set amount of product during electrolysis. *(7 marks)*

(Note: in order to provide a clear explanation for students some answers are longer than would be required for an answer in the exam.)

Question 36—Forensic Chemistry

(a) HPLC separates components of a mixture on the basis of their relative affinity for the stationary and mobile phases as they pass through the chromatography column. Comparison of the chromatogram of a mixture with chromatography standards (produced under the same conditions) allows substances to be identified. In this case, the chromatograms indicate that the suspect's equipment contains compounds with the same retention times as *C* and *D* on the standard. The identification of these substances as methcathinone and MDMA should be confirmed using a technique such as mass spectroscopy to compare their molecular mass and fragmentation patterns with those of the reference compounds. *(3 marks)*

(b) (i) *X* is starch. *Y* is glycogen. *Z* is cellulose. *(2 marks)*

(ii) Cellulose is a condensation polymer of β-glucose and as such is a polysaccharide. In each polymer chain about 10 000 β-glucose molecules are joined by β-1,4-glycoside linkages, resulting in flat, ribbon-like, unbranched polymer chains. This structure allows the polymer chains to form extensive hydrogen bonding with adjacent chains, resulting in the strength and rigidity that makes cellulose suitable for its function as a plant structural material. *(3 marks)*

(c) (i) The river water contains mercury. The river water sample spectrum contains all of the spectral lines of mercury but only some lines corresponding to those of cadmium and lead. *(2 marks)*

(ii) A flame test can be used to observe the emission spectrum of an element such as copper. In order to do this, a platinum loop attached to a glass rod is cleaned by dipping into concentrated HCl, then is heated in a blue Bunsen flame until no colour is observed. The loop is then dipped into a sample of copper(II) chloride and returned to the blue Bunsen flame. The individual emission spectral lines of the blue-green flame can be observed using a spectroscope. Concentrated HCl is highly corrosive, so rubber gloves should be worn to avoid skin contact. Safety glasses should also be worn to avoid eye contact with the HCl. Flammable materials should be kept away from the Bunsen to avoid a fire. *(3 marks)*

(d) (i) In order to confirm that a compound is organic, a sample can be burned in air and any gas produced bubbled into limewater. If the sample is organic carbon dioxide will be produced, the presence of which is confirmed by the production of a white precipitate in the limewater. *(2 marks)*

(ii) The following elimination sequence of tests can be carried out to distinguish between an alkanoic acid, an alkene and an alkane.

1. Place 1 mL samples of each compound in three separate test tubes. Add 1 mL of saturated sodium carbonate solution to each sample. If gas bubbles are observed then the sample is an alkanoic acid (acid + carbonate $\rightarrow$ salt + water + carbon dioxide).

2. Place 1 mL of a fresh sample of each of the two remaining compounds in a test tube, then add two drops of bromine water. Stopper and shake the tube. If the bromine water decolourises (orange → colourless) immediately under ordinary laboratory lighting then the sample is an alkene (undergoing an addition reaction with the halogen).

3. Place the remaining sample with bromine water (from the second test) in the sunlight. If the bromine water slowly decolourises then the sample is an alkane (undergoing a substitution reaction in the presence of UV light).

Because hydrocarbons are toxic by all routes of exposure, the experiment should be done in a fume cupboard to avoid inhalation of fumes. Rubber gloves should be used to avoid skin contact. Alkanoic acids are corrosive to eyes and skin, so safety glasses should also be worn. Hydrocarbons are flammable so the experiment should not be carried out near naked flames or ignition sources. *(3 marks)*

(e) Electrophoresis exploits different migration rates of ions in an electric field, and is commonly used in the analysis of biological molecules. A typical application of electrophoresis is the separation of proteins in a mixture. A solution of the mixture is applied to a solid matrix (e.g. gel or buffer-soaked paper) which is placed between the electrodes of a DC source. Charged species in the solution migrate towards one or the other electrode (depending on their overall charge) at a characteristic rate. The migration rate depends on the applied voltage, the nature of the solid matrix, the total charge on the species, and the molecular mass. The total charge depends on the pH of the buffer system used. After the species in the mixture have been separated they are stained so that they can be seen. Electrophoresis of a reference mixture of known proteins on the same gel or paper allows identification of the proteins in the unknown sample. This technique can be very useful for identifying the origin of plant or animal products in foods, since different species have characteristic protein profiles.

However, electrophoresis would not be suitable to analyse explosive debris, which is a mixture of organic and inorganic material. The chemical components of the debris depend on the explosive used, but may include metals (such as iron, magnesium and/or aluminium), inorganic compounds (such as NH_4NO_3, sodium and/or potassium chlorate and/or perchlorate, e.g. $NaClO_3$ and $NaClO_4$), and nitrated organic compounds (such as trinitrotoluene, TNT). These species are unable to be identified by electrophoresis because their very low molecular mass and fixed charge (or lack of charge), which cannot be manipulated by changing buffer solution, means that they are not separated under the conditions used in electrophoresis.

Given that the explosive residue found at the crime scene is likely to be present in trace amounts, and that it could contain a range of different compounds (including those listed above), chromatography is a much better technique for its analysis than electrophoresis, which requires larger sample sizes and typically analyses a group of the same type of compound such as proteins or DNA fragments. There are many different types of chromatography. Gas-liquid chromatography (GLC) and high-performance liquid chromatography (HPLC) are the most common chromatographic techniques in forensic science.

In GLC a small quantity of the sample is dissolved in an appropriate solvent, injected onto the column (the stationary phase) and vaporised. The column is typically 1 to 4 m long and is packed with a solid such as silica or alumina. The mobile phase is an inert gas such as nitrogen or helium. As the gas travels through the column, the temperature of the column is raised slowly from about 50 to 250 °C. Compounds in the mixture are separated on the basis of their boiling points and the temperature of the column, and their affinity for the stationary phase and the mobile phase. Separated substances exit the stationary phase at different times and are detected. Substances can be identified using the chromatographs of reference samples analysed under the same conditions and/or by coupling the GLC with a mass spectrometer, which analyses each compound as it exits the GLC. GLC is extremely sensitive and can detect minute amounts of compounds. It can be used to analyse a wide variety of substances.

In HPLC the solutes are migrated through the stationary phase using a liquid mobile phase under pressure, rather than a gas at increasing temperature. For this reason HPLC is preferred for the analysis of explosive debris as it is conducted at room temperature rather than an elevated temperature, which may cause safety issues. *(7 marks)*

(Note: in order to provide a clear explanation for students some answers are longer than would be required for an answer in the exam.)

CHAPTER 12

2014 HIGHER SCHOOL CERTIFICATE EXAMINATION

Chemistry

General Instructions

- Reading time – 5 minutes
- Working time – 3 hours
- Write using black or blue pen
  Black pen is preferred
- Draw diagrams using pencil
- Board-approved calculators may be used
- A data sheet and a Periodic Table are provided at the back of this paper
- Write your Centre Number and Student Number where required

Total marks – 100

Section I

75 marks

This section has two parts, Part A and Part B

Part A – 20 marks

- Attempt Questions 1–20
- Allow about 35 minutes for this part

Part B – 55 marks

- Attempt Questions 21–31
- Allow about 1 hour and 40 minutes for this part

Section II

25 marks

- Attempt ONE question from Questions 32–36
- Allow about 45 minutes for this section

Section I
75 marks

Part A – 20 marks
Attempt Questions 1–20
Allow about 35 minutes for this part

Use the multiple-choice answer sheet for Questions 1–20.

---

1 In which layer of the atmosphere is ozone considered a pollutant?

(A) Mesosphere

(B) Stratosphere

(C) Thermosphere

(D) Troposphere

2 What is the IUPAC name of the following compound?

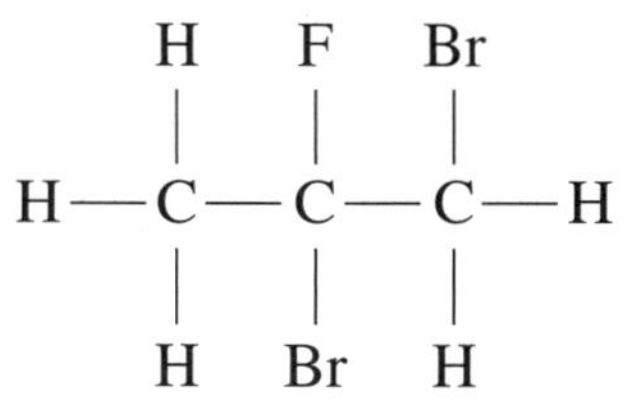

```
     H   F   Br
     |   |   |
 H — C — C — C — H
     |   |   |
     H   Br  H
```

(A) 1,2-dibromo-2-fluoropropane

(B) 2,3-dibromo-2-fluoropropane

(C) 2-fluoro-2,3-dibromopropane

(D) 2-fluoro-1,2-dibromopropane

3 Which row of the table correctly matches the scientist(s) with their theory of acids?

	Scientist(s)	*Theory*
(A)	Arrhenius	Acids contain oxygen
(B)	Brönsted and Lowry	Acids are proton donors
(C)	Davy	Acids are able to produce hydrogen ions in water
(D)	Lavoisier	Acids contain hydrogen

4 Which of the following equations correctly represents catalytic cracking of a petroleum fraction?

(A) $C_{15}H_{32}(g) \xrightarrow{AlSi_2O_6} C_{15}H_{32}(s)$

(B) $nC_2H_4(g) \xrightarrow{AlSi_2O_6} -(CH_2-CH_2)_n-(s)$

(C) $C_{15}H_{32}(g) \xrightarrow{AlSi_2O_6} C_7H_{16}(g) + 4C_2H_4(g)$

(D) $C_7H_{16}(g) + 4C_2H_4(g) \xrightarrow{AlSi_2O_6} C_{15}H_{32}(g)$

5 Which row of the table correctly matches the reactant and the product of an *addition reaction*?

	Reactant	*Product*
(A)	$CH_3 - CH_2 - CH_2 - CH_2 - OH$	$CH_3 - CH_2 - CH = CH_2$
(B)	$CH_3 - CH_2 - CH_2 - CH(OH) - CH_3$	$CH_3 - CH_2 - CH = CH - CH_3$
(C)	$CH_3 - CH = CH - CH_2 - CH_3$	$CH_3 - CH_2 - CH(Cl) - CH_2 - CH_3$
(D)	$CH_3 - C(=O) - OH$	$CH_3 - C(=O) - O - CH_3$

6 Drinking water is regularly tested to ensure that it is safe for consumption.

Which of the following test results indicates the highest drinking-water quality?

	Dissolved oxygen (mg/L)	*Nitrate* (mg/L)	*Total dissolved solids* (mg/L)	*Turbidity* (NTU)
(A)	2	0.1	50	50
(B)	8	0.1	50	2
(C)	2	2	200	2
(D)	8	2	200	50

7 This table contains information on three indicators.

Indicator	*pH range*	*Colour (lower pH – higher pH)*
Methyl orange	3.1–4.4	red – yellow
Methyl red	4.4–6.2	pink – yellow
Phenolphthalein	8.3–10.0	colourless – pink

A substance is tested with each of the indicators and the results are recorded below.

Indicator	*Colour*
Methyl orange	yellow
Methyl red	yellow
Phenolphthalein	colourless

Which of the following substances will produce these results?

(A) Lemonade pH 2.9

(B) White wine pH 4.2

(C) Tap water pH 7.2

(D) Ammonia pH 11.2

8 The graph shows the pH of a solution of a weak acid, HA, as a function of temperature.

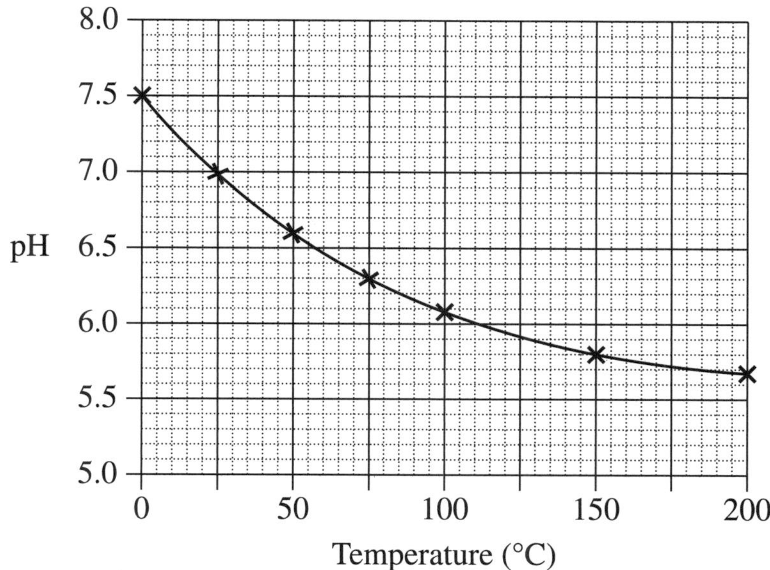

What happens as the temperature decreases?

(A) HA becomes less ionised and the H^+ concentration increases.

(B) HA becomes less ionised and the H^+ concentration decreases.

(C) HA becomes more ionised and the H^+ concentration increases.

(D) HA becomes more ionised and the H^+ concentration decreases.

9 Four compounds, W, X, Y and Z, are represented below.

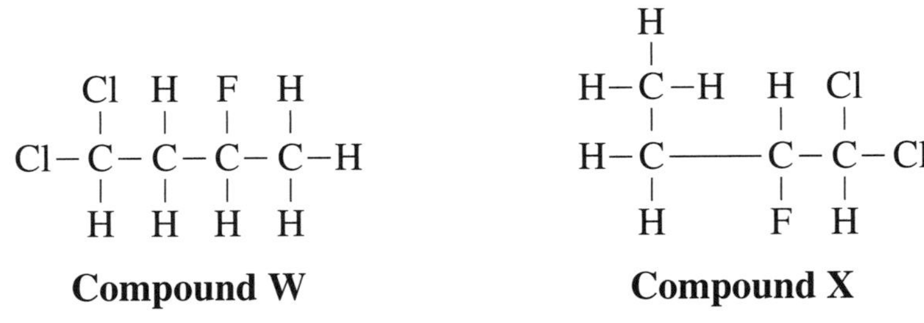

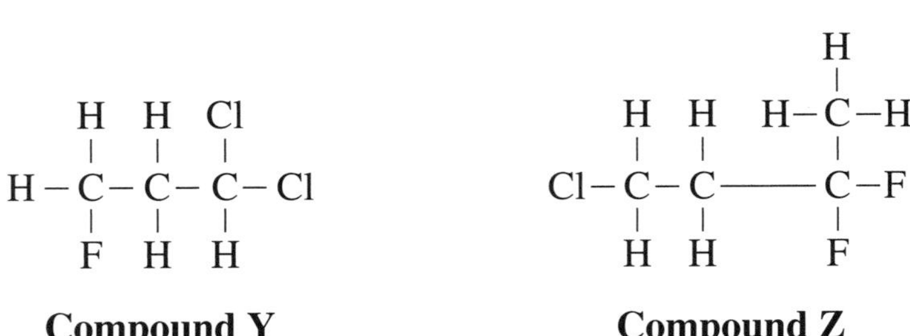

Which of the following is a pair of isomers?

(A) W and X

(B) W and Y

(C) X and Y

(D) Y and Z

10 The following equation represents a chemical system in equilibrium:

$$OCl^-(aq) + H_2O(l) \rightleftharpoons HOCl(aq) + OH^-(aq)$$

Which of the following is an acid/base conjugate pair?

(A) H_2O / HOCl

(B) HOCl / OH^-

(C) HOCl / OCl^-

(D) OCl^- / H_2O

11 Which of the following does NOT represent the formation of a coordinate covalent bond?

(A) $H:\overset{H}{\overset{\cdot\cdot}{\underset{H}{\underset{\cdot\cdot}{C}}}}\cdot \quad + \quad \cdot\overset{\cdot\cdot}{\underset{\cdot\cdot}{Cl}}: \longrightarrow H:\overset{H}{\overset{\cdot\cdot}{\underset{H}{\underset{\cdot\cdot}{C}}}}:\overset{\cdot\cdot}{\underset{\cdot\cdot}{Cl}}:$

(B) $H:\overset{\cdot\cdot}{\underset{H}{\underset{\cdot\cdot}{O}}}: \quad + \quad H^+ \longrightarrow \left[H:\overset{\cdot\cdot}{\underset{H}{\underset{\cdot\cdot}{O}}}:H\right]^+$

(C) $H:\overset{\cdot\cdot}{\underset{H}{\underset{\cdot\cdot}{N}}}:H \quad + \quad H^+ \longrightarrow \left[H:\overset{H}{\overset{\cdot\cdot}{\underset{H}{\underset{\cdot\cdot}{N}}}}:H\right]^+$

(D) $\dot{\dot{:}}O::O\dot{\dot{:}} \quad + \quad \overset{\cdot\cdot}{\underset{\cdot\cdot}{O}}: \longrightarrow \dot{\dot{:}}O::\underset{\cdot\cdot}{O}:\overset{\cdot\cdot}{\underset{\cdot\cdot}{O}}:$

12 The diagram shows the pH values of some substances.

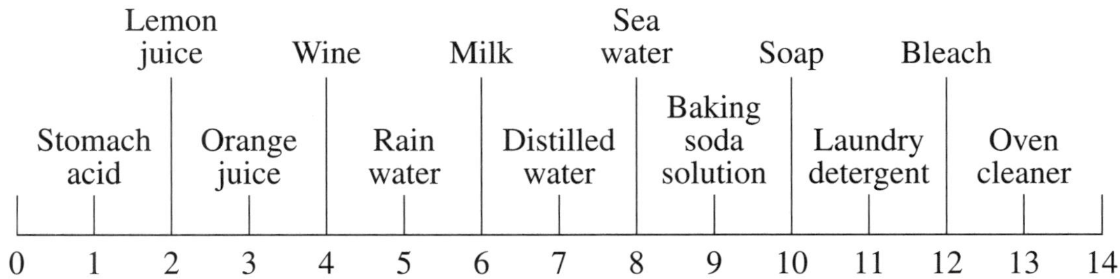

Based on the information provided, which of the following statements about the relative concentration of hydrogen ions is correct?

(A) It is 2 times higher in bleach than in milk.

(B) It is 10 times lower in stomach acid than in soap.

(C) It is 1000 times lower in distilled water than in wine.

(D) It is 100 times higher in laundry detergent than in baking soda solution.

13 This equation shows an equilibrium established in the synthesis of ammonia from its component gases:

$$N_2(g) + 3H_2(g) \rightleftharpoons 2NH_3(g)$$

If the volume of the reaction chamber is suddenly halved at time T, which of the following best depicts changes in the concentration of ammonia over time?

(A)

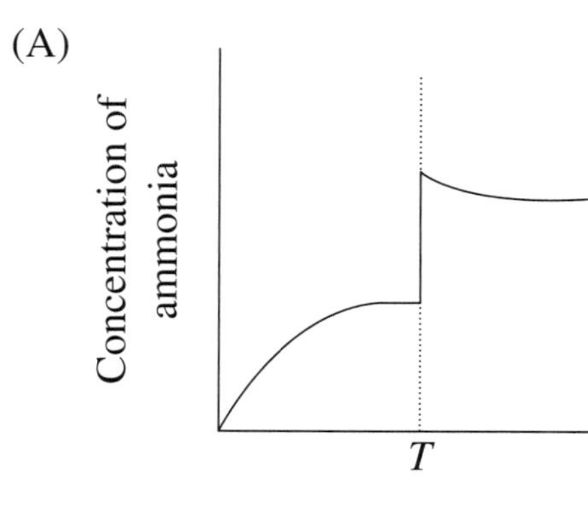

(B)

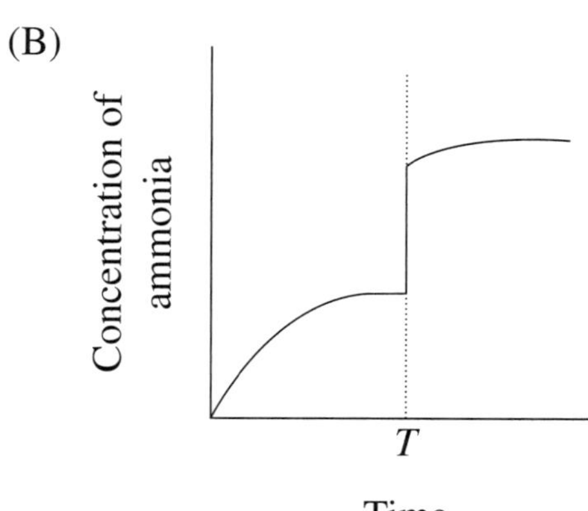

(C)

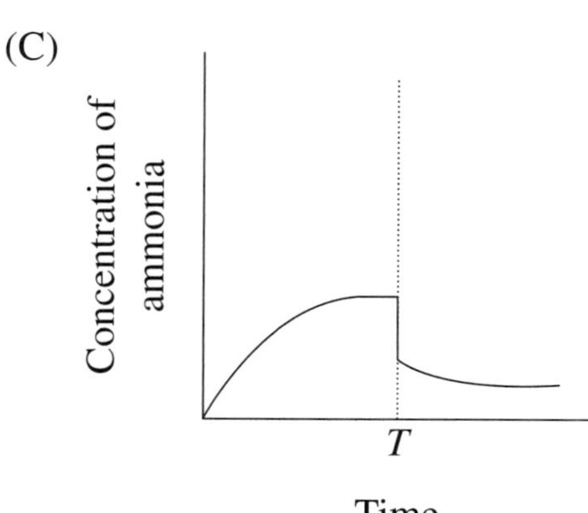

(D)

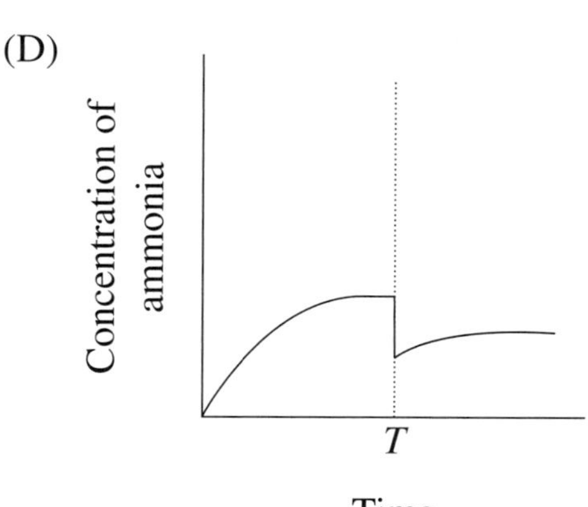

14 What is the pH of a 0.018 mol L^{-1} solution of hydrochloric acid?

(A) 0.74

(B) 0.96

(C) 1.04

(D) 1.74

15 If exactly one gram of each of the following compounds is treated with excess hydrochloric acid, which would release the greatest volume of $CO_2(g)$ at 25°C and 100 kPa?

(A) K_2CO_3

(B) $KHCO_3$

(C) Na_2CO_3

(D) $NaHCO_3$

16 In which of the following reactions is the metal species reduced?

(A) $2FeCl_2(aq) + Cl_2(g) \rightarrow 2FeCl_3(aq)$

(B) $CuS(s) + O_2(g) \rightarrow Cu(s) + SO_2(g)$

(C) $2Al(OH)_3(aq) \rightarrow Al_2O_3(s) + 3H_2O(l)$

(D) $Ca(s) + 2HCl(aq) \rightarrow CaCl_2(aq) + H_2(g)$

17 What is the standard cell potential for the reaction of 1.0 mol L^{-1} acidified potassium dichromate $(K_2Cr_2O_7(aq))$ with aqueous sulfur dioxide $(SO_2(aq))$ under standard conditions?

(A) 1.20 V

(B) 1.52 V

(C) 2.24 V

(D) 3.20 V

18 This is a representation of a segment of the polymer nylon 6,6.

$$-\underset{\underset{O}{\|}}{C}-(CH_2)_4-\underset{\underset{O}{\|}}{C}-\underset{\underset{H}{|}}{N}-(CH_2)_6-\underset{\underset{H}{|}}{N}-\underset{\underset{O}{\|}}{C}-(CH_2)_4-\underset{\underset{O}{\|}}{C}-\underset{\underset{H}{|}}{N}-(CH_2)_6-\underset{\underset{H}{|}}{N}-$$

Which of the following represents the two monomers that are used to produce nylon 6,6?

(A) $HO-\underset{\underset{O}{\|}}{C}-(CH_2)_4-NH_2$ and $H_2N-(CH_2)_6-NH_2$

(B) $HO-\underset{\underset{O}{\|}}{C}-(CH_2)_4-\underset{\underset{O}{\|}}{C}-OH$ and $H_2N-(CH_2)_4-\underset{\underset{O}{\|}}{C}-OH$

(C) $HO-\underset{\underset{O}{\|}}{C}-(CH_2)_6-\underset{\underset{O}{\|}}{C}-OH$ and $H_2N-(CH_2)_4-NH_2$

(D) $HO-\underset{\underset{O}{\|}}{C}-(CH_2)_4-\underset{\underset{O}{\|}}{C}-OH$ and $H_2N-(CH_2)_6-NH_2$

19 An experimental car using ethanol as a fuel source requires 2270 kJ of energy for every kilometre travelled.

Given that the heat of combustion of ethanol is 1360 kJ mol^{-1}, what is the maximum distance that the car can travel on 1.0 kilogram of ethanol?

(A) 1.7 km

(B) 13 km

(C) 28 km

(D) 36 km

20 This graph represents the yield of an equilibrium reaction at different temperature and pressure conditions inside a reaction vessel.

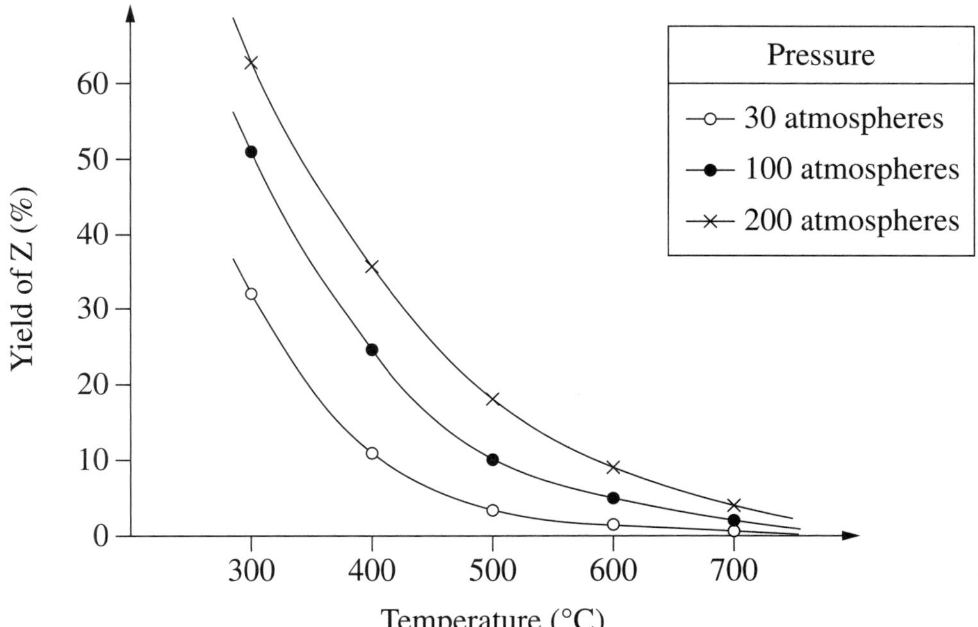

Which of the following reactions could produce the trends shown in the graph?

(A) $X(g) + Y(g) \rightleftharpoons 3Z(g)$ $\Delta H = +100$ kJ

(B) $X(g) + Y(g) \rightleftharpoons 2Z(g)$ $\Delta H = -100$ kJ

(C) $2X(g) + 2Y(g) \rightleftharpoons Z(g)$ $\Delta H = +100$ kJ

(D) $4X(g) + 2Y(g) \rightleftharpoons 3Z(g)$ $\Delta H = -100$ kJ

2014 HIGHER SCHOOL CERTIFICATE EXAMINATION

Chemistry

Centre Number

Student Number

Section I (continued)

Part B – 55 marks
Attempt Questions 21–31
Allow about 1 hour and 40 minutes for this part

Answer the questions in the spaces provided. These spaces provide guidance for the expected length of response.

Show all relevant working in questions involving calculations.

Question 21 (3 marks)

The transuranic artificial element curium–242 can be produced in two ways. The nuclear equations for the two processes are given below. **3**

Process *A*

$$^{241}_{95}\mathrm{Am} + ^{1}_{0}n \rightarrow ^{242}_{95}\mathrm{Am} \rightarrow ^{242}_{96}\mathrm{Cm} + ^{0}_{-1}e$$

Process *B*

$$^{239}_{94}\mathrm{Pu} + ^{4}_{2}\mathrm{He} \rightarrow ^{242}_{96}\mathrm{Cm} + ^{1}_{0}n$$

Compare these two processes of production for the element curium–242.

....

....

....

....

....

....

....

....

Question 22 (6 marks)

A student performed a first-hand investigation to determine the quantitative relationship between heat of combustion and molecular mass of alkanols. The student did this by burning different alkanols to heat water as shown in the diagram below. The calculated heats of combustion for four of the alkanols are given in the table.

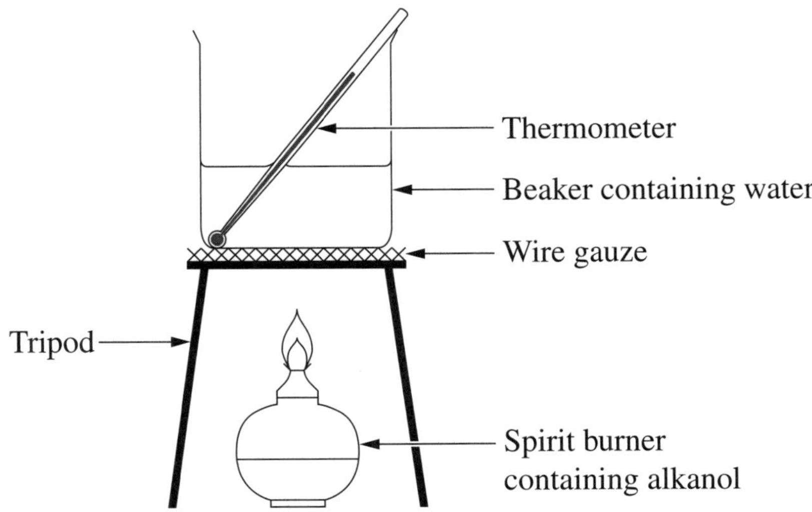

Alkanol	*Molecular mass* (g mol^{-1})	*Calculated heat of combustion* (kJ mol^{-1})	*Theoretical heat of combustion* (kJ mol^{-1})
methanol	32	150	726
ethanol	46	950	1367
propan-1-ol	60	1500	2021
butan-1-ol	74	2250	2676

Question 22 continues

Question 22 (continued)

(a) On the grid below, graph both the calculated and the theoretical heat of combustion against the molecular mass of the alkanols. **3**

Heat of combustion versus molecular mass

Heat of combustion ($kJ\ mol^{-1}$)

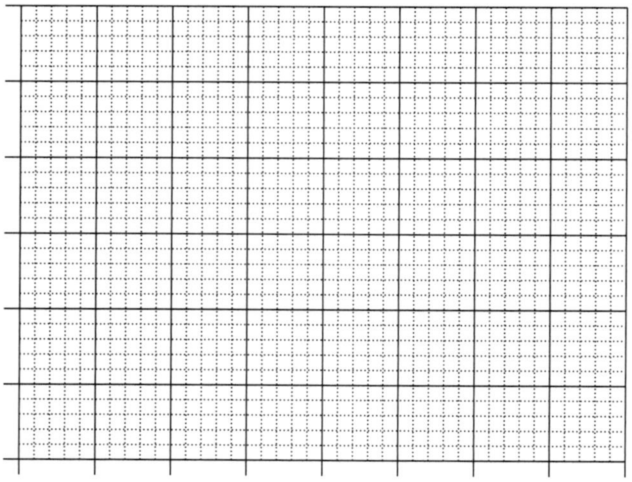

Molecular mass ($g\ mol^{-1}$)

(b) Discuss the validity of the student's investigation. **3**

....

....

....

....

....

....

....

....

End of Question 22

Question 23 (3 marks)

This diagram shows a town situated near agriculture and industry.

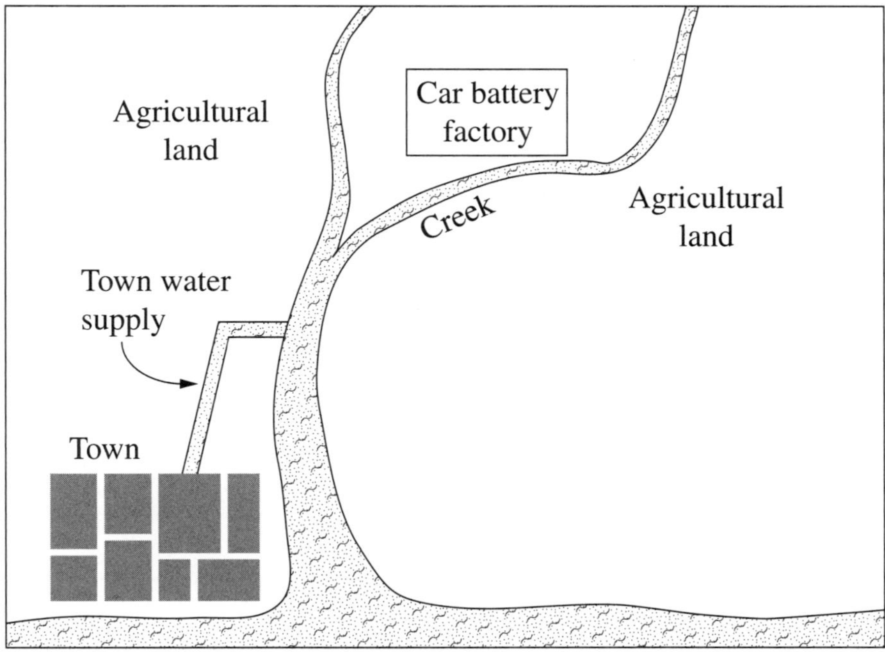

The town relies on the river for its water supply.

(a) Identify ONE chemical species that could be a contaminant of the water supply. **1**

....

(b) Explain the need to monitor the levels of a contaminant in water supplies. **2**

....

....

....

2014 HIGHER SCHOOL CERTIFICATE EXAMINATION

Chemistry

Centre Number

Section I – Part B (continued)

Student Number

Question 24 (5 marks)

A solution contains carbonate, chloride and sulfate ions. **5**

Describe a sequence of tests that could be used to confirm the presence of each of these ions. Include ONE relevant chemical equation.

Question 25 (4 marks)

Under conditions of low oxygen levels, octane can undergo incomplete combustion according to the following chemical equation:

$$2C_8H_{18}(l) + 17O_2(g) \rightarrow 6C(s) + 4CO(g) + 6CO_2(g) + 18H_2O(l)$$

(a) Explain the need to monitor this process. **2**

....

....

....

....

(b) Calculate the mass of soot $(C(s))$ produced if 4.2 moles of octane are combusted in this way. **2**

....

....

....

....

2014 HIGHER SCHOOL CERTIFICATE EXAMINATION

Chemistry

Section I – Part B (continued)

Centre Number

Student Number

Question 26 (6 marks)

A first-hand investigation to produce an ester is to be carried out in a school laboratory, using an alkanol, an alkanoic acid and a suitable catalyst.

(a) Name an ester that could be produced in a school laboratory. **1**

....

(b) Describe how potential hazards associated with the three chemicals required for this investigation could be addressed. **5**

....

....

....

....

....

....

....

....

....

....

Question 27 (6 marks)

Limestone ($CaCO_3$) contributes to the hardness of water by releasing Ca^{2+} ions in water. The chemical equation for this exothermic reaction is shown.

$$CaCO_3(s) + H_2O(l) + CO_2(g) \rightleftharpoons Ca^{2+}(aq) + 2HCO_3^-(aq)$$

(a) Explain why increasing the temperature of hard water would reduce its hardness. **2**

....

....

....

....

(b) Describe how atomic absorption spectroscopy (AAS) could be used to measure the effectiveness of heating water to reduce its hardness. **4**

....

....

....

....

....

....

....

....

2014 HIGHER SCHOOL CERTIFICATE EXAMINATION

Chemistry

Section I – Part B (continued)

Centre Number

Student Number

Question 28 (5 marks)

A galvanic cell has been constructed as shown in the diagram.

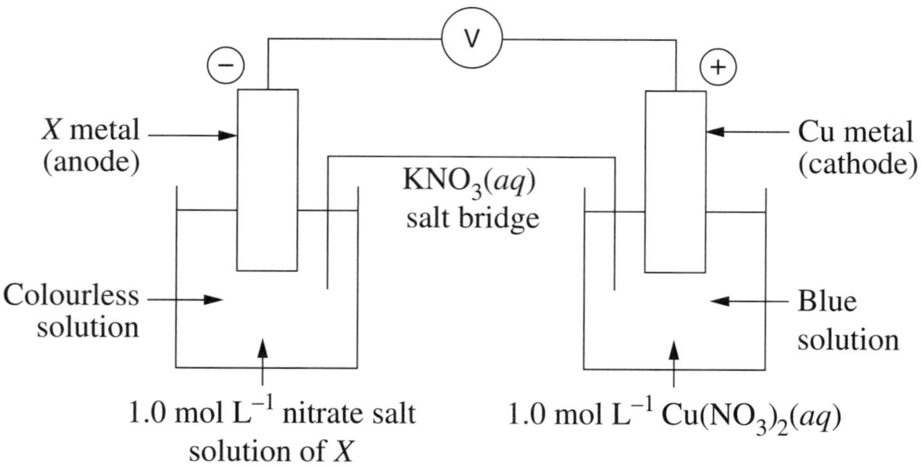

V

⊖ ⊕

X metal (anode)

Cu metal (cathode)

$KNO_3(aq)$ salt bridge

Colourless solution

Blue solution

1.0 mol L^{-1} nitrate salt solution of *X*

1.0 mol L^{-1} $Cu(NO_3)_2(aq)$

(a) Explain the colour change in the copper half-cell as the reaction proceeds. **2**

....

....

....

....

(b) The theoretical standard potential for this galvanic cell is 2.02 V. **3**

Identify metal *X* and justify your answer.

....

....

....

....

....

....

Question 29 (5 marks)

An experiment was performed to model the formation of acid rain. **5**

A sample of sulfur was burned on a metal spoon. While alight, it was placed in a gas jar with some dry litmus paper.

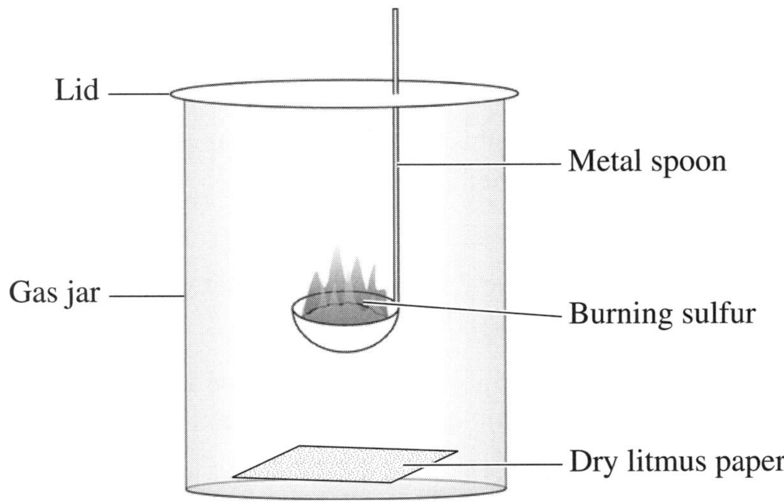

When a fine mist of water was sprayed into the jar, the litmus paper turned red.

Assess the suitability of this experiment as a model for the formation of acid rain.

....

....

....

....

....

....

....

....

....

....

2014 HIGHER SCHOOL CERTIFICATE EXAMINATION

Chemistry

Section I – Part B (continued)

Centre Number

Student Number

Question 30 (5 marks)

A batch of dry ice (solid CO_2) was contaminated during manufacture. To determine its purity, the following steps were carried out.

Step 1: A 0.616 gram sample of the contaminated dry ice was placed in a clean, dry flask.

Step 2: 50.00 mL of 1.00 mol L^{-1} sodium hydroxide was added to the flask. The sodium hydroxide was in excess.

Step 3: The flask was sealed to prevent loss of carbon dioxide gas and the reaction allowed to reach completion, according to this equation:

$$2NaOH(aq) + CO_2(s) \rightarrow Na_2CO_3(aq) + H_2O(l)$$

Step 4: The remaining sodium hydroxide was titrated against a 1.00 mol L^{-1} solution of hydrochloric acid. The average volume of HCl used was 27.60 mL.

(a) Calculate the number of moles of NaOH added in Step 2. **1**

....

....

(b) Calculate the percentage purity by mass of this batch of dry ice. **4**

....

....

....

....

....

....

....

....

Question 31 (7 marks)

With reference to the underlying chemistry and with relevant equations, assess the impacts on society of TWO uses of ethanol. 7

....

....

....

....

....

....

....

....

....

....

....

....

....

....

....

....

....

....

2014 HIGHER SCHOOL CERTIFICATE EXAMINATION

Chemistry

Section II

25 marks
Attempt ONE question from Questions 32–36
Allow about 45 minutes for this section

Answer parts (a)–(c) of the question in Section II Answer Booklet 1.
Answer parts (d)–(e) of the question in Section II Answer Booklet 2.
Extra writing booklets are available.

Show all relevant working in questions involving calculations.

Question 32	Industrial Chemistry
Question 33	Shipwrecks, Corrosion and Conservation
Question 34	The Biochemistry of Movement *(Not included in this reproduction)*
Question 35	The Chemistry of Art *(Not included in this reproduction)*
Question 36	Forensic Chemistry

Question 32 — Industrial Chemistry (25 marks)

Answer parts (a)–(c) in Section II Answer Booklet 1.

(a) This diagram illustrates some of the properties of sulfuric acid. **3**

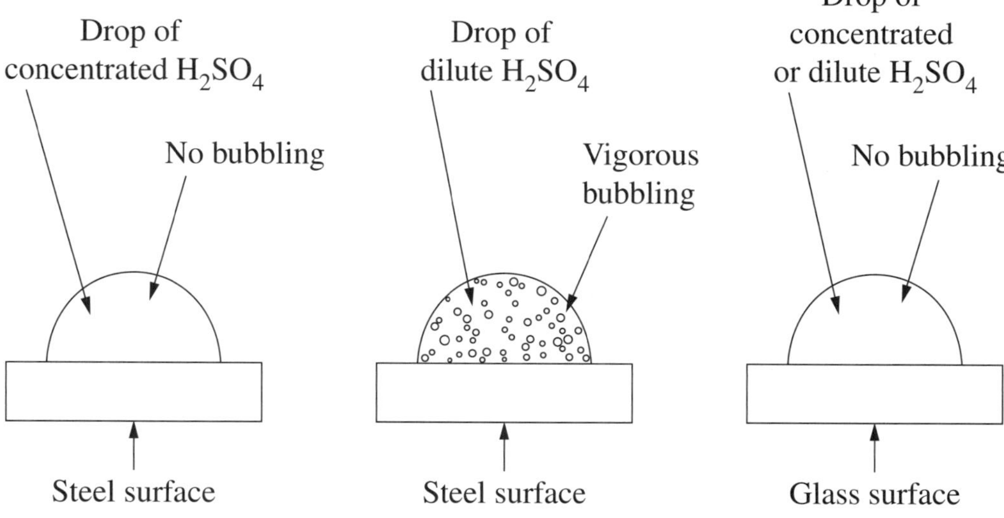

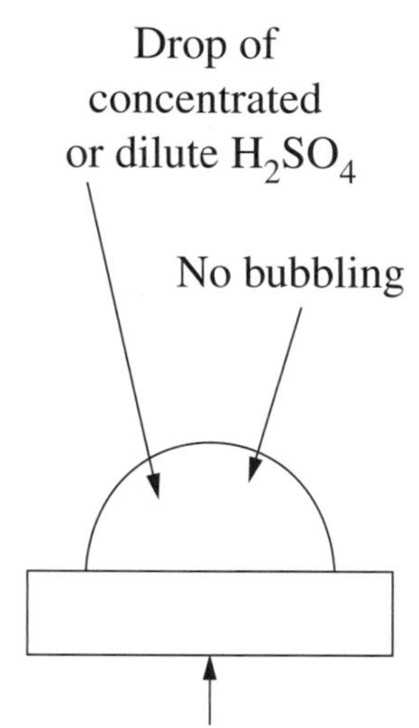

Explain, with reference to the diagram, how sulfuric acid should be transported.

(b) Nitrosyl chloride is introduced into an empty container. It then dissociates into nitric oxide and chlorine according to the equation:

$$2NOCl(g) \rightleftharpoons 2NO(g) + Cl_2(g)$$

The reaction is endothermic.

(i) Explain the effect on the yield of NO(g) if the temperature is increased. **2**

(ii) The equilibrium constant, K, for the reaction is 0.028. **3**

Calculate the equilibrium concentration of NOCl(g) if the equilibrium concentration of $Cl_2(g)$ is 0.17 mol L^{-1}.

Question 32 continues

Question 32 (continued)

(c) Electrolysis is used in the industrial production of sodium hydroxide.

(i) Contrast the energy transformations in galvanic and electrolytic cells. **2**

(ii) The table shows the products of three different electrolytic cells involving aqueous or molten sodium chloride. **3**

Cell	*Anode*	*Cathode*
X	$O_2(g)$	$H_2(g)$
Y	$Cl_2(g)$	$H_2(g)$
Z	$Cl_2(g)$	$Na(l)$

Explain which of the three electrolytic cells from the table is used for the industrial production of sodium hydroxide.

Answer parts (d)–(e) in Section II Answer Booklet 2.

(d) An investigation is to be conducted to model a chemical step involved in the Solvay process.

(i) Outline a valid procedure that could be used to carry out the investigation. **2**

(ii) Describe TWO limitations associated with the procedure outlined in part (i). **3**

(e) Explain how the differences in the structure and composition of soaps and detergents determine their uses and their impacts on the environment. **7**

End of Question 32

Question 33 — Shipwrecks, Corrosion and Conservation (25 marks)

Answer parts (a)–(c) in Section II Answer Booklet 1.

(a) This diagram shows the various layers of a pipe. **3**

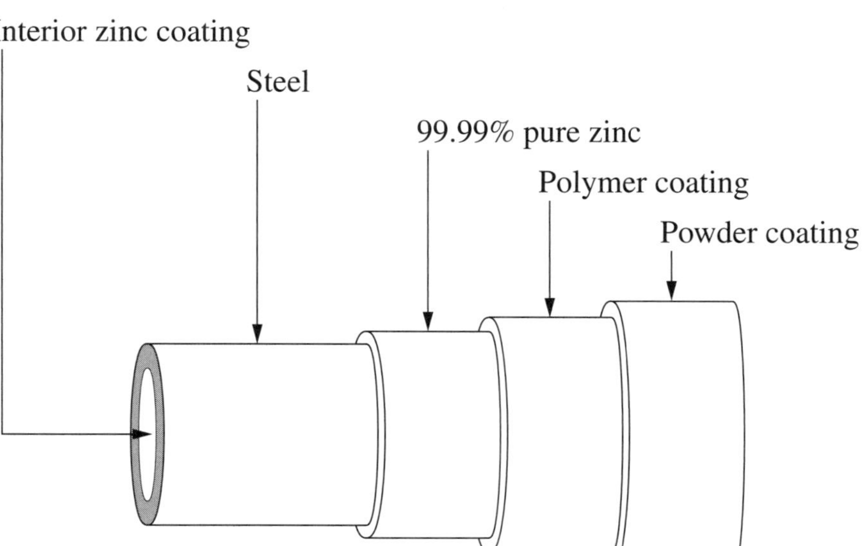

Outline TWO reasons why this pipe would be resistant to corrosion.

(b) Iron corrodes differently under acidic and neutral conditions.

(i) Write an equation to represent the process of rusting under neutral conditions. **2**

(ii) Explain why iron will corrode faster under acidic conditions than under neutral conditions. **3**

Question 33 continues

Question 33 (continued)

(c) A wooden artefact was recovered from a shipwreck. After it was removed from the ocean, it was placed in a tank of distilled water. The water was changed weekly and the chloride ion concentration was monitored. After 10 weeks, the artefact was dried slowly in a controlled environment.

The graph shows the chloride ion concentration in the tank in the first five weeks.

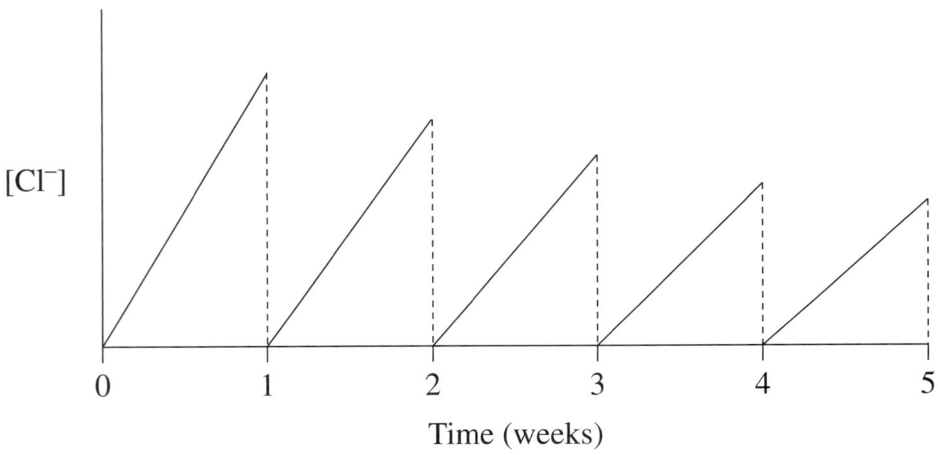

(i) Account for the shape of this graph. **2**

(ii) Explain the possible effect on the artefact if it had been simply left to dry instead of undergoing the procedure described above. **3**

Answer parts (d)–(e) in Section II Answer Booklet 2.

(d) An investigation is to be set up in a school laboratory to determine the rate of corrosion of iron in different oxygen concentrations.

(i) Identify the variables that need to be kept constant in this investigation. **2**

(ii) Describe TWO limitations in making qualitative and/or quantitative observations in this investigation. **3**

(e) Explain why a range of factors should be considered when using electrolysis to clean and stabilise a metal artefact. **7**

End of Question 33

Question 36 — Forensic Chemistry (25 marks)

Answer parts (a)–(c) in Section II Answer Booklet 1.

(a) This picture shows a forensic scientist collecting a blood sample from a crime scene. **3**

Explain TWO errors that this scientist is making while collecting the blood sample.

(b) Hydrolysis can be used to break down proteins into amino acids.

(i) This equation represents the hydrolysis of a dipeptide. **2**

$$H_2N-\underset{\substack{|\\ CH_3}}{CH}-\overset{\substack{O\\ ||}}{C}-\overset{\substack{H\\ |}}{N}-CH_2-CH_2-C\begin{matrix}\diagup O \\ \\ \diagdown OH\end{matrix} + H_2O \rightarrow \ldots\ldots + \ldots\ldots$$

In your answer booklet, draw the TWO products of the hydrolysis of the dipeptide.

(ii) Describe how protein hydrolysis is used in forensic analysis. **3**

Question 36 continues

Question 36 (continued)

(c) (i) An unknown substance was collected. It was analysed and its mass spectrum collected. **2**

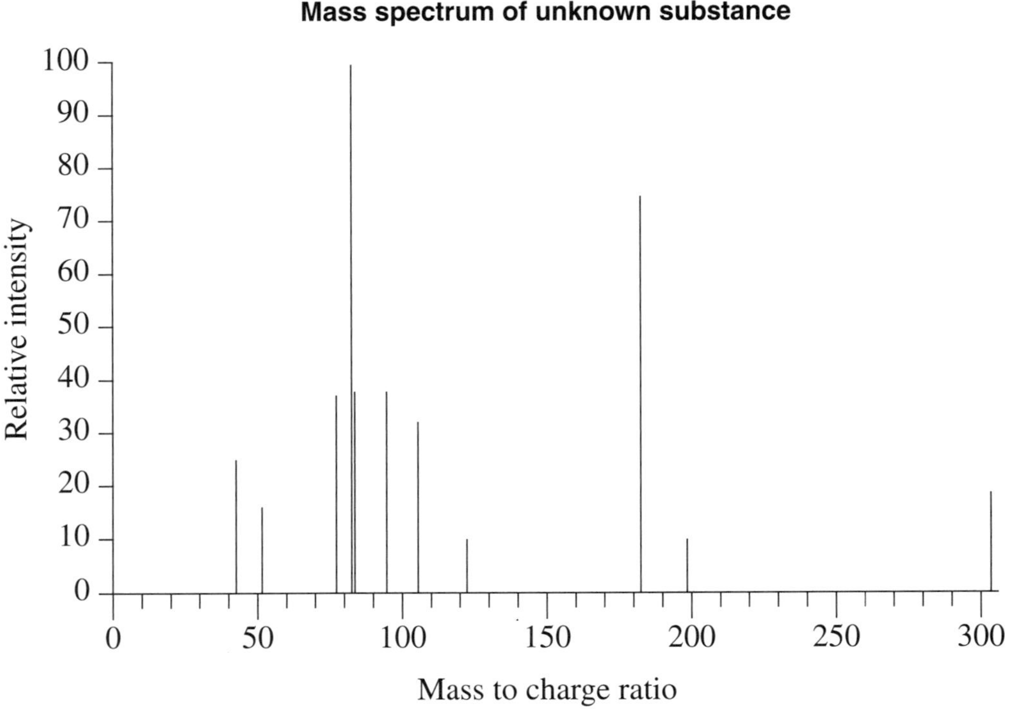

This table shows the mass to charge ratio of a selection of fragments in the mass spectrum for three compounds of interest in forensic investigations.

Compound name	*Significant fragments (mass to charge ratio)*
Caffeine	67 109 194
Cocaine	82 94 182
Paracetamol	43 109 151

Using the information in the table, identify the unknown substance and justify your choice.

(ii) Describe how mass spectrometry can be useful for analysing forensic evidence. **3**

Question 36 continues

Question 36 (continued)

Answer parts (d)–(e) in Section II Answer Booklet 2.

(d) An investigation is to be conducted in a school laboratory to separate organic compounds using chromatography.

(i) Outline a valid procedure that could be used in a school laboratory to carry out the investigation. **2**

(ii) Describe TWO limitations of carrying out the procedure outlined in part (i). **3**

(e) Explain why DNA evidence may be challenged when used in court cases. **7**

End of Question 36

End of paper

2014 HSC Examination Paper

Sample Answers

Section I Part A

(Total 20 marks)

1 D Pollution is defined as the introduction of harmful substances into the environment. Ozone is introduced to the troposphere by man and is harmful to human health.

2 A

3 B

4 C Cracking must involve splitting the molecule into at least two smaller products.

5 C Addition reactions involve adding atoms across a double bond.

6 B High quality water looks clear and does not contain microbes or traces of dissolved substances that could affect the taste. Turbidity refers to the clarity. Dissolved solids affect its taste and high levels of dissolved oxygen are an indicator of low levels of microbes.

7 C If methyl red is yellow the pH is above 6.2 and if phenolphthalein is colourless the pH is below 8.3.

8 B

9 A Isomers have different structural formulae but the same molecular formula. Compounds W and X have the formula $C_4H_7Cl_2F$. In compound W the fluorine atom is found on carbon number 3 while in compound X it is found on carbon number 2.

10 C

11 A In a coordinate covalent bond the electron pair is derived from the one atom. The electrons in the covalent bond in (A) are derived from the carbon and the chlorine atoms.

12 C A decrease in pH by one point indicates an increase in hydrogen ions by 10 times. Wine is three pH points lower than distilled water and therefore the concentration of hydrogen ions is $10 \times 10 \times 10$ higher in wine.

13 B When the volume of the chamber is halved the concentration of ammonia will instantly be doubled. The reaction will then shift to the right to partially counteract the increase in the overall pressure, increasing the concentration of ammonia.

14 D $pH = -\log_{10}[H^+] = -\log_{10}[0.018]$

15 D The volume of CO2(g) produced is dependent upon the number of moles of reactant. In each case, one mole of the compound forms one mole of $CO_2(g)$. The reactant with the smallest molar mass will have the greatest number of moles and will produce the greatest volume.

16 B The oxidation state of copper drops from +2 to 0 and is therefore reduced.

17 A $K_2Cr_2O_7$ is reduced with a reduction potential of 1.36 V. SO_2 is oxidised with an oxidation potential of −0.16 V.

$1.36 - 0.16 = 1.20$ V

18 D

19 B $n(\text{ethanol}) = \dfrac{1000}{46} = 21.74$ mol

Heat produced by burning = 21.74 × 1360 kJ = 29 565.2 kJ

Distance travelled $= \dfrac{29\,565.2}{2270} = 13.0$ km

20 D As temperature rises the yield of Z decreases. Therefore the forward reaction must be exothermic. As the pressure rises the yield of Z increases. Therefore there must be more moles of gaseous reactant than product.

Section I Part B

21 Process A occurs in a nuclear reactor where a target of americium-241 is introduced into the reactor and is bombarded by neutrons. Process B occurs in a particle accelerator where alpha particles are accelerated at very high speeds into plutonium-239 targets. In Process A beta particles are emitted, whereas in Process B a neutron is emitted. In both cases a small particle is being assimilated into the nucleus of a large atom, increasing the number of protons and therefore creating an element with a higher atomic number than the target.

(3 marks)

22 (a)

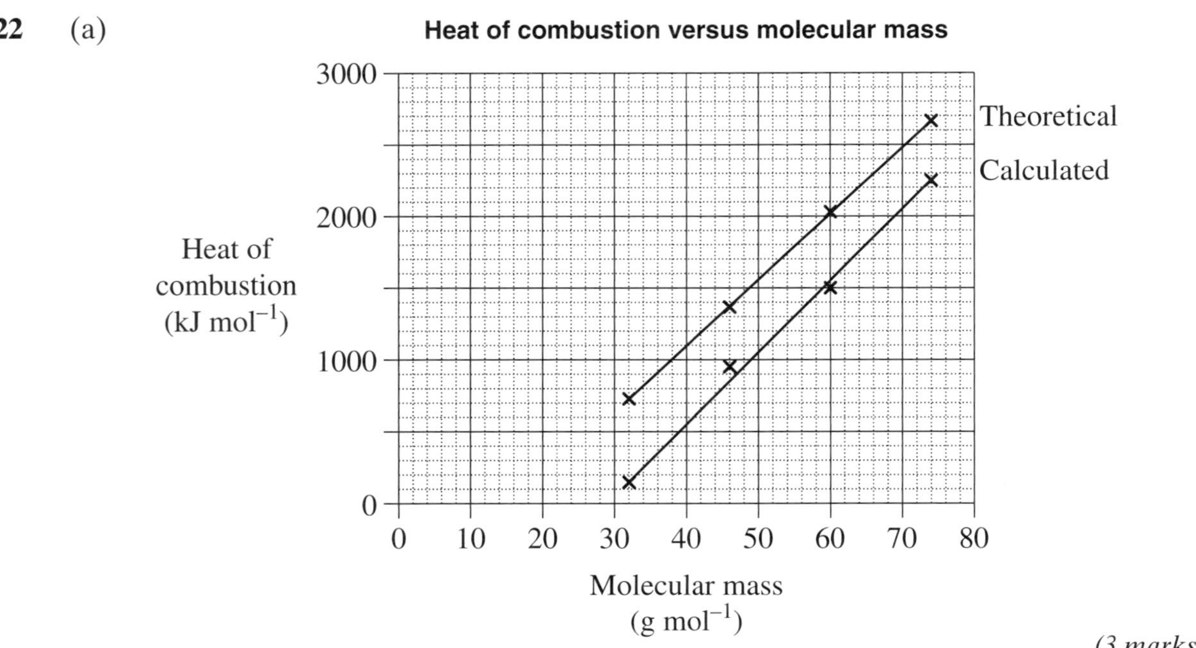

(3 marks)

(b) This is a valid procedure as it addresses the aim of determining the relationship between heat of combustion and molecular mass. Though the calculated heats of combustion are low, the general trend reflected in the results is similar to the theoretical value, as the amount of heat lost for each fuel is fairly consistent. However, this is not a valid procedure to determine heat of combustion because there are too many opportunities for heat to not be absorbed by the water in the beaker and lost from the system. The flame from the burner is not as close as it could be to the beaker, resulting in heat loss to the surroundings. A wire gauze is used which will direct heat away from the beaker. The water is contained in a glass beaker, which is not insulated. It will radiate heat to the surroundings. The beaker is not covered, increasing the heat loss from the surface of the water to the surroundings. The combustion is incomplete because insufficient oxygen is present at the wick, producing less heat than if combustion was complete. The different values for the calculated and theoretical heat of combustion reflect this invalid procedure. *(3 marks)*

23 (a) lead from the car battery factory *(1 mark)*

(b) A contaminant such as lead needs to be monitored because it can be absorbed into the human body through drinking water and from other sources such as breathing air and eating contaminated food. Once in the body, contaminants such as lead accumulate over time and can reach toxic levels. At these levels, a contaminant such as lead can cause neurological problems, behavioural disorders in children and disruption to enzyme systems. *(2 marks)*

24 1. Add a few drops of 1 mol L^{-1} nitric acid. Effervescence (fizzing) indicates the presence of carbonate. Add more nitric acid until effervescence ceases.

$2H^+(aq) + CO_3^{2-}(aq) \rightarrow CO_2(g) + H_2O(l)$

2. Add a few drops of 0.1 mol L^{-1} silver nitrate to the previously acidified solution. A white precipitate, which darkens when exposed to sunlight, indicates the presence of chloride.

$Ag^+(aq) + Cl^-(aq) \rightarrow AgCl(s)$

3. Filter the previous mixture and test the filtrate. Add a few drops of 1 mol L^{-1} barium nitrate. A fine white precipitate indicates the presence of sulfate ions.

$Ba^{2+}(aq) + SO_4^{2-}(aq) \rightarrow BaSO_4(s)$ *(5 marks)*

25 (a) This process produces particulate carbon (soot) which, when breathed in over a period of time, can cause cancer. Carbon monoxide also needs monitoring because it is toxic to humans. It combines with haemoglobin in the blood, reducing the capacity of the blood to absorb oxygen. Poor combustion of fuels also generates less energy than complete combustion. Monitoring allows this problem to be identified and corrected. *(2 marks)*

(b) $n(C) = 3 \times n(C_8H_{18}) = 4.2 \times 3 = 12.6$ mol soot

$m(C) = 12.6 \times 12.01 = 151.33 \text{ g} = 150 \text{ g}$ *(2 marks)*

26 (a) Octyl ethanoate (or octyl acetate) *(1 mark)*

(b) The alkanol required for this investigation is octan-1-ol. It is volatile and flammable. This hazard can be addressed by heating the mixture in a hot water bath on a hot plate rather than by exposing the alkanol to the naked flame of a Bunsen burner.

The alkanoic acid required is glacial acetic acid (ethanoic acid). It is corrosive and may produce tissue damage, particularly on the mucous membranes of eyes, mouth and respiratory tract. To address this risk to the eyes, safety glasses should be worn.

The catalyst used is 18 mol L^{-1} sulfuric acid. It is highly corrosive and very hazardous in case of skin contact and may produce burns. It should only be handled by a qualified chemistry teacher; however, students may use concentrated acid in small quantities (less than 5 mL) after appropriate safety warnings. Latex gloves should be worn to protect the skin from contact with this acid. *(5 marks)*

27 (a) When temperature is increased the equilibrium will shift to the left to counteract the change and favour the endothermic reaction. In the process Ca^{2+} ions will be precipitated. *(2 marks)*

(b) Hardness is due to the concentration of Ca^{2+} ions. Atomic absorption spectroscopy (AAS) uses the principle of selective absorption of light by metal ions to determine concentration of very small amounts. The greater the absorption, the higher the ion concentration. Heated and unheated samples of water can be tested by AAS and the concentrations of calcium compared. A calcium cathode lamp is used as a source of light. It emits a characteristic spectrum through a vaporised sample of water containing the ion being tested. Standardised solutions are used to obtain various concentration absorptions. The sample is sprayed into the flame and calcium ions absorb selective wavelengths of light passing through the sample. The emerging light passes through a monochromator containing a narrow slit and the intensity of a single wavelength beam is measured by a photomultiplier. This measures light intensity and converts it to an electrical signal. The amount of light absorbed relative to the standards is converted into a concentration figure. If the concentration of calcium in the heated water was fo und to be lower than unheated water, the heating method would be described as effective. *(4 marks)*

28 (a) Blue copper ions are being reduced to copper metal at the cathode. Therefore the concentration of copper ions is dropping and the intensity of the blue-coloured solution is diminishing.

$Cu^{2+}(aq) + 2e^- \rightarrow Cu(s)$ *(2 marks)*

(b) Metal X is aluminium, as aluminium has an oxidation potential of 1.68 V. This oxidation potential, when added to the reduction potential of copper ions (0.34 V), is equal to the theoretical potential of 2.02 V.

$Cu^{2+}(aq) + 2e^- \rightarrow Cu(s) \qquad E^0 = 0.34\ V$

$X(s) \rightarrow X^{3+}(aq) + 3e^- \qquad E^0 = x$

$$0.34 + x = 2.02\ V$$
$$x = 1.68\ V$$

(3 marks)

29 Acid rain is rain with a lower pH than normal owing to man-made acidic oxide gases dissolving and reacting with water in the atmosphere. In this experiment, sulfur is burnt in air forming sulfur dioxide.

$$S(s) + O_2(g) \rightarrow SO_2(g)$$

When the fine mist of water is sprayed into the jar the SO_2 will react with the water forming sulfurous acid.

$$SO_2(g) + H_2O(l) \rightarrow H_2SO_3(aq)$$

Sulfurous acid will ionise, forming H^+ ions that turn litmus red, indicating the presence of acid.

This experiment models one way in which acid rain can form in industrial areas. Sulfur minerals (rather than pure sulfur) are burnt in power stations (as they are present in coal), forming SO_2 which goes on to react with moisture in the atmosphere forming sulfurous acid.

Sulfur dioxide is also produced by smelting metal sulphides, and naturally by volcanoes and decomposition of rotting organic wastes. These processes can also produce acid rain.

ASSESSMENT

Weighing up the pros and cons, this is a suitable model to show one way that acid rain forms because it reacts with one of the same non-metals that on combustion go on to form acid rain in nature, and it shows that the acids produced result in a falling pH as would occur in the atmosphere. *(5 marks)*

30 (a)

$$\begin{aligned} n(\text{NaOH}) &= C \times V \\ &= 1.00 \times 0.05000 \\ &= 0.0500 \text{ mol} \end{aligned}$$

(1 mark)

(b)

$$\begin{aligned} n(\text{HCl}) \text{ reacted} &= C \times V \\ &= 1.00 \times 0.02760 \\ &= 0.02760 \text{ mol} \end{aligned}$$

$n(\text{NaOH})$ present at completion $= n(\text{HCl})$ reacting $= 0.02760$ mol

$$\begin{aligned} n(\text{NaOH}) \text{ reacting with } CO_2 &= \text{initial moles} - \text{final moles} \\ &= 0.0500 - 0.02760 \\ &= 0.0224 \text{ mol} \end{aligned}$$

$$n(CO_2) \text{ reacting} = \frac{n(\text{NaOH})}{2} = 0.0112 \text{ mol}$$

$$m(CO_2) = 0.0112 \times 44.01 = 0.493 \text{ g}$$

$$\% \text{ purity} = \left(\frac{m_{CO_2}}{m_{\text{dry ice}}}\right) \times 100 = \frac{0.493}{0.616} \times 100 = 80.0\%$$

(4 marks)

31 Answers may vary. Following are two possibilities.

Use 1: Ethanol as a fuel

Ethanol is produced by fermentation of sugary crops, such as sugar cane and corn, and as such it is a renewable resource.

$$C_6H_{12}O_6(aq) \rightarrow 2C_2H_5OH(aq) + 2CO_2(g)$$

When combusted, ethanol forms carbon dioxide and water and releases heat, and so is a useful fuel.

$$C_2H_5OH(l) + 3O_2(g) \rightarrow 2CO_2(g) + 3H_2O(l)$$

Therefore, theoretically, the carbon dioxide released when ethanol is combusted is balanced by that absorbed during photosynthesis. Carbon dioxide is thought to be a main cause of global warming, which will have many impacts on society, such as costs associated with preparing for and dealing with the increasing frequency of natural disasters and their impacts on people and structures. Because ethanol is considered to be carbon neutral, in that it does not lead to a net increase in carbon dioxide in the atmosphere, using ethanol as a fuel has a significant positive impact on society.

However, fossil fuels are used in the farming process and in the distillation and distribution of ethanol, so ethanol is not as clean a fuel as first imagined. Ethanol does lead to a net increase in atmospheric carbon dioxide levels.

Ethanol does not release as much energy as petrol when combusted. However, because it burns more efficiently than petrol (being a small molecule), the net energy available is about the same.

The impact on society of clearing land for wholesale growth of sugar crops should aso be considered. This is arable land that would otherwise have been used to grow food crops.

ASSESSMENT

Weighing up the pros and cons, using ethanol as a fuel blend would appear to have a positive impact on society owing to the reduced production of carbon while not requiring modifications to car engines nor causing significant reductions in energy output.

Use 2: Ethanol as a solvent

Ethanol is a good solvent for both polar and non-polar substances. The alcohol, OH group, interacts with other polar molecules such as water, forming hydrogen bonds. The non-polar ethyl end, C_2H_5, attracts non-polar solutes such as grease and oils by forming dispersion forces, making it a useful cleaning agent.

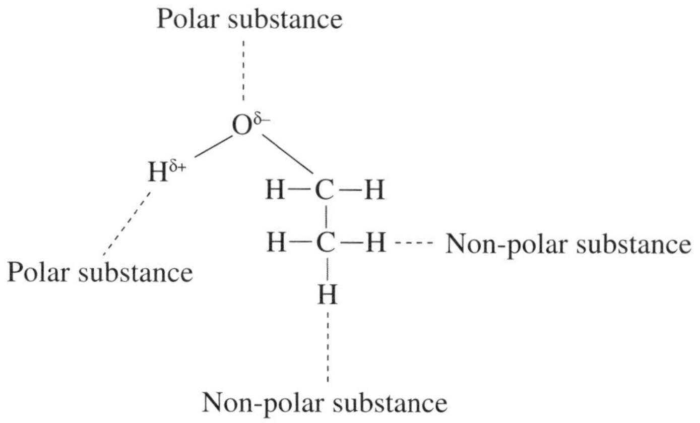

Polar substance

$O^{\delta-}$

$H^{\delta+}$

H—C—H

H—C—H ---- Non-polar substance

Polar substance

H

Non-polar substance

Therefore ethanol is used as a solvent in cosmetics, perfume and toiletries and in industry in paints, resins and lacquers.

The impact on society of using ethanol for this purpose is positive as lives are made easier with these products being available to mankind.

On the negative side, industrial ethanol is derived from the cracking of petroleum to ethylene and its hydration. The mining of petroleum is environmentally damaging. Ethanol is also a volatile, flammable chemical. Its storage can pose a risk to society if it comes into contact with heat or flames.

ASSESSMENT

Weighing up the pros and cons, using ethanol as a solvent for cleaning and in manufacturing and industry would appear to outweigh a few negative impacts.

(7 marks)

Section II—Options

Question 32—Industrial Chemistry

(a) The transportation of sulfuric acid depends on its concentration. The diagram shows that concentrated sulfuric acid does not react with a steel surface, so large volumes of concentrated sulfuric acid can be transported in steel tanks by land or sea, but not by air. The tanks should be labelled with the appropriate hazardous chemical (HAZCHEM) labels. Small volumes of concentrated sulfuric acid can also be transported in glass bottles, with which it does not react, as shown in the diagram. However, glass cannot be used for large volumes because it is fragile. The diagram shows that dilute sulfuric acid does react with steel. As a result, dilute sulfuric acid must be transported in relatively small volumes, in glass bottles. When steel reacts with dilute sulfuric acid the predominant reaction is:

$2Fe(s) + 3H_2SO_4(aq) \rightarrow Fe_2(SO_4)_3(aq) + 3H_2(g)$ *(3 marks)*

(b) (i) The forward reaction is endothermic, so increasing the temperature (adding heat energy) will shift the equilibrium position in the forward direction, because Le Chatelier's principle predicts that the equilibrium position will shift in the direction that consumes the added heat energy. As a result, the yield of $NO(g)$ will increase. *(2 marks)*

(ii)

	NOCl	Cl_2	NO
Equilibrium concentration (mol L^{-1})	x	0.17	0.34

$$K = \frac{[NO]^2[Cl_2]}{[NOCl]^2} = 0.028$$

$$0.028 = \frac{(0.34)^2 \times (0.17)}{x^2}$$

$$x^2 = \frac{0.019652}{0.028} = 0.70185$$

Therefore $x = 0.84$ mol L^{-1} *(3 marks)*

(c) (i) Electrolytic cells convert electrical energy into chemical potential energy, while galvanic cells convert chemical potential energy into electrical energy. This means that electrolytic cells require a constant electric current to operate, while a working galvanic cell produces an electric current. *(2 marks)*

(ii) The industrial production of sodium hydroxide uses brine (aqueous sodium chloride) as a starting material. In the electrolytic process, Cl^- ions are oxidised at the anode to produce chlorine gas and water is reduced at the cathode to produce hydrogen gas and hydroxide ions.

$2Cl^-(aq) \rightarrow Cl_2(g) + 2e^-$

$2H_2O(l) + 2e^- \rightarrow 2OH^-(aq) + H_2(g)$

Sodium ions migrate from the anolyte to the catholyte in the membrane process to produce aqueous sodium hydroxide. The overall reaction that occurs is:

$2NaCl(aq) + 2H_2O(l) \rightarrow 2NaOH(aq) + Cl_2(g) + H_2(g)$

This reaction corresponds to Cell Y.

Cell X is used for the electrolysis of water and does not produce sodium hydroxide. Cell Z involves the electrolysis of molten sodium chloride, and sodium metal is produced rather than sodium hydroxide. *(3 marks)*

(d) (i) Answers may vary. Following is one example.

To model the formation of sodium hydrogen carbonate, mix 100 mL each of saturated $NH_3(aq)$ and saturated $NaCl(aq)$ in a 2 L conical flask (in a fume cupboard to avoid inhalation of $NH_3(g)$, which is toxic). Swirl the mixture while adding lumps of $CO_2(s)$ (dry ice) with tongs. Wear insulating gloves to avoid skin contact with the dry ice because it causes skin burns. Continue to add the dry ice with swirling until a precipitate of $NaHCO_3$ forms. *(2 marks)*

(ii) While only two limitations are required for this answer, three have been given here as examples.

A limitation of this procedure is that it requires the use of a fume cupboard, and so may not be possible in all school laboratories. Saturated ammonia solution (required to produce the ammoniacal brine) is toxic by inhalation, and is very pungent. Doing the experiment in a laboratory with open windows would not be sufficient to eliminate all risk of illness. Another limitation is that the $CO_2(s)$ sublimes very rapidly and escapes from the reaction mixture as $CO_2(g)$. For this reason a large quantity of $CO_2(s)$ must be added, slowly and with constant swirling. This increases both the time and the cost of the modelling experiment. An additional limitation is the difficultly in fully saturating the salt with ammonia because ammonia evaporates readily. *(3 marks)*

(e) Soaps and detergents are composed of a hydrophilic ionic 'head', which may be polar or ionic, and a hydrophobic 'tail', which is a long-chain, non-polar, hydrocarbon segment. It is this combination of structural features that gives rise to the cleaning action of soaps and detergents. The non-polar tail is soluble in oil, while the polar/ionic head dissolves in water. When soap or detergent is added to a mixture of oil and water, and the mixture is agitated, the soap/detergent particles surround the grease droplet. The hydrophobic end dissolves in the grease droplet, and the hydrophilic end dissolves in the water and is surrounded by water molecules. The resulting spherical particle is called a micelle, and the mixture becomes an emulsion of oil dispersed throughout the water, which can be washed away. The hydrophobic tail is a non-polar hydrocarbon chain, typically C_{12}–C_{15}, for both soap and synthetic detergents. It is the hydrophilic head that shows structural variation, giving rise to different uses and effectiveness in hard water.

A typical soap anion structure is $CH_3(CH_2)_{14}COO^-$, in which the head is the anionic carboxylate group. Soaps are typically used for personal hygiene, but they are of limited use in hard water because of the presence of Mg^{2+} and Ca^{2+} ions, which precipitate with soap anions. Soaps are easily decomposed by organisms in water and sewage (i.e. they are biodegradable) and as a result their use does not pose any grave environmental concerns.

Anionic detergents have the benzene sulfonate group as the hydrophilic head. Unlike soaps, these detergents do not precipitate with Ca^{2+} or Mg^{2+}, so are excellent for use in hard water. Because they are powerful surfactants, they tend not to be used for personal hygiene; rather, they are found in products such as washing-up liquids and laundry detergents. Early versions of anionic detergents contained branched-chain hydrophobic regions. These were not very biodegradable. As a result, many rivers and dams suffered from a build-up of foam and detergent, which caused considerable environmental damage. The development of linear alkylbenzene sulfonates eliminated this problem because the linear molecules are much more easily broken down in the environment.

Cationic detergents contain a quaternary ammonium salt as the hydrophilic head. The cationic head is attracted to glass and some fibres, so the detergent particles remain attached to the surface of glass objects, some fabrics and hair after washing. To avoid slippery glassware they are not used as dishwashing liquids, but are used as fabric softeners and hair conditioners, where their use leaves clothes and fibres feeling soft. Cationic detergents also have antiseptic properties, so they are used in nappy washes and domestic disinfectants. Their cationic nature prevents interaction with Mg^{2+} and Ca^{2+} ions, so they are effective in hard water. At low concentrations, cationic detergents are biodegradable and do not result in significant environmental damage. However, at high concentration (i.e. if they are overused) their antiseptic properties could result in the death of bacteria in sewage works that are responsible for the breakdown of sewage.

Non-ionic detergents tend to be polyethers. They are very low-frothing detergents, so find application in front-loading washing machines and dishwashers. Their non-ionic nature also makes them effective in hard water. They do not pose significant environmental issues.

While modern synthetic detergents themselves do not pose significant environmental issues, up until relatively recently their commercial formulations have included phosphate compounds as 'builders' to increase the detergent's effectiveness. The release of phosphates into waterways greatly increases nutrient concentrations, leading to algal blooms and eutrophication—a serious, negative impact on water quality and ecosystems. The use of phosphate builders has been greatly reduced, and in many detergents, eliminated. As a result, modern synthetic detergents do not cause algal blooming.

Anionic detergent

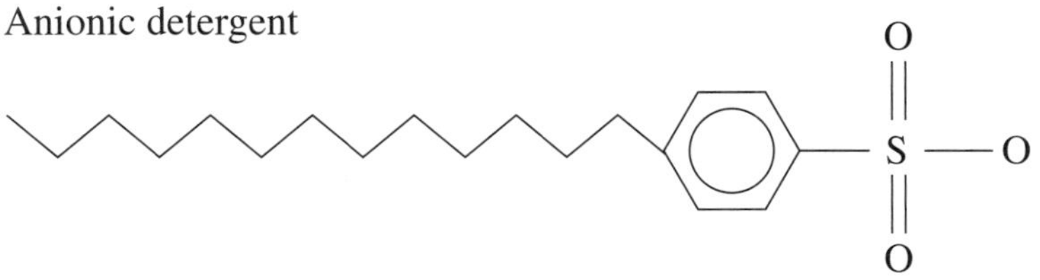

Cationic detergent

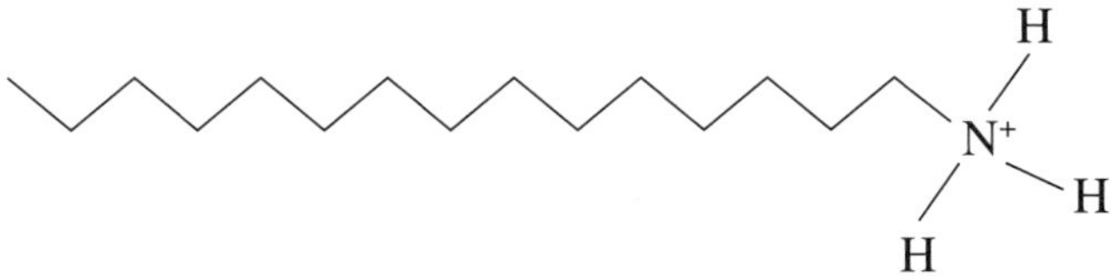

Non-ionic detergent

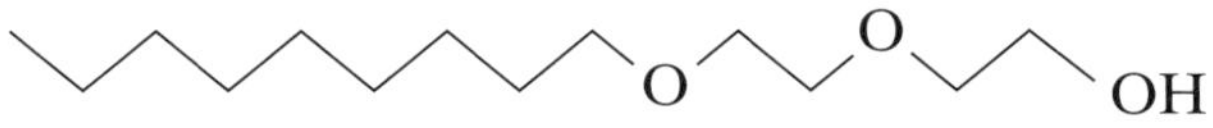

(7 marks)

Question 33—Shipwrecks, Corrosion and Conservation

(a) Reason 1: The interior zinc coating over the steel surface will act as a sacrificial anode. The zinc has a higher oxidation potential than iron so will corrode in preference to it, protecting the iron.

$Fe(s) \rightarrow Fe^{2+}(aq) + 2e^-$ $\quad E^0 = 0.44\ V$

$Zn(s) \rightarrow Zn^{2+}(aq) + 2e^-$ $\quad E^0 = 0.76\ V$

Reason 2: The polymer coating layer acts as a physical barrier that prevents oxygen and water (the necessary conditions for corrosion) from the outside of the pipe coming into contact with the steel layer. *(3 marks)*

(b) (i) $4Fe(s) + 3O_2(g) + 4H_2O(l) \rightarrow 2Fe_2O_3.2H_2O(s)$ *(2 marks)*

(ii) Under neutral conditions oxygen and water are reduced.

$\frac{1}{2}O_2(g) + H_2O(l) + 2e^- \rightarrow 2OH^-(aq)$ $\quad E^0 = 0.4\ V$

Under acidic conditions an alternative reduction reaction will occur with a higher reduction potential.

$\frac{1}{2}O_2(g) + 2H^+(aq) + 2e^- \rightarrow H_2O(l)$ $\quad E^0 = 1.23\ V$

The higher potential will result in a faster reaction rate. *(3 marks)*

(c) (i) In the first week the chloride ions diffused out of the wood cells into the water tank at a constant rate, owing to the diffusion gradient that existed between the highly concentrated cell interior and the outside distilled water in the container. Over each of the next four weeks the concentration of chloride ions in the water returned to zero when the salty distilled water was replaced with fresh water. It then rose as chloride ions continued to diffuse out of the wood. The concentration of chloride in the tank at the end of each week was progressively less because the diffusion gradient was less as the wood cells began with a lower internal concentration of salt. *(2 marks)*

(ii) If the artefact had been left to dry, the water within the wood cells would have evaporated, leaving the salt within the cells to crystallise. The salt crystals would have grown and eventually caused the cell walls to crack. The wood would lose its solid structure and become brittle. Deformation of the artefact would result. *(3 marks)*

(d) (i) The pieces of iron in each test would need to be the same type and have the same surface area, size and shape. Temperature, water volume and water type or concentration (if salty) would also need to be kept constant, as would the time between recording results. *(2 marks)*

(ii) Answers will vary. The question only requires two limitations. Three examples are provided here.

One limitation in making qualitative observations is that it is difficult to compare the level of corrosion under different conditions when the rusty precipitate does not always remain attached to the metal. It slips off and becomes suspended in the solution and falls to the bottom as sediment. It remains loosely attached to the metal in other conditions and hence the metal may appear to have corroded to a greater extent than it has.

One limitation in making quantitative observations is that if the corrosion rate is to be determined by the change in mass of the metal object over the given time, then the metal needs to be removed from the container, washed, dried and weighed. It is difficult to maintain consistency with this process and the result may be that more vigorous washing of one piece of metal may result in a greater mass loss than with another.

Another limitation is controlling the oxygen levels in each experiment. Low oxygen levels in the water can be obtained by boiling the water and allowing it to cool in a closed tube, to prevent oxygen re-entering the water. When the tube is opened to place the iron in the water some oxygen may re-enter the water. *(3 marks)*

(e) Submerged iron artefacts are likely to be impregnated with chloride. Electrolysis works by driving electrons into the artefact and making it the cathode. Some of the iron ions are returned to iron metal, cleaning the surface of corrosion products and restoring the metal.

$Fe(OH)_2(s) + 2e^- \rightarrow Fe(s) + 2OH^-(aq)$ and

$Fe(OH)Cl(s) + 2e^- \rightarrow Fe(s) + OH^-(aq) + Cl^-(aq)$

This also prevents any further corrosion, thereby stabilising the artefact. Making the artefact cathodic and negatively charged also helps the cleaning process by removing ions from small cracks throughout the metal surface by repelling anions such as Cl^- and SO_4^{2-}.

It is evident that voltage, polarity, the nature of electrolyte and electrodes, and the placement of circuit components are critical considerations contributing to the effective cleaning and stabilisation of the metal artefact.

Voltage needs to be maintained at a low level. Lower voltage over longer periods is a more effective way of returning metal ions to metal because the atoms form a regular lattice, not clumps.

The surface and type of anode needs consideration. An inert, stainless steel mesh is often used as the anode, and water is oxidised to oxygen gas on its surface. When a high surface area anode is used rather than a solid piece of metal the bubbles of oxygen provide a soft abrasion that helps to clean the surface of the artefact of any loose material without damaging the original metal.

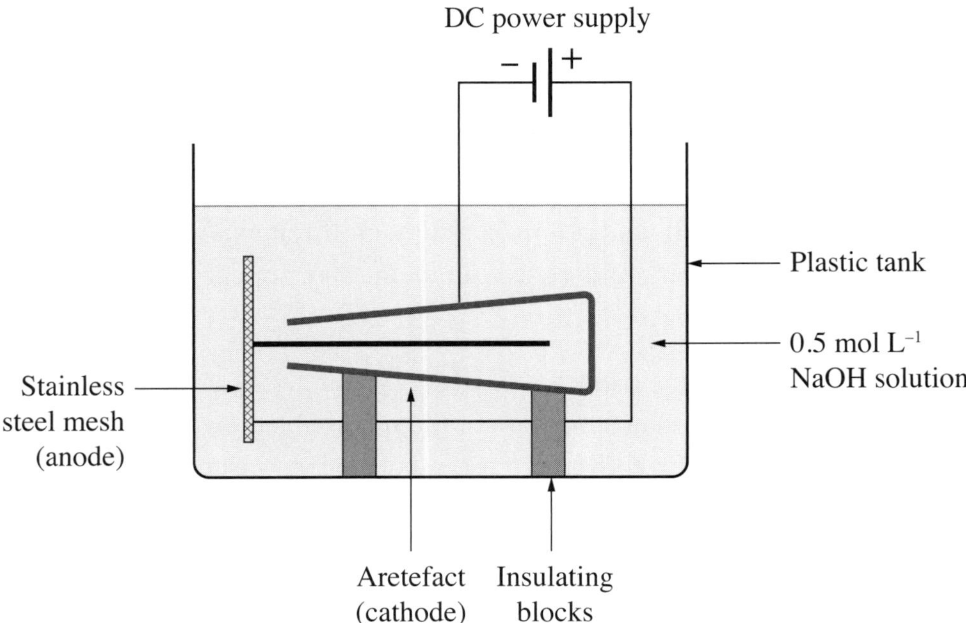

It is vitally important that the artefact be maintained as the cathode at all times during electrolysis. If not, the artefact itself will be the anode and will further corrode.

The nature of the electrolyte and the composition of the artefact need consideration. A 5% sodium hydroxide solution is a suitable electrolyte for most metals, including iron. It provides a conducting solution but also prevents the surrounding solution from becoming acidic and accelerating corrosion. However, it is not suitable for lead, copper and its alloys. A sodium carbonate or sodium hydrogen carbonate electrolyte is best for these as the lead, tin and zinc in copper alloys can dissolve in high pH solutions.

In order to avoid further corrosion, the conserved object should be coated with an appropriate lacquer or wax. *(7 marks)*

Question 36—Forensic Chemistry

(a) Answers will vary. Two examples are given below.

The scientist is not wearing protective clothing. There are fluids on the floor, which could be infectious bodily fluids, and the scientist could be exposing herself to harmful pathogens. Also, by not wearing gloves, and by having her hair uncovered, she risks contaminating the samples with her own biological material. This would reduce the credibility of the evidence, and of her as a professional if she is required to defend her evidence in court.

There is no sign of a record sheet, or labelling of the sample that she is collecting. As a result the chain of evidence will be invalid. This will also reduce the credibility of her evidence as well as her professional credibility. *(3 marks)*

(b) (i)

$$H_2N-\underset{\displaystyle H_3C}{\overset{|}{CH}}-C\begin{matrix}/\!\!/O\\ \backslash OH\end{matrix} \qquad H_2N-CH_2-CH_2-C\begin{matrix}/\!\!/O\\ \backslash OH\end{matrix}$$

(2 marks)

(ii) Hydrolysis reactions are used to split proteins into small polypeptide fragments, or individual amino acids. Restriction enzymes, which cleave the amino acid chain at specific sites, are used to carry out the reaction. Following hydrolysis a mixture of amino acids or polypeptides of different mass and overall charge is produced. This mixture can be separated into its components using a chromatographic technique or electrophoresis. In electrophoresis the sample is placed in the centre of a polyacrylamide gel in a buffer solution, between the electrodes of a DC power source. The amino acids or polypeptides will migrate through the gel at a rate that depends on their mass and overall charge. Pure samples of known proteins can be treated in the same way and run alongside the mixture being analysed. This combination of techniques allows for the identification of proteins in samples that might be useful in the analysis of plant and/or animal material, such as in the determination of its origin or species. *(3 marks)*

(c) (i) The sample is cocaine. Peaks corresponding to all three fragments (with mass to charge ratios of 82, 94 and 182) are present in the spectrum. None of the significant peaks for caffeine are present, and only one of the three for paracetamol is possibly present (at 43). *(2 marks)*

(ii) In mass spectrometry a very small sample (only microgram quantities are required) containing a pure substance is injected into the spectrometer where it is vaporised. The sample is then bombarded with electrons so that it fragments. These fragments are accelerated in a magnetic field, and are detected at the end of the spectrometer, producing a graph of mass to charge ratio versus peak intensity. The fragmentation pattern of different molecules is characteristic, so the spectrum of the unknown in the sample can be compared with those of known substances to determine the sample's identity. Mass spectroscopy can also be used in conjunction with gas chromatography and high performance liquid chromatography (HPLC). In these instances, multiple substances present in a sample are first separated by chromatography, and then each separated component is transferred automatically to the mass spectrometer for identification. This could be used to determine the source of paint found at a crime scene, or the oil found at an oil spill for example.

(3 marks)

(d) (i) The following procedure outlines how paper chromatography can be used to separate the components of leaf extract.

Grind three healthy, green leaves from a hibiscus plant in a mortar and pestle in 5 mL of ethanol. Carefully spot the resulting green solution onto a line drawn near the bottom of a strip of filter paper. Place the strip of filter paper into a 250 mL beaker containing a small amount of ethanol, such that the solvent is not higher in the beaker than the line drawn near the bottom of the filter paper strip. Place a watch glass on the beaker as a lid. Stop the experiment when the solvent has travelled to 1 cm from the top of the strip. Remove the strip and draw a second line across it, marking the solvent front. The presence of chlorophyll is indicated by a discrete green spot on the resulting chromatograph. In order to avoid the inhalation of toxic ethanol vapours, the experiment should be performed in a fume cupboard, and sources of ignition avoided to eliminate the risk of a fire (ethanol is flammable).

(2 marks)

(ii) Answers will vary. Two limitations of the experiment outlined in part (i) are given below.

The technique is not quantitative, so it is not possible to determine how much chlorophyll was extracted from the leaf.

It is difficult to put a very small but concentrated spot onto the line on the filter paper strip. This means that as the solvent front moves up the paper, the spot spreads out more than it should, and the compounds may not be separated effectively. In order to avoid this a very fine needle should be used to make the spot rather than a plastic pipette. *(3 marks)*

(e) DNA evidence is extremely useful in forensic science for many reasons, but it may also be challenged when presented in court. Chromosomes contain sections of DNA that are not genes. These sections of DNA are as unique to an individual as their fingerprint. They can be analysed in a process called DNA fingerprinting, which allows samples found at a crime scene to be compared with samples from suspects to help either confirm guilt or innocence.

DNA must first be obtained from crime scene samples by breaking the cell membranes, and centrifuging to separate the DNA from the other cell contents and cytoplasm. The DNA present is then copied many times in a process known as the polymerase chain reaction (PCR). In this process the DNA double helix is separated by heating and suitable enzymes are added to catalyse the process of copying the DNA. The PCR process relies on the complementary base-pairing of DNA and results in the production of large numbers of exact copies of the original DNA. As a result, extremely small samples can be analysed using this technique.

The DNA strands are then cut into fragments using restriction enzymes, and these are separated by electrophoresis. The separated fragments are dyed so that they can be seen, and are then compared to standards (e.g. the DNA from a suspect, to help confirm guilt or innocence in a crime, or from a suspected parent in a paternity case). If 13 intron tandem repeat fragments are used for DNA fingerprinting then there is only a 1-in-10 billion chance of two fragment patterns being identical.

DNA analysis is very useful because DNA is present in almost all cells, which means that different types of samples can be compared (e.g. skin left at a crime scene can be compared with a blood or saliva sample). The PCR process allows extremely small samples to be analysed. However, DNA evidence can be challenged. For example, closely related individuals (e.g. family members, parents and children) will have more similarities in their DNA fingerprints than unrelated or more distantly related individuals. Fingerprinting results could be challenged, particularly if an identical twin is suspected of a crime. DNA analysis could also be challenged because of contamination of evidence, or if the correct chain of custody procedures are not followed. *(7 marks)*

CHAPTER 13

2015 HIGHER SCHOOL CERTIFICATE EXAMINATION

Chemistry

General Instructions

- Reading time – 5 minutes
- Working time – 3 hours
- Write using black pen
- Draw diagrams using pencil
- Board-approved calculators may be used
- A data sheet and a Periodic Table are provided at the back of this paper

Total marks – 100

Section I

75 marks

This section has two parts, Part A and Part B

Part A – 20 marks

- Attempt Questions 1–20
- Allow about 35 minutes for this part

Part B – 55 marks

- Attempt Questions 21–30
- Allow about 1 hour and 40 minutes for this part

Section II

25 marks

- Attempt ONE question from Questions 31–35
- Allow about 45 minutes for this section

Section I
75 marks

Part A – 20 marks
Attempt Questions 1–20
Allow about 35 minutes for this part

Use the multiple-choice answer sheet for Questions 1–20.

---

1 In which layer of the atmosphere does ozone absorb the most UV radiation?

(A) Mesosphere

(B) Stratosphere

(C) Thermosphere

(D) Troposphere

2 Which type of glassware is used in a titration to deliver an accurate volume of a solution to a known volume of another solution?

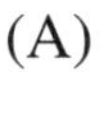
(A)

(B)

(C)
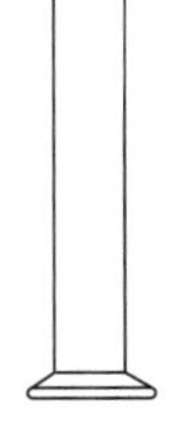

(D)

3 What flame colour do copper ions produce when heated?

(A) Brick red

(B) Blue-green

(C) Pale purple

(D) Yellow-orange

4 What happens to Fe^{2+} in the following reaction?

$$Sn^{4+} + Fe^{2+} \rightarrow Sn^{3+} + Fe^{3+}$$

(A) It undergoes oxidation and gains electrons.

(B) It undergoes reduction and gains electrons.

(C) It undergoes oxidation and loses electrons.

(D) It undergoes reduction and loses electrons.

5 The oxides CaO, CO_2, Na_2O and N_2O_4 are placed in water to form four separate solutions.

Which row of the table correctly indicates the solutions with pH less than 7 and the solutions with pH greater than 7?

	Solutions			
	pH less than 7		*pH greater than 7*	
(A)	CO_2	N_2O_4	CaO	Na_2O
(B)	CaO	N_2O_4	CO_2	Na_2O
(C)	CaO	Na_2O	CO_2	N_2O_4
(D)	CO_2	Na_2O	CaO	N_2O_4

6 Which of the following is the most suitable replacement for CFCs in terms of reducing their environmental impact?

(A) CH_4

(B) CH_2F_2

(C) CH_2ClF

(D) $CHCl_2CCl_2F$

7 A diagram of a simple cell is shown.

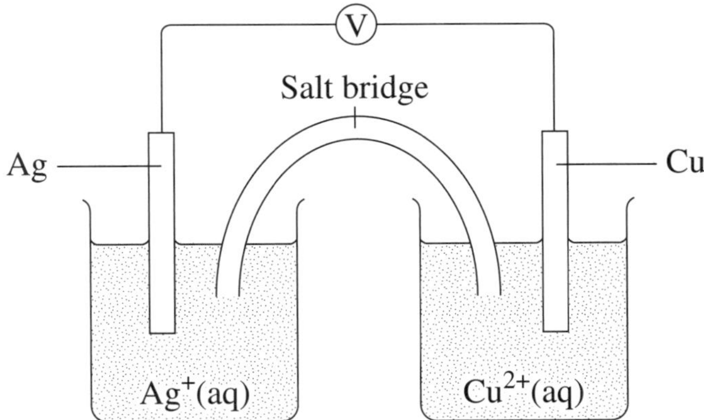

Which of the following occurs when the cell is in operation?

(A) Silver ions are formed in solution.

(B) The copper electrode loses electrons.

(C) Electrons travel through the electrolyte.

(D) The copper electrode increases in mass.

8 Which of the following statements best explains the solubility of ethanol in octane?

(A) Ethanol and octane are both non-polar.

(B) Ethanol forms hydrogen bonds with octane.

(C) Ethanol forms dispersion forces with octane.

(D) Ethanol forms dipole-dipole bonds with octane.

9 What are the reactants used to make this compound?

$$CH_3CH_2CH_2-\overset{\overset{\large O}{\|}}{C}-O-CH_2CH_2CH_3$$

(A) Butan-1-ol and butanoic acid

(B) Butan-1-ol and propanoic acid

(C) Propan-1-ol and butanoic acid

(D) Propan-1-ol and propanoic acid

10 Which of the equations correctly describes incomplete combustion?

(A) $C_2H_5OH(l) + 2O_2(g) \rightarrow 2CO(g) + 3H_2O(l)$

(B) $C_2H_5OH(l) + \frac{7}{2}O_2(g) \rightarrow 2CO_2(g) + 3H_2O(l)$

(C) $C_2H_5OH(l) + 3O_2(g) \rightarrow 2CO_2(g) + 3H_2O(l)$

(D) $C_2H_5OH(l) + 2O_2(g) \rightarrow C(s) + CO(g) + 3H_2O(l)$

11 Two monomers are shown.

$$HO-CH_2-CH_2-OH \qquad HO-\overset{O}{\overset{\|}{C}}-(CH_2)_2-\overset{O}{\overset{\|}{C}}-OH$$

Which of the following shows a condensation polymer that could be formed from the monomers?

(A) $$\left[CH_2-CH_2-\overset{O}{\overset{\|}{C}}-(CH_2)_2-\overset{O}{\overset{\|}{C}} \right]_n$$

(B) $$HO-CH_2-CH_2-O-\overset{O}{\overset{\|}{C}}-(CH_2)_2-\overset{O}{\overset{\|}{C}}-OH$$

(C) $$\left[C-CH_2-CH_2-O-O-\overset{O}{\overset{\|}{C}}-(CH_2)_2-\overset{O}{\overset{\|}{C}}-O \right]_n$$

(D) $$\left[O-CH_2-CH_2-O-\overset{O}{\overset{\|}{C}}-(CH_2)_2-\overset{O}{\overset{\|}{C}} \right]_n$$

12 A transuranic element can be produced in a nuclear reactor according to this equation:

$$^{239}_{94}\text{Pu} + 2\text{X} \rightarrow \, ^{241}_{94}\text{Pu} \rightarrow \, ^{241}_{95}\text{Am} + \text{Y}$$

Which row of the table correctly identifies X and Y?

	X	Y
(A)	Neutron	Electron
(B)	Proton	Neutron
(C)	Neutron	Proton
(D)	Proton	Electron

13 Which of the following solutions has the highest pH?

(A) 1.0 mol L^{-1} acetic acid

(B) 0.10 mol L^{-1} acetic acid

(C) 1.0 mol L^{-1} hydrochloric acid

(D) 0.10 mol L^{-1} hydrochloric acid

14 The graph shows the changes in pH during a titration.

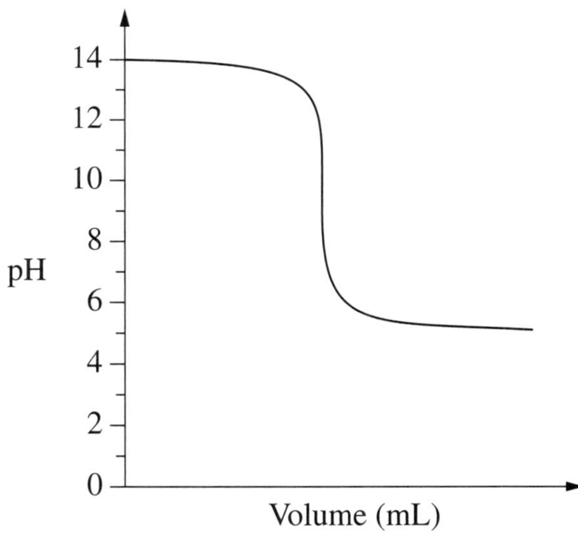

Which pH range should an indicator have to be used in this titration?

(A) 3.1–4.4

(B) 5.0–8.0

(C) 6.0–7.6

(D) 8.3–10.0

15 Part of a water catchment is shown in the diagram.

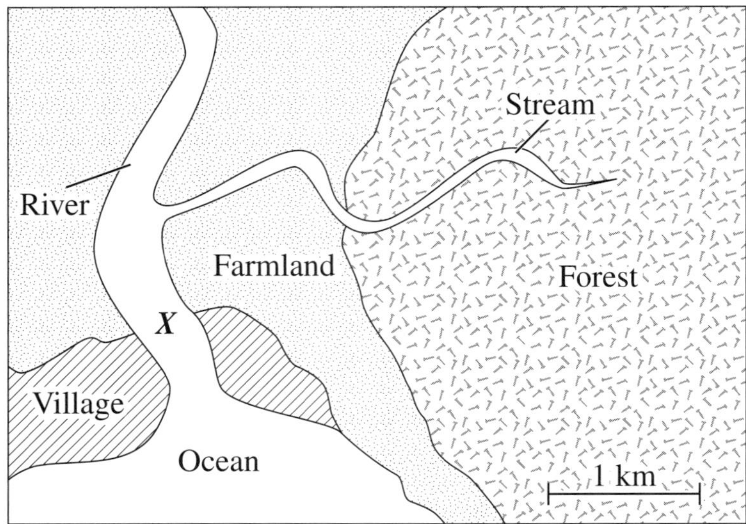

A sample of river water taken from point ***X*** is analysed.

Which row of the table shows the most likely results?

	Results of water analysis at ***X***			
	Turbidity (NTU)	*BOD* (ppm)	*pH*	*Total dissolved solids* (ppm)
(A)	400	18	6.5	22 000
(B)	22	3	8.5	17
(C)	5	18	6.5	22 000
(D)	400	3	8.5	17

16 The equation describes an equilibrium reaction occurring in a closed system.

$$X(g) + Y(g) \rightleftharpoons 4Z(g) \qquad \Delta H = +58 \text{ kJ}$$

Under which set of conditions would the highest yield of Z(*g*) be obtained?

	Temperature (°C)	*Pressure* (kPa)
(A)	50	100
(B)	50	200
(C)	300	100
(D)	300	200

17 What volume of carbon dioxide will be produced if 10.3 g of glucose is fermented at 25°C and 100 kPa?

(A) 1.30 L

(B) 1.42 L

(C) 2.57 L

(D) 2.83 L

Use this information to answer Questions 18–19.

A sample of pond water from a contaminated site was analysed to determine the concentration of lead ions using the following procedure.

- A measuring cylinder was used to collect a 50 mL sample from the pond.
- The sample was placed in a clean dry beaker.
- 25.0 mL of 0.200 mol L^{-1} sodium chloride solution was added to the sample.
- The precipitate of lead(II) chloride that formed was filtered, dried and weighed. It had a mass of 0.13 g.

18 How could the reliability of the analysis of the pond water be improved?

(A) Analyse more samples from the same pond

(B) Use 50 mL of distilled water as a control sample

(C) Analyse samples from different ponds on the site

(D) Remove other contaminants from the sample before the analysis

19 What was the concentration of lead ions in the sample?

(A) 5.0×10^{-3} mol L^{-1}

(B) 5.8×10^{-3} mol L^{-1}

(C) 9.3×10^{-3} mol L^{-1}

(D) 10.7×10^{-3} mol L^{-1}

20 The table shows the heat of combustion of four straight chain alkanols.

Number of C atoms in straight chain alkanol	*Heat of combustion* ($kJ\ mol^{-1}$)
1	726
3	2021
5	3331
7	4638

What is the mass of water that could be heated from 20°C to 45°C by the complete combustion of 1.0 g of heptan-1-ol?

(A) 0.032 kg

(B) 0.044 kg

(C) 0.36 kg

(D) 0.38 kg

2015 HIGHER SCHOOL CERTIFICATE EXAMINATION

Chemistry

Centre Number

Student Number

Section I (continued)

Part B – 55 marks
Attempt Questions 21–30
Allow about 1 hour and 40 minutes for this part

Answer the questions in the spaces provided. These spaces provide guidance for the expected length of response.

Show all relevant working in questions involving calculations.

Extra writing space is provided on pages 27 and 28. If you use this space, clearly indicate which question you are answering.

Write your Centre Number and Student Number at the top of this page.

Please turn over

Question 21 (4 marks)

(a) Outline a suitable method to prepare a natural indicator. **2**

....

....

....

....

....

(b) How could a natural indicator be tested? **2**

....

....

....

....

....

Question 22 (7 marks)

The table shows data for ozone concentrations over 50 years in the upper atmosphere above Antarctica.

Year	*Ozone Concentration* (Dobson Units)
1955	320
1960	300
1970	300
1980	260
1995	130
2000	130
2005	150

(a) Draw a line graph of the data on the grid provided. **4** 

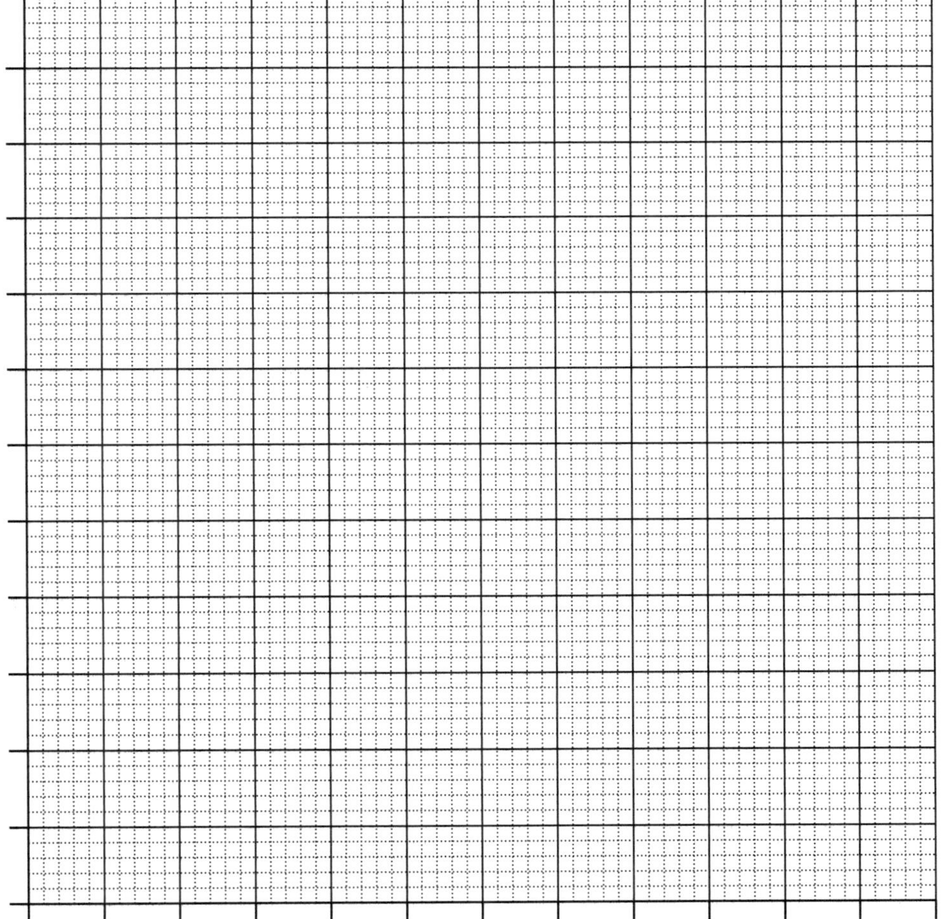

Question 22 continues

Question 22 (continued)

(b) Describe a method by which this data could have been measured. **3**

....

....

....

....

....

....

....

....

End of Question 22

Question 23 (4 marks)

Explain how the structure and chemistry of ONE of the following cells determines its cost and practicality. **4**

- button cell
- fuel cell
- vanadium redox cell
- lithium cell
- liquid junction photovoltaic device (eg the Gratzel cell)

Name of cell: ....

....

....

....

....

....

....

....

....

....

....

....

....

Question 24 (5 marks)

(a) Explain why the salt, sodium acetate, forms a basic solution when dissolved in water. Include an equation in your answer. **2**

....

....

....

....

(b) A solution is prepared by using equal volumes and concentrations of acetic acid and sodium acetate. **3**

Explain how the pH of this solution would be affected by the addition of a small amount of sodium hydroxide solution. Include an equation in your answer.

....

....

....

....

....

Question 25 (7 marks)

(a) Describe the steps involved in the process of *addition polymerisation*. **3**

....

....

....

....

....

....

....

....

(b) Explain the uses of polyethylene and polystyrene in terms of their structures and properties. **4**

....

....

....

....

....

....

....

....

....

....

Question 26 (7 marks)

A sodium hydroxide solution was titrated against citric acid ($C_6H_8O_7$) which is triprotic.

(a) Draw the structural formula of citric acid (2–hydroxypropane–1,2,3–tricarboxylic acid). **1**

(b) How could a computer-based technology be used to identify the equivalence point of this titration? **2**

....

....

....

....

....

Question 26 continues

Question 26 (continued)

(c) The sodium hydroxide solution was titrated against 25.0 mL samples of 0.100 mol L^{-1} citric acid. The average volume of sodium hydroxide used was 41.50 mL. **4**

Calculate the concentration of the sodium hydroxide solution.

....

....

....

....

....

....

....

....

....

....

End of Question 26

Question 27 (5 marks)

Name a radioisotope used in a non-medical industry and discuss its use in that industry in terms of its properties. **5**

....

....

....

....

....

....

....

....

....

....

....

....

....

....

....

Question 28 (3 marks)

The equipment shown is set up. After some time a ring of white powder is seen to form on the inside of the glass tube.

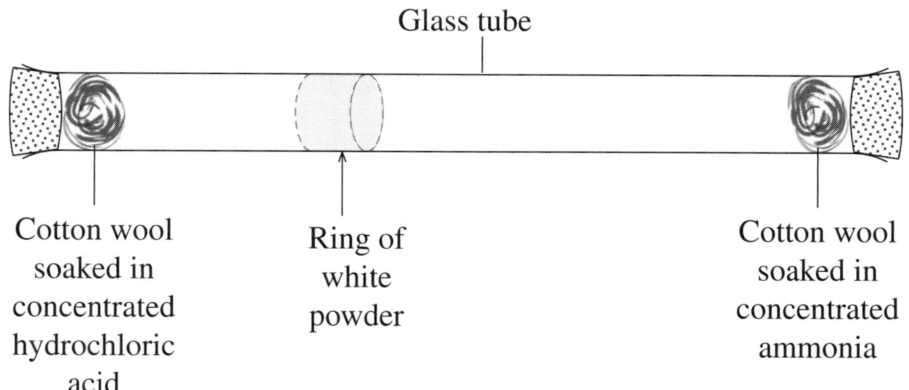

(a) Why would this NOT be an acid–base reaction according to Arrhenius? **1**

....

....

(b) Explain why this would be considered a Brönsted–Lowry acid–base reaction. Include an equation in your answer. **2**

....

....

....

....

....

....

Question 29 (7 marks)

The procedure of a first-hand investigation conducted in a school laboratory to determine the percentage of sulfate in a lawn fertiliser is shown.

- 2.00 g of a sample of fertiliser was ground up and placed in a beaker.
- It was dissolved in about 200 mL of 0.1 mol L^{-1} hydrochloric acid, stirred and filtered.
- Excess barium chloride solution was quickly added to this beaker and a precipitate formed.
- The precipitate was then allowed to settle, filtered using filter paper and the residue collected.
- The residue was dried and weighed and had a mass of 2.23 g.

(a) Suggest modifications that could be made to the procedure to improve the results of this investigation. Justify your suggestions. **4**

....

....

....

....

....

....

....

....

....

....

....

....

Question 29 continues

Question 29 (continued)

(b) Calculate the percentage of sulfate in the original fertiliser sample. **3**

....

....

....

....

....

....

....

....

End of Question 29

Question 30 (6 marks)

The graph shows the percentage yield of ammonia produced from nitrogen and hydrogen at different temperatures and pressures. **6**

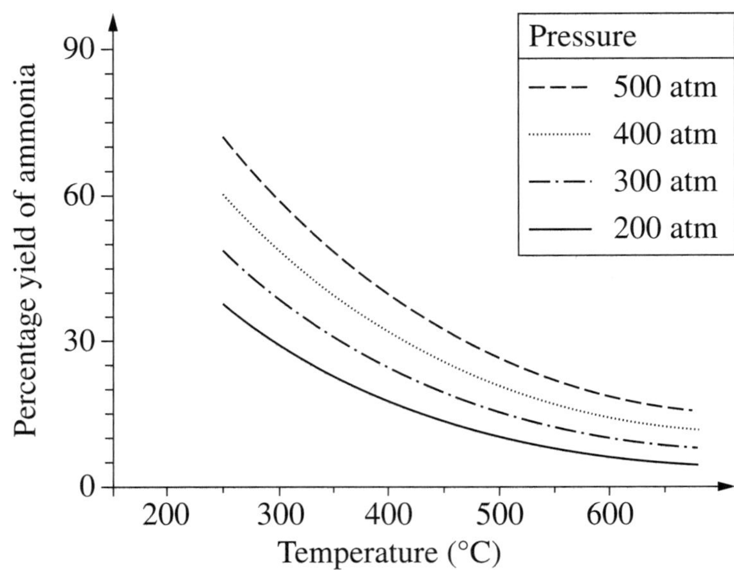

The Haber process is the main industrial procedure for the production of ammonia. Explain the conditions used in the Haber process with reference to the graph.

....

....

....

....

....

....

....

....

....

....

....

....

....

....

....

....

....

....

2015 HIGHER SCHOOL CERTIFICATE EXAMINATION

Chemistry

Section II

25 marks
Attempt ONE question from Questions 31–35
Allow about 45 minutes for this section

Answer parts (a)–(d) of one question in the Section II Writing Booklet. Extra writing booklets are available.

Show all relevant working in questions involving calculations.

Question 31 Industrial Chemistry

Question 32 Shipwrecks, Corrosion and Conservation

Question 33 The Biochemistry of Movement *(Not included in this reproduction)*

Question 34 The Chemistry of Art *(Not included in this reproduction)*

Question 35 Forensic Chemistry

Question 31 — Industrial Chemistry (25 marks)

Answer parts (a) and (b) of the question on pages 2–4 of the Section II Writing Booklet. Start each part of the question on a new page.

(a) At temperatures above 100°C, hydrogen and carbon monoxide react to form methanol gas in this reversible reaction.

$$2H_2(g) + CO(g) \rightleftharpoons CH_3OH(g)$$

A mixture of hydrogen, carbon monoxide and methanol is placed in a container with a volume that can be changed. The mixture is allowed to reach equilibrium.

(i) The initial volume of the container is 1.00 L. **2**

Account for any changes in the concentration of hydrogen gas when the volume of the container is rapidly increased to 2.00 L.

(ii) The initial mixture placed in the container had 0.50 mol of hydrogen, 1.00 mol of carbon monoxide and 2.50 mol of methanol. Once the volume of the container had been increased to 2.00 L and equilibrium had been re-established, the number of moles of hydrogen in the mixture had changed by 0.36 mol. **3**

Calculate the equilibrium constant for this reaction.

(b) (i) Describe how saponification can be safely carried out as part of a first-hand investigation. **3**

(ii) Explain the chemistry related to the cleaning properties of the product of saponification. **4**

Question 31 continues

Question 31 (continued)

Answer parts (c) and (d) of the question on pages 5–8 of the Section II Writing Booklet. Start each part of the question on a new page.

(c) The diagram shows part of the Solvay process for producing sodium carbonate.

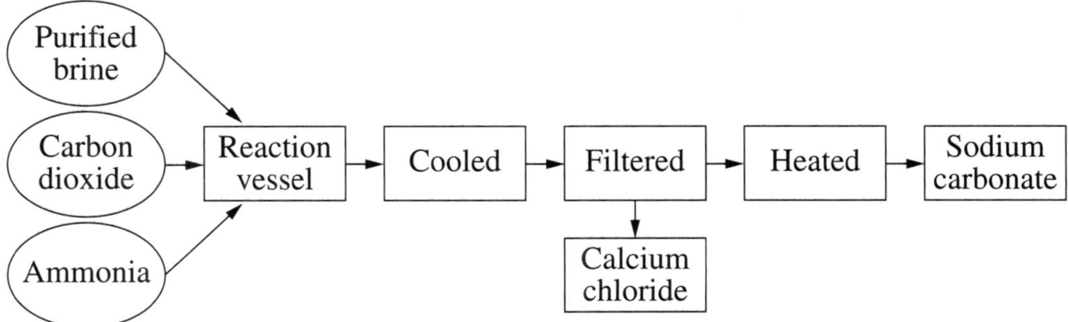

(i) Outline the chemistry of the production of sodium carbonate in the process shown. Include equations in your answer. **3**

(ii) By making specific reference to the diagram, justify the requirements for the location of a Solvay process plant. **3**

(d) Compare the membrane cell method with ONE other method used in the industrial production of sodium hydroxide in terms of technical and environmental issues. **7**

End of Question 31

Question 32 — Shipwrecks, Corrosion and Conservation (25 marks)

Answer parts (a) and (b) of the question on pages 2–4 of the Section II Writing Booklet. Start each part of the question on a new page.

(a) (i) Outline the limitations of using paint to protect ships that are in constant use. **2**

(ii) Explain the chemical principles involved in the use of a sacrificial anode. Include relevant chemical equations in your answer. **3**

(b) (i) Describe a valid and reliable first-hand investigation that can be used to compare the rates of corrosion of iron with ONE named form of steel. **3**

(ii) Explain how the percentage composition of steel can determine its properties, with reference to TWO types of steel. **4**

Question 32 continues

Question 32 (continued)

Answer parts (c) and (d) of the question on pages 5–8 of the Section II Writing Booklet. Start each part of the question on a new page.

(c) (i) The equipment in the photograph was used in an attempt to plate a metal spoon with magnesium using an electrolytic cell containing a solution of magnesium sulfate. **3**

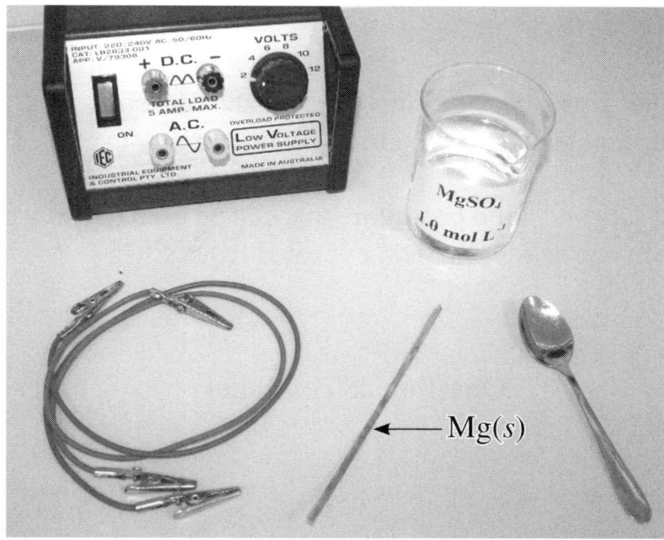

Draw a labelled scientific diagram of the electrolytic cell. Include the cathode, anode, direction of electron flow and polarity of the electrodes.

(ii) Explain how Davy's work increased our understanding of electron transfer reactions. **3**

(d) Two identical ships are sunk in seawater. One is sunk in shallow water (60 m) and the other in deep water (4000 m). Explain how the rusting processes differ in these two ships. Include equations in your answer. **7**

End of Question 32

Question 35 — Forensic Chemistry (25 marks)

Answer parts (a) and (b) of the question on pages 2–4 of the Section II Writing Booklet. Start each part of the question on a new page.

(a) (i) Compare the composition of glycogen with that of cellulose. **2**

(ii) Relate the differences in composition of glycogen and cellulose to their different structures. **3**

(b) (i) Identify the structure of amino acids and describe the relationship between amino acids and proteins. **3**

(ii) Describe a safe and valid procedure that can be used to show the presence of protein in egg white. Include expected results. **4**

Question 35 continues

Question 35 (continued)

Answer parts (c) and (d) of the question on pages 5–8 of the Section II Writing Booklet. Start each part of the question on a new page.

(c) (i) George and Linda have one child together. Each of them also has one child from a previous relationship. A schematic representation of their DNA profiles is shown below. **3**

DNA profiles obtained from George, Linda and the three children

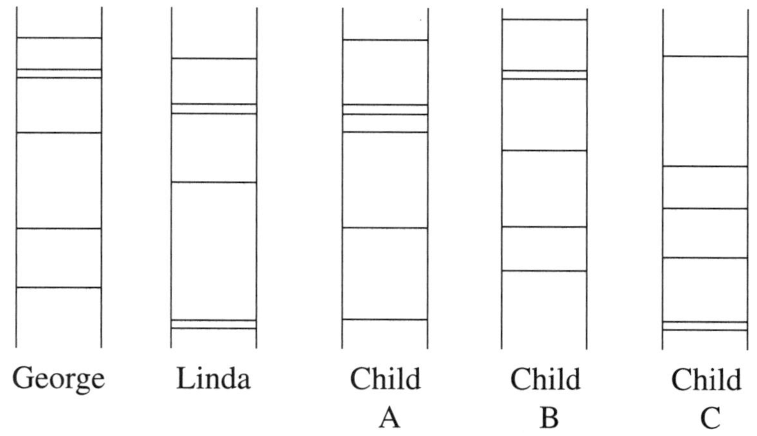

Use the information in the DNA profiles to identify the relationships of Child A, Child B and Child C to George and Linda. Justify your answer.

(ii) Describe the benefits of maintaining DNA data banks. **3**

(d) Name ONE chromatography technique and assess its use in the analysis of forensic evidence. **7**

End of paper

2015 HSC Examination Paper

Sample Answers

Section I Part A
(Total 20 marks)

1 B

2 D A burette

3 B

4 C

5 A

6 B Difluoromethane. Although hydrocarbons are used as CFC replacements, methane (CH_4, option A) is not used. It is a greenhouse gas so it has a negative environmental impact. Options C and D contain chlorine, which is a destructive halogen and should not be used as a CFC replacement.

7 B

8 C

9 C

10 A

11 D

12 A

13 B Dilute weak acid

14 D The indicator needs to change colour completely at the equivalence point which corresponds to the centre of the vertical part of the graph.

15 A You would expect the total dissolved solids to be high as point *X* is estuarine (so will contain seawater salts) and it is downstream from farmland (run-off contains fertiliser ions). Farm run-off also contains clays and leads to high turbidity.

16 C The forward reaction is endothermic. A high temperature will favour the forward reaction, as this absorbs heat, producing a higher yield of *Z*. A low pressure will also favour the forward reaction, as the forward reaction will increase the pressure by converting 2 moles of gas into 4 moles.

17 D $C_6H_{12}O_6 \rightarrow 2CO_2 + 2C_2H_5OH$

$n(C_6H_{12}O_6) = 10.3/180.2 = 0.057$ mol

$n(CO_2) = 2 \times 0.057 = 0.114$ mol

$V(CO_2) = 0.114 \times 24.79$ L $= 2.83$ L

18 A

19 C $n(PbCl_2) = 0.13/278.1 = 4.67 \times 10^{-4}$ mol

$n(Pb^{2+}) = 4.67 \times 10^{-4}$ mol

$c(Pb^{2+}) = 4.67 \times 10^{-4}/0.05 = 9.3 \times 10^{-3}$ mol L^{-1}

20 D Heat absorbed by calorimeter $= |\Delta H| = m\,C\,\Delta T$

$$m = \frac{\Delta H}{C\,\Delta T}$$

$$= \frac{4638 \times 10^3}{(4.18 \times 10^3) \times 25}$$

= 4.38 J per 116.2 g heptan-1-ol

= 44.38/116.2

= 0.38 kg

Note: The units of C (4.18×10^3) are in J kg^{-1} K^{-1}, so the heat has to be converted to joules from kilojoules. The mass of water will then be in kg on calculation. Also, the heat of combustion is the absolute value of the enthalpy of combustion. This removes the negative sign in the formula.

Section I Part B

21 (a) 1. Chop a small red cabbage leaf into pieces.

2. Grind up the pieces in a mortar and pestle.

3. Boil for a few minutes in 100 mL distilled water.

4. Decant the cabbage extract into a clean 100 mL beaker. *(2 marks)*

(b) Determine the colour of the indicator over the full pH range from 1 to 14.

In order to do this, dilute a 1.0 mol L^{-1} HCl solution until the pH is 1 when measured with a pH probe. Place 2 mL of this solution into a test tube. Continue to dilute the HCl to produce solutions with pH of 3 and 5. Do the same with 1.0 mol L^{-1} NaOH to produce solutions with pH 14, 12 and 9. Measure 2 mL of distilled water to make a solution of pH ~7 (pure water is slightly acidic due to dissolved carbon dioxide). Add 10 drops of indicator to each solution and record the colour for each pH. A useful indicator should produce a distinctly different colour for each pH. *(2 marks)*

22 (a)

Ozone concentrations in the upper air above Antarctica

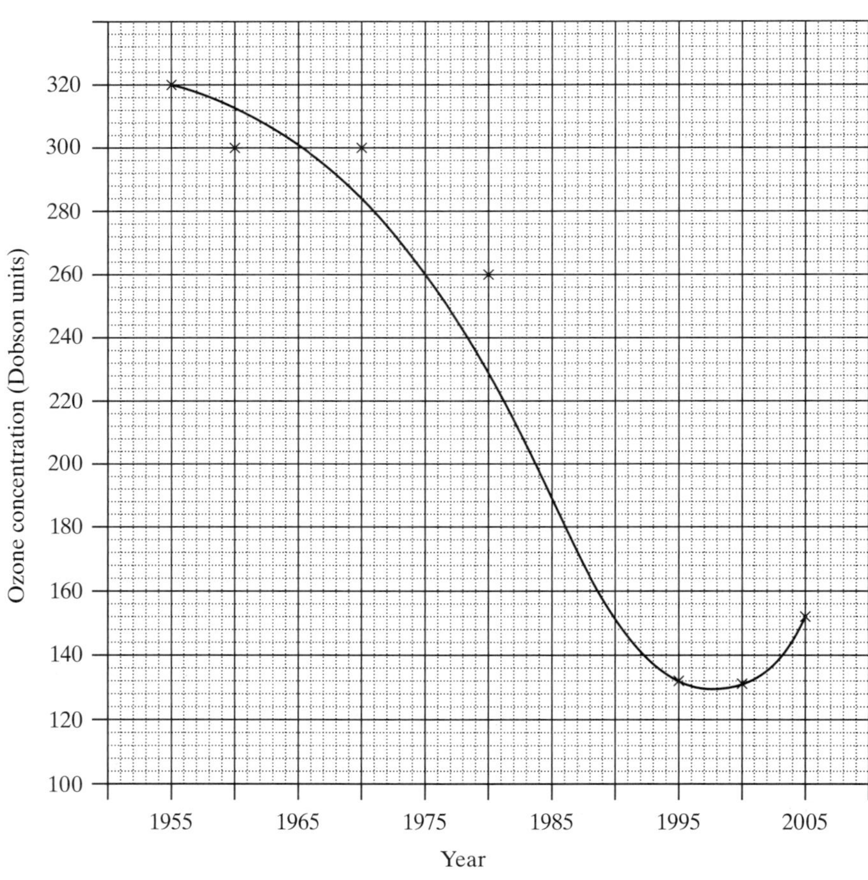

(4 marks)

(b) The concentration of ozone in the stratosphere was measured in the 1980s and 1990s using the Total Ozone Mapping Spectrometer on board a satellite. It compared incident solar radiation to radiation backscattered from the atmosphere to determine ozone levels. *(3 marks)*

23 Answers may vary. Following is one example.

Name of cell: fuel cell

The most successful fuel cell to date uses oxygen gas and hydrogen gas as reactants and produces water.

Hydrogen is supplied to the anode chamber and oxygen to the cathode chamber. The gases diffuse through the porous electrodes which are made of platinum and act as catalysts. It is the platinum that is responsible for most of the cost of a fuel cell, as this is a relatively expensive metal. The gases react with the acidic electrolyte as follows:

Anode reaction: $H_2(g) \rightarrow 2H^+(aq) + 2e^-$

Cathode reaction: $O_2(g) + 4H^+(aq) + 4e^- \rightarrow 2H_2O(l)$

The overall reaction is: $2H_2(g) + O_2(g) \rightarrow 2H_2O(l)$

The theoretical emf is 1.23 V; however, this is rarely attained due to the slow migration of H^+ ions from anode to cathode.

Fuel cells are small in size, provide constant voltage, and produce water as a reaction product. Space shuttles used fuel cells for electricity and drinking water. However, the need for a constant fuel supply is not very practical. The major problem with fuel cell usage on Earth is the storage of hydrogen gas. Hydrogen is explosive and could not be stored in large amounts or carried by cars.

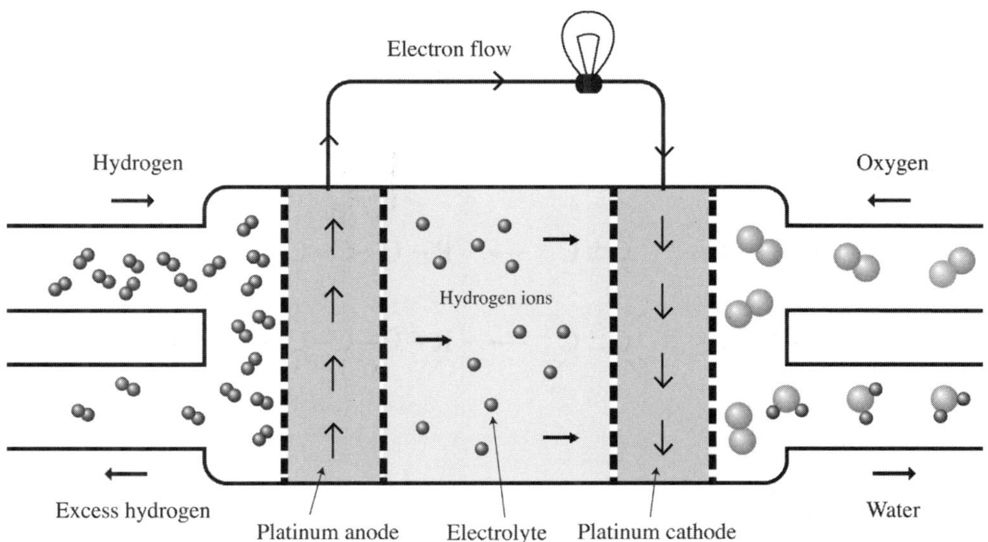

(4 marks)

24 (a) Sodium acetate is a basic salt. Sodium is a neutral conjugate of a strong base. Acetate is the strong base of a weak acid, acetic acid. A strong base will react with water, accepting a proton and producing hydroxide.

$CH_3COO^-(aq) + H_2O(l) \rightleftharpoons CH_3COOH(aq) + OH^-(aq)$ *(2 marks)*

(b) The pH of the solution will not be affected by adding sodium hydroxide. The acetic acid/sodium acetate solution will act as a pH buffer.

$CH_3COOH(aq) + H_2O(l) \rightleftharpoons CH_3COO^-(aq) + H_3O^+(aq)$

When hydroxide is added to the buffer, it will neutralise the hydronium ion, reducing its concentration. The equilibrium will shift to the right, according to Le Chatelier's principle, to partially counteract this change and restore the pH to what it was originally. *(3 marks)*

25 (a) The monomers in addition polymerisation are small alkene molecules, such as ethylene and propylene.

Step 1. A peroxide initiator is heated so that it splits into radicals, and added to the monomers.

Step 2. The double bond in the monomers break as they are activated by the initiator, forming free radicals.

Step 3. The free radical monomers combine by sharing unbonded electrons, propagating the polymer (i.e. growing the polymer chain).

Step 4. An inhibitor is added to halt the polymer growth.

Initiation

R–O–O–R → 2R–O•

R–O• + C=C → R–O–C–C•

Propagation

R–O–C–C• + C=C → R–O–C–C–C–C•

R–O–C–C–C–C• + C=C → R–O–C–C–(C–C)–C–C•

Termination

R–O–C–C–(C–C)–C–C• + •C–C–(C–C)–C–C–O–R

↓

R–O–C–C–(C–C)–C–C–C–C–(C–C)–C–C–O–R

(3 marks)

(b) Low-density polyethylene is composed of long carbon chains with numerous branches. The branches prevent the carbon chains from packing closely together. These long chains have weak dispersion forces between them, resulting in a transparent, soft and flexible plastic. This makes the plastic useful for cling wrap, as it can be stretched over containers of food. It will cling to the container, keeping the food fresh. Polyethylene allows oxygen to pass through but is resistant to moisture.

Polyethylene

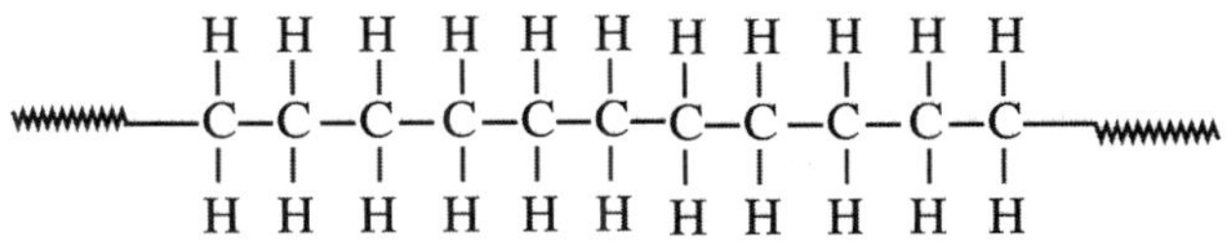

```
     H H H H H H H H H H H
     | | | | | | | | | | |
~~~~-C-C-C-C-C-C-C-C-C-C-C-~~~~
     | | | | | | | | | | |
     H H H H H H H H H H H
```

Polystyrene is composed of long carbon chains with phenyl group branches attached to every second carbon atom. Steam or air is pumped into the melted plastic to form an expanded, water resistant product trade named as Styrofoam. The trapped gases in the plastic give the material excellent heat insulating properties, so Styrofoam is used to make disposable drink containers which hold heat well. The trapped gas in the polystyrene also provides shock absorbance, so the plastic is used to pack white goods and electrical products for transport, protecting them from damage and water spoilage.

Polystyrene

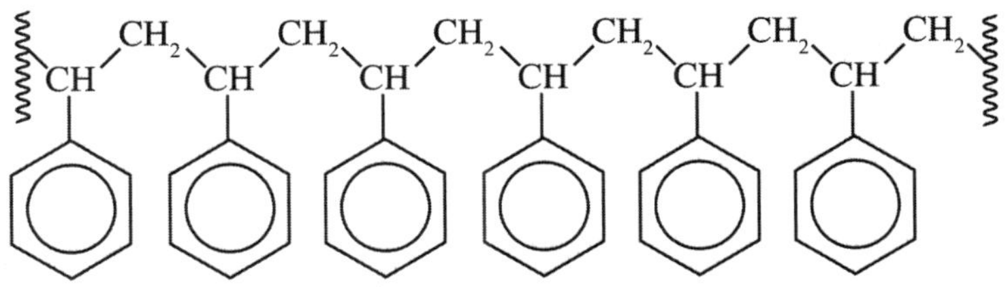

(4 marks)

26 (a)

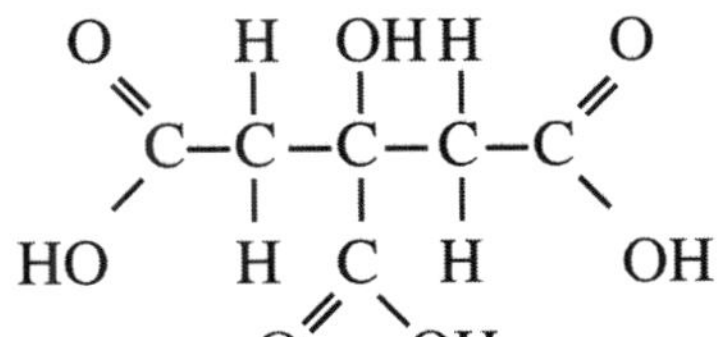

```
 O     H  OH H    O
  \\   |  |  |   //
   C - C - C - C - C
  /    |   |   |    \
HO     H   C   H     OH
          // \
         O    OH
```

(1 mark)

(b) The equivalence point is reached when the stoichiometric amounts of acid and base have been combined. This point can be determined using a chemical indicator. Alternatively, a pH probe can be placed in the reaction flask attached to a data logger. As the sodium hydroxide is added to the citric acid in the flask, a graph is plotted showing the pH changing with volume of base added. The equivalence point is identified when a steep rise in pH is indicated on the graph. *(2 marks)*

(c) $C_6H_8O_7(aq) + 3NaOH(aq) \rightarrow 3H_2O(l) + C_6H_5O_7Na_3(aq)$

$n(C_6H_8O_7) = 0.100 \times 0.0250 = 0.00250$ mol

$n(NaOH)$ at equivalence point $= 0.00250 \times 3 = 0.00750$ mol

$c(NaOH) = 0.00750/0.04150 = 0.18072$ mol L^{-1}

$= 0.181$ mol L^{-1} (to 3 significant figures) *(4 marks)*

27 Answers may vary. Following is one example.

Americium-241 is an artificially produced radioisotope used in smoke detectors. Smoke detectors are used in factories and homes to warn against fires, and are important safety devices because of their obvious potential to save lives and property.

Americium-241 has two significant properties that make it useful for smoke detectors: it has a long half-life of 432 years, and it decays by emitting alpha particles.

$^{241}_{95}Am \rightarrow ^{237}_{93}Np + ^{4}_{2}He$

The long half-life means that the small amount of americium-241 in smoke detectors (<35 kBq) never needs replacing.

Americium-241 emits alpha particles and low energy gamma rays. The alpha particles are absorbed within the detector, or in a few centimetres of air, so they do not pose a health hazard. Most of the gamma rays produced by the americium-241 escape harmlessly.

The alpha particles emitted by the americium-241 collide with oxygen and nitrogen molecules in the air in the detector's ionisation chamber to produce oxygen and nitrogen ions. A low-voltage electric current is applied across two electrodes in the ionisation chamber. The ions conduct the current. When smoke enters the space between the electrodes, the alpha radiation is absorbed by the smoke particles. This causes the electric current to fall, which sets off an alarm *(5 marks)*

28 (a) Arrhenius defined an acid as a substance that produces hydrogen ions when dissolved in water and a base as a substance producing hydroxide ions when dissolved in water. In this example the acid is reacting as a gas and not dissolved in water and the base (ammonia) does not contain hydroxide ions. *(1 mark)*

(b) A Brönsted–Lowry acid donates a proton and a base accepts a proton. In this example the HCl molecule donates a proton to the NH_3 molecule, so the HCl is an acid and NH_3 is a base.

$HCl(g) + NH_3(g) \rightarrow NH_4Cl(s)$ *(2 marks)*

29 (a) Modification 1: The excess barium chloride should be added slowly to increase the likelihood that all of the sulfate ions come in contact with barium and are precipitated.

Modification 2: The precipitate should not be allowed to settle prior to filtering. If it settles there is a chance that the solid will be difficult to extract from the beaker. The mixture should be stirred and slowly added to the filter. Additional water should be added progressively to ensure complete transfer.

Modification 3: Filtering barium sulfate using filter paper can be problematic as the precipitate is very fine and, depending on the grade of paper used, can pass through the filter paper. Instead, a sintered glass funnel can be used with a vacuum pump to increase the chances of collecting the barium sulfate residue.

Modification 4: The residue should be washed a few times with distilled water to remove any soluble substances that would otherwise dry out with the residue and add to the final mass.

Modification 5: Following the initial drying and weighing of the residue, it should be reweighed after further drying to ensure that it is completely dry. Only when the mass is constant can you be certain that all moisture has evaporated and is not a part of the final mass.

Modification 6: The whole procedure should be repeated five or more times and the results averaged to obtain a more reliable result. *(4 marks)*

(b) % mass SO_4^{2-} in $BaSO_4$ = 96.07/233.37 × 100 = 41.166%

mass SO_4^{2-} in residue = 0.41166 × 2.23 = 0.918 g

% mass SO_4^{2-} in fertiliser = 0.918/2.00 × 100 = 45.9% *(3 marks)*

30 The graph shows a decreasing yield of ammonia with an increasing temperature at all pressures. This occurs because the reaction to produce ammonia from hydrogen and nitrogen is exothermic.

$3H_2(g) + N_2(g) \rightleftharpoons 2NH_3(g) \quad \Delta H = -92\ \text{kJ mol}^{-1}$

Lower temperatures favour the forward reaction to produce ammonia. However, low temperatures cause a slow rate of reaction. The temperature used to maximise the yield at a suitable rate is 400 to 500 °C using an iron oxide catalyst.

The graph shows that increasing the pressure increases the yield of ammonia. This is because there are 4 moles of reactant producing 2 moles of product. At higher pressures, the reaction will favour the forward reaction as, according to Le Chatelier's principle, the equilibrium will shift to partially counteract the imposed change. A forward shift will counteract the high pressure and in the process, increase the yield of ammonia.

The highest yield results from a pressure of 500 atm. However, high pressures can put a strain on the machinery, causing a safety risk. For this reason, pressures of 250 atm are commonly used. Even though the yield is less and the rate of reaction is slower, it is safer and less costly to run at this pressure.

In addition to temperature and pressure there are other conditions that improve the process. The most efficient reactant ratio is 3:1 H_2:N_2. Removing the ammonia produced by liquefaction also causes the equilibrium to shift toward the product, increasing the yield of ammonia. *(6 marks)*

Section II—Options

Question 31—Industrial Chemistry

(a) (i) Doubling the volume rapidly will initially halve the concentration of hydrogen gas in the system. It will also halve the pressure. According to Le Chatelier's principle, this will result in a shift in the equilibrium position favouring the production of gas because it will offset the imposed pressure reduction. In this case that means a shift to the left-hand side. As a result the concentration of hydrogen gas will increase, but not back to its original value. (Note that the total number of hydrogen molecules in the system at equilibrium will have increased.) *(2 marks)*

(ii)

	$2H_2(g)$ +	$CO(g)$	$CH_3OH(g)$
Initial number of mol in 1 L	0.50	1.00	2.50
Change: increase in $n(H_2)$ 0.36 mol	+0.36	+0.18	–0.18
Final number of mol in 2 L	0.86	1.18	2.32
Final concentration	0.43	0.59	1.16

$$K = \frac{[CH_3OH]}{[H_2]^2.[CO]}$$

$$= \frac{1.16}{(0.43)^2 \times 0.59}$$

$$= 10.6$$

$= 11$ (Note that the limiting number of significant figures is 2.) *(3 marks)*

(b) (i) The following steps are necessary in order to safely carry out saponification in the laboratory.

1. Put on safety glasses and rubber gloves because both 4 mol L^{-1} NaOH (corrosive) and saturated NaCl solution (irritant) are harmful to eyes and skin.
2. Use a measuring cylinder to transfer 50 mL of 4 mol L^{-1} $NaOH(aq)$ to a large evaporating basin.
3. Add 50 mL of coconut oil using a measuring cylinder. Place a large clock glass over the basin to avoid splashing.
4. Gently boil the mixture for 30 minutes, using a Bunsen burner.
5. Carefully add water as necessary from a wash bottle to maintain approximately 100 mL of liquid in the evaporating basin.
6. When white solids have been formed, allow the mixture to cool.
7. Once cool, add 50 mL of saturated $NaCl(aq)$ and gently stir.
8. Separate the white solid by filtration.
9. Wash the residue with cold, deionised water, and allow it to dry. The residue is the desired product, soap. *(3 marks)*

(ii) The product of saponification is soap. Soap is composed of the sodium salts of fatty acids, sodium carboxylates. When dissolved in water the salts dissociate to form sodium ions and carboxylate ions. The latter have an anionic, hydrophilic 'head' (the carboxylate ion) and a hydrophobic carbon-chain tail. When mixed with grease or oil and agitated, the anionic head interacts via ion-dipole attractions with the water dipoles. The hydrophobic tail dissolves in the fat/oil droplets by dispersion forces and as a result, spherical micelles of fat/soap form and the fat/oil particles are dispersed throughout the water. The surface anionic groups on the micelle prevent the fat particles from joining back together. This forms an oil-water emulsion with the soap acting as the surfactant. *(4 marks)*

(c) (i) When carbon dioxide is bubbled into the ammoniacal brine solution, carbonic acid forms:

$$CO_2(g) + H_2O(l) \rightleftharpoons H_2CO_3(aq)$$

Carbonic acid reacts with the weak base ammonia to produce ammonium and hydrogen carbonate ions, resulting in a mixture of Na^+, HCO_3^-, Cl^- and NH_4^+. When cooled, $NaHCO_3$ crystallises and the solid is obtained simply by filtration. When heated, $NaHCO_3(s)$ produces the product sodium carbonate:

$$2NaHCO_3(s) \rightarrow Na_2CO_3(s) + H_2O(g) + CO_2(g).$$

The filtrate is retained. The CO_2 is recycled, to add to the CO_2 produced by the heating of $CaCO_3(s)$ before the first step shown:

$$CaCO_3(s) \rightarrow CaO(s) + CO_2(g)$$

The $CaO(s)$ from this step is added to water to create an alkaline suspension called 'milk of lime', $Ca(OH)_2$. The filtrate from above is mixed with the milk of lime. An acid-base reaction occurs:

$$Ca(OH)_2(aq) + 2NH_4Cl(aq) \rightarrow 2NH_3(g) + 2H_2O(l) + CaCl_2(aq).$$

The ammonia is recycled and added back into the reaction vessel. The calcium chloride must be disposed of as a waste product, along with excess heat produced by the process. *(3 marks)*

(ii) The ideal location for a Solvay plant is by the sea. This is because the sea is an abundant source of brine (saturated NaCl), which is a required feedstock. A coastal location also means that the waste product, $CaCl_2(aq)$, can be disposed of by discharge into the ocean because the addition of Ca^{2+} and Cl^- ions to sea water has negligible effect on their concentrations. The third advantage of a coastal location is that the product is more cheaply shipped to international markets if it is close to a harbour. *(3 marks)*

(d) Two methods for the production of sodium hydroxide are the membrane cell process and the mercury process, with the former being the most current technology. Both processes are electrolytic, so require an anode, cathode and electrical energy.

The cathode in the mercury process is liquid mercury. Sodium ions are reduced here:

$Na^+ + e^- \rightarrow Na(s)$

The Na(s) dissolves in the liquid mercury cathode to form an amalgam. This is pumped into a separate compartment where it is sprayed into water to give NaOH(aq):

$2Na(s) + 2H_2O(l) \rightarrow 2NaOH(aq) + H_2(g)$.

The anode in this process is an inert electrode such as titanium. Here, chloride ions are oxidised to chlorine:

$2Cl^- \rightarrow Cl_2(g) + 2e^-$.

The removal of the amalgam in the flowing cathode results in high purity NaOH(aq), and ensures that products are kept separate.

The membrane cell process is a much more recent method, in which a selectively permeable membrane is used to keep anode and cathode compartments separated. This means that flowing mercury is not required to separate the products. Chloride is oxidised to chlorine gas at an inert anode in this system too, and water is reduced at a steel mesh cathode:

$2H_2O(l) + 2e^- \rightarrow H_2(g) + 2OH^-(aq)$

The selectively permeable membrane allows Na^+ ions to pass through into the cathode compartment containing OH^- ions, resulting in very high purity NaOH(aq).

The technical issues associated with the mercury process revolve around the need to avoid mercury leaks. This has proved to be very difficult. Losses of mercury to the environment have occurred, and the impacts on the environment have been catastrophic. Mercury is a highly poisonous neurotoxin which bio-accumulates. Mercury that leaked from electrolysis plants has poisoned people, killed other organisms and destroyed habitats. The mercury process also has high, ongoing energy requirements since it involves electrolysis. This energy is generally provided by burning fossil fuels, which releases CO_2 into the atmosphere, contributing to climate change.

While the membrane process still has an energy requirement, so is responsible for $CO_2(g)$ emissions, its technical requirements and environmental impacts are significantly lower than for the mercury process. No toxic heavy metals are involved in the membrane process, so the technical difficulties associated with preventing their release into the environment is non-existent. The development of suitable, inexpensive and long-lasting ion-selective membranes which allowed only the sodium ions to migrate to the cathode was a technical issue that has been overcome. The membrane is important because it eliminates a separation step which would increase the costs and energy requirements of the process. *(7 marks)*

Question 32—Shipwrecks, Corrosion and Conservation

(a) (i) Paint is used to cover the steel hull of a ship because it forms a physical barrier. The layer of paint protects the steel from coming into contact with oxidants, water and oxygen. The limitation of paint is that it can be scratched, exposing the underlying steel to these oxidants and causing corrosion. *(2 marks)*

(ii) A sacrificial anode consists of a metal that is more reactive than the metal it is being attached to. It will corrode in preference to the metal it is attached to and will become the anode. The other metal becomes cathodic and is protected from corrosion.

Zinc is often used as a sacrificial anode as it has a slightly higher oxidation potential than iron.

$Zn(s) \rightarrow Zn^{2+}(aq) + 2e^-$ $E^0 = 0.76$ V

$Fe(s) \rightarrow Fe^{2+}(aq) + 2e^-$ $E^0 = 0.44$ V

In this example the zinc will corrode and become anodic and the iron will remain protected from corrosion because it becomes the cathode. Fe^{2+} ions are reduced. Overall:

$Zn(s) + Fe^{2+}(aq) \rightarrow Zn^{2+}(aq) + Fe(s)$ *(3 marks)*

(b) (i) Brush clean five pieces of iron. Weigh each, then place them into five separate test tubes. Half submerge the iron pieces in salt water. Brush clean and weigh five pieces of structural steel, of the same surface area as the iron pieces, and set them up in exactly the same way (i.e. with the same volume of water and salt concentration).

In order to determine the extent of corrosion, weigh each piece of metal once a day for five days, taking it out of the water, cleaning and drying it before weighing. Record the weight. The greater the corrosion, the greater the weight loss.

This is a valid experiment as the method used tests the aim (to compare the corrosion rates) and the variables have been controlled. The reliability of the experiment is improved by having five samples of each metal. Ensure the samples are of the same size and shape (i.e. surface area) so that a valid comparison of the average percentage mass lost for both metal types can be made. *(3 marks)*

(ii) The percentage of carbon in steel has a direct effect on its properties. The more carbon, the harder and less malleable the steel becomes. Mild steel is predominantly composed of iron, with a small amount of carbon (< 0.2%) and manganese (0–0.4%). This gives a relatively soft and malleable steel that can be easily worked and welded. Tool steel contains considerably more carbon (0.6–1.3%) and manganese (0.3–0.9%). This gives a much harder steel that has low malleability and is brittle. It is strong and resistant to wear. *(4 marks)*

(c) (i)

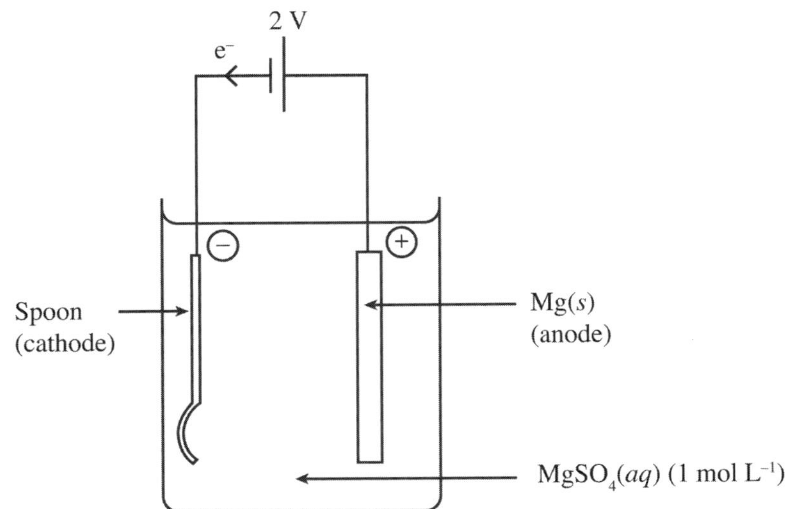

(3 marks)

(ii) Davy reasoned that the chemical combination occurred between substances of opposite charge and that chemical action was the cause of the electricity produced in simple electrolytic cells. He constructed the largest battery ever built and used it to electrolyse aqueous salt solutions to produce hydrogen and oxygen.

He built simple molten salt electrolytic cells and used them to isolate previously undiscovered elements such as potassium, sodium, calcium and strontium. He suggested that zinc attached to copper could prevent its corrosion. *(3 marks)*

(d) Corrosion is the process whereby a metal breaks down through oxidation, causing the metal atom to turn into ions. Corrosion of iron is also known as rusting.

The current choice of metal for ship construction is iron. Iron will rust in the presence of water and oxygen according to the following equations:

Oxidation: $Fe(s) \leftrightarrow Fe_{2+}(aq) + 2e^-$

Reduction: $O_2(g) + 2H_2O(l) + 4e^- \leftrightarrow 4OH^-(aq)$

Net equation: $2Fe(s) + O_2(g) + 2H_2O(l) \leftrightarrow 2Fe^{2+}(aq) + 4OH^-(aq)$

The depth at which a ship sinks can affect the rusting process owing to differences in factors such as temperature, pH, oxygen concentration and the presence of sulfate-reducing bacteria.

Temperature: With an increase in temperature comes an increase in reaction rate. The above reactions will achieve activation energy faster in warmer waters at 60 m than in the cold depths at 4000 m. You would therefore expect iron to rust at a greater rate at 60 m than at 4000 m.

pH: Gas solubility will increase with a decrease in temperature. At 4000 m there is likely to be greater solubility of carbon dioxide, which will result in a higher concentration of dissolved carbon dioxide from the respiration of living things. This can react with water to produce carbonic acid.

$H_2O(l) + CO_2(g) \rightarrow H_2CO_3(aq)$

Iron will rust faster in acidic conditions owing to a different reduction reaction with a higher reduction potential.

$\frac{1}{2}O_2(g) + 2H^+ + 2e^- \rightarrow H_2O$

You would therefore expect rusting to occur at a greater rate at depth where gas solubility is higher and pH is lower.

Oxygen concentration: Gas solubility will increase with a decrease in temperature; however, the concentration of O_2 will be higher at 60 m depth than at 4000 m because oxygen enters at the surface of the ocean and is progressively used by living things as it diffuses downwards. As O_2 is a reactant in the reduction reaction, an increase in concentration will result in faster corrosion. Therefore you would expect rusting of iron to be faster in the shipwreck at 60 m, where dissolved oxygen concentrations are higher, than in a wreck at 4000 m.

Sulfate-reducing bacteria: Anaerobic bacteria can be found in areas low in dissolved oxygen, such as at a depth of 4000 m. These bacteria promote the corrosion of the iron hulls of ships by reducing sulfate ions in the ocean and using the electrons from the oxidation of iron.

Oxidation: $Fe(s) \rightarrow Fe^{2+}(aq) + 2e^-$

Reduction: $SO_4^{2-}(aq) + 10H^+(aq) + 8e^- \rightarrow H_2S(aq) + 4H_2O(l)$

In order to obtain the electrons for this reduction, the bacteria oxidise the iron at corrosion sites like shipwrecks. Although these sulfate-reducing bacteria are present in low concentrations in the open ocean, they are present in very high concentrations on submerged iron objects at depth. You would therefore expect rusting to be more extensive on deep-sea shipwrecks compared to those at 60 m.

Taking into account all these factors, you would expect rusting of iron to be more extensive in the shallower 60 m waters, compared to deep 4000-m waters. This is because the access to dissolved oxygen is likely to be the most significant factor affecting rusting, as oxygen is the usual oxidant in this reaction. *(7 marks)*

Question 35—Forensic Chemistry

(a) (i) Glycogen and cellulose are polymers of glucose. In cellulose the β-glucose molecules are connected via β-glycosidic bonds, while in glycogen the α-glucose monomers are linked by α-glycosidic bonds. *(2 marks)*

(ii) The β-glycosidic bonds between glucose molecules in cellulose result in a highly linear chain, with hydroxyl groups on both sides of the polymer chain. This results in very strong hydrogen bonding between polymer chains, giving cellulose its strength. The α-glycosidic bonds in glycogen result in an entirely different, highly branched structure. Glycogen is found in the form of granules inside many cell types, particularly muscle and liver cells. *(3 marks)*

(b) (i) Amino acids contain an amine and an alkanoic acid functional group. Their structure is shown in the following diagram.

Amine functional: H_2N—C(H)(R)—C(=O)OH :Alkanoic acid functional

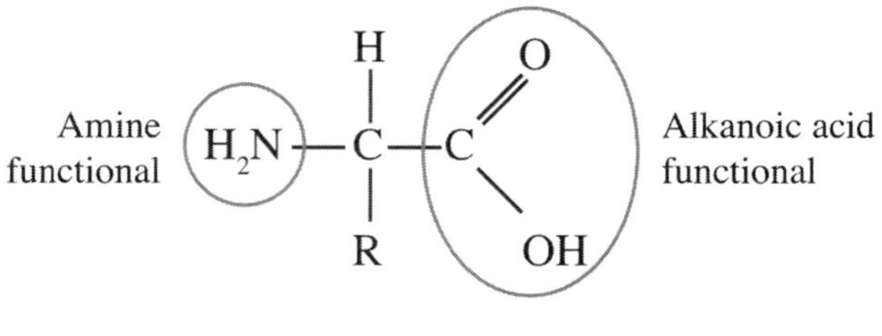

Amino acids vary in the R group, and 20 amino acids are used as monomers in the synthesis of proteins.

Proteins are polymers in which amino acid groups are bonded together via amide (or peptide) bonds to form a polypeptide. The alkanoic acid functional group of one molecule reacts with the amine group of a second molecule (eliminating a water molecule), and the reaction continues at the acid end of the second molecule, and so on. The structure of the growing polypeptide chain is shown in the following diagram.

```
  H   H   O   H   H   O   H   H   O
  |   |   ||  |   |   ||  |   |   ||
—N———C———C———N———C———C———N———C———C—
      |           |           |
      R           R           R
```

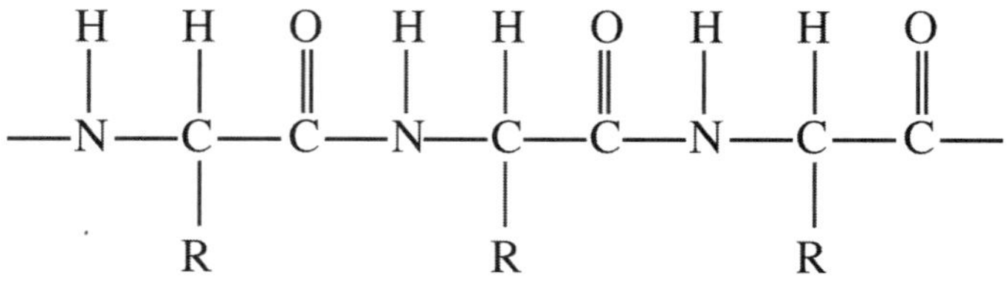

(3 marks)

(ii) Protein in food such as eggwhite can be detected using Biuret reagent. The reagent contains a mixture of $NaOH(aq)$, $CuSO_4(aq)$ and aqueous potassium sodium tartrate to prevent the Cu^{2+} precipitating in the alkaline conditions. The presence of $Cu^{2+}(aq)$ means the colour is initially pale blue, changing on addition of protein to purple (or violet-blue). The following steps are necessary to carry out the test.

1. Transfer 1 mL of eggwhite to a test tube.
2. Transfer 1 mL of deionised water to a second test tube.
3. Add 10 drops of Biuret reagent to both test tubes.
4. Stopper both test tubes and mix gently, observing any colour changes.

The test tube containing the eggwhite should change colour from blue to purple (or violet-blue). The test tube containing water (the control, to ensure a valid result) remains blue.

Safety glasses should be worn to avoid eye contact because NaOH(*aq*) is corrosive and Cu^{2+} salts are eye irritants. *(4 marks)*

(c) (i) The DNA profiles shown would be from the 'fingerprint' region of the DNA chosen for analysis. This region does not code for genes as such, but is as individual as a person's fingerprint. Similarities in DNA from this region can help determine paternity, because there will be more similarity between closely related individuals than between other individuals. Child A's DNA is similar to DNA from both George and Linda, so Child A is their child. On the basis of similarities with one or other parent, Child B is George's child and Child C is Linda's child. *(3 marks)*

(ii) DNA data banks are controversial, with many people concerned about privacy issues and the abuse of the information contained therein. Supporters of DNA data banks cite many benefits. For example, having the DNA of every citizen in a country on file would make solving crimes much more efficient because suspects could be eliminated or identified very quickly and accurately. It would make the current system fairer: people arrested for an offence, whether they are guilty or not, generally have their DNA recorded. This system is selective and unfair, and results in ethnic minorities and young people being over represented. The maintenance of DNA data banks would make it easier to track the movements of terrorists, who are currently able to travel from country to country using false documentation. DNA data banks may also help identify victims from accidents, such as plane crashes. *(3 marks)*

(d) A number of answers are possible. One example is given below.

Forensic samples known to contain proteins can be analysed using electrophoresis, a chromatographic technique that separates molecules on the basis of their size and overall charge. In this technique, the two ends of a solid support matrix, such as a wet filter paper or a polyacrylamide gel, are immersed in a buffer solution, and the electrodes of a DC power source are connected. This applies an electric field across the support matrix. The sample is applied to the middle of the support, and the components of the mixture in the sample migrate towards the electrodes at different rates, depending on their size and charge.

Different proteins in the mixture have specific masses, and are composed of a different sequence of amino acids (their primary structure). The amino acids in the protein have acidic, basic or neutral side chains and as a result their interaction with the buffer produces species with different mass:charge ratios. This causes them to migrate to one or the other electrode, at different rates. The bands are stained after separation so that the proteins' positions on the matrix can be visualised. Running a sample of known proteins on the same support under the same conditions allows a reference to be obtained. Proteins in the mixture can be identified by comparing the bands in the separated sample with those of the reference results. Different animal species have different protein compositions so the resulting protein profile can be used to identify the origins of a sample.

This chromatographic technique can be used to distinguish between the different meat products used in food preparation. For example, in the 1990s electrophoresis was used to determine that some Australian restaurants were using ling, a cheaper fish, in dishes claiming to be expensive barramundi. While there are other more useful techniques in forensic chemistry, such as DNA analysis, electrophoresis is extremely useful in some situations because it is rapid and inexpensive compared to other techniques. In cases involving protein analysis, such as that described above, it is the most appropriate technique. *(7 marks)*

2016 HIGHER SCHOOL CERTIFICATE EXAMINATION

Chemistry

General Instructions

- Reading time – 5 minutes
- Working time – 3 hours
- Write using black pen
- Draw diagrams using pencil
- Board-approved calculators may be used
- A data sheet and a Periodic Table are provided at the back of this paper

Total marks – 100

Section I

75 marks

This section has two parts, Part A and Part B

Part A – 20 marks

- Attempt Questions 1–20
- Allow about 35 minutes for this part

Part B – 55 marks

- Attempt Questions 21–30
- Allow about 1 hour and 40 minutes for this part

Section II

25 marks

- Attempt ONE question from Questions 31–35
- Allow about 45 minutes for this section

Section I
75 marks

Part A – 20 marks
Attempt Questions 1–20
Allow about 35 minutes for this part

Use the multiple-choice answer sheet for Questions 1–20.

---

1 What is the name of this compound?

```
H           H
 \         /
  C   =   C
 /         \
H           Cl
```

(A) Styrene

(B) Ethylene

(C) Chloroethane

(D) Vinyl chloride

2 Which of the following metal ions would NOT cause heavy metal pollution if released in high concentrations?

(A) Copper

(B) Lead

(C) Mercury

(D) Sodium

3 What is the molecular formula of pentanoic acid?

(A) C_5H_9O

(B) $C_5H_{10}O$

(C) $C_5H_{10}O_2$

(D) $C_5H_{11}O_2$

4 Which row of the table correctly identifies an application of polystyrene and the reason for its suitability for that application?

	Application	*Reason for suitability*
(A)	Shopping bags	Rigidity
(B)	Shopping bags	Flexibility
(C)	Screwdriver handles	Rigidity
(D)	Screwdriver handles	Flexibility

5 Which of the following diagrams best represents the bonding between molecules of water and ethanol?

```
                        H
                        |
        H   H           O—H
        |   |         ⋰
(A)  H—C—C—O
        |   |     \
        H   H      H
```

```
                        H
                         \
                          O
                         /
        H   H          H
        |   |        ⋰
(B)  H—C—C—O
        |   |     \
        H   H      H
```

```
                    O
                   / \
                 H    H
               ⋰
        H   H
        |   |
(C)  H—C—C—O
        |   |     \
        H   H      H
```

```
        H   H
        |   |
(D)  H—C—C—O
        |   |     \
        H   H      H
              ⋱
                O—H
                |
                H
```

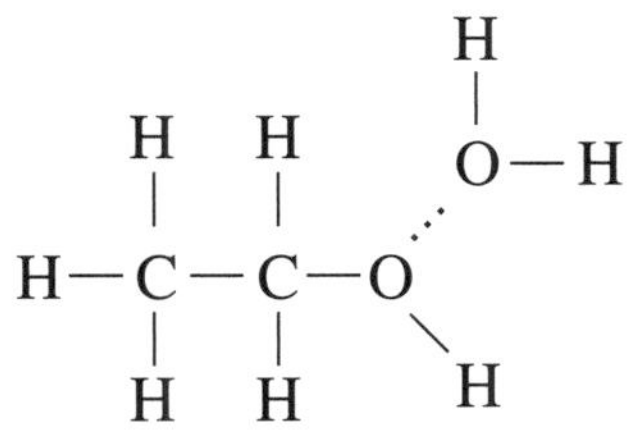

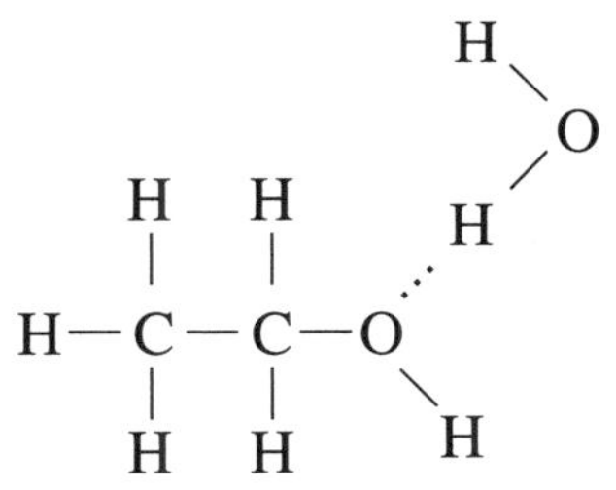

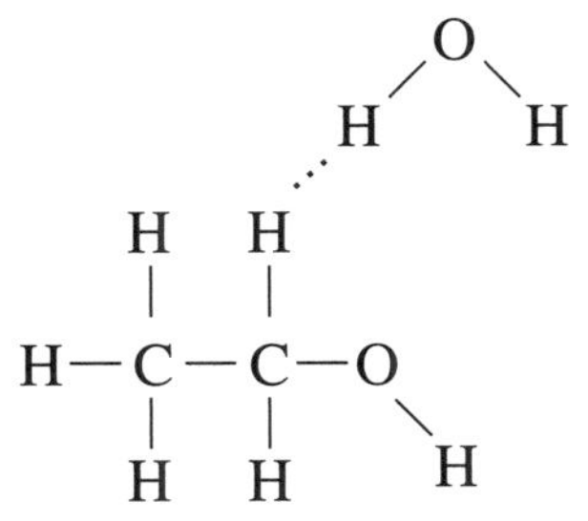

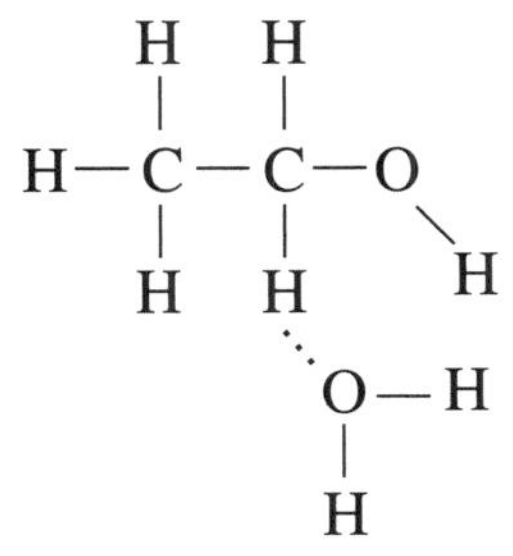

6 Which combination of equimolar solutions would produce the most basic mixture?

(A) Acetic acid and barium hydroxide

(B) Acetic acid and sodium carbonate

(C) Sulfuric acid and barium hydroxide

(D) Sulfuric acid and sodium carbonate

7 Which indicator in the table would be best for distinguishing between lemon juice (pH = 2.3) and potato juice (pH = 5.8)?

	Indicator	*Colour at different pH*	
(A)	Crystal violet	0.2 – yellow	1.8 – blue
(B)	Methyl orange	3.2 – red	4.4 – yellow
(C)	Bromothymol blue	6.0 – yellow	7.6 – blue
(D)	Phenolphthalein	8.2 – colourless	10.0 – pink

8 The following procedure was used to test water hardness.

- 5.0 mL of hard water was placed in a test tube.
- 0.1 mL of liquid soap was added to the test tube.
- The sample was shaken for 30 seconds.
- The height of bubbles was measured.

What would be a suitable control to use with this procedure?

(A) Not adding any soap to the test tube

(B) Not placing any water in the test tube

(C) Using a second sample of the hard water

(D) Replacing the hard water with distilled water

9 Curium is produced according to this equation.

$$^{239}_{94}\text{Pu} + X \longrightarrow {}^{242}_{96}\text{Cm} + {}^{1}_{0}\text{n}$$

What is X in the equation?

(A) A proton

(B) A neutron

(C) A beta particle

(D) An alpha particle

10 Which of the following is the conjugate base of the $H_2PO_4^-$ ion?

(A) H_3PO_4

(B) H_3PO_3

(C) HPO_4^{2-}

(D) HPO_3^{2-}

11 What is the IUPAC name of the following compound?

```
      F   Cl
      |   |
  F — C — C — H
      |   |
      F   Br
```

(A) 1-bromo-1-chloro-2,2,2-trifluoroethane

(B) 1-chloro-1-bromo-2,2,2-trifluoroethane

(C) 2-chloro-2-bromo-1,1,1-trifluoroethane

(D) 2-bromo-2-chloro-1,1,1-trifluoroethane

12 Which of the following could be added to 100 mL of 0.01 mol L^{-1} hydrochloric acid solution to change its pH to 4?

(A) 900 mL of water

(B) 900 mL of 0.01 mol L^{-1} hydrochloric acid

(C) 9900 mL of water

(D) 9900 mL of 0.01 mol L^{-1} hydrochloric acid

13 The flow chart shows the steps used to identify a sample of a substance.

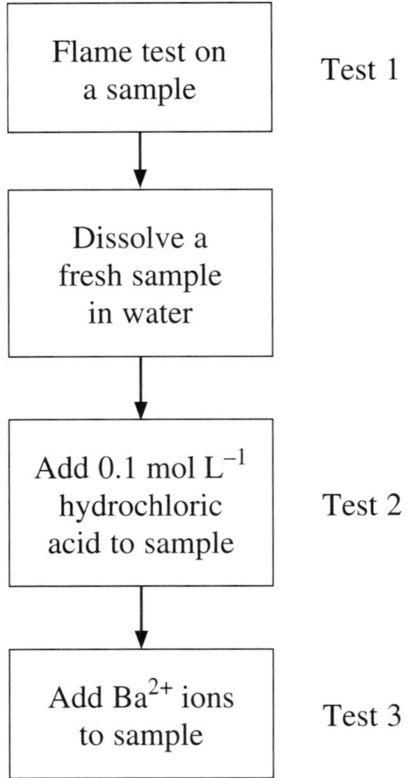

If the substance is sodium sulfate, what should have been observed in Tests 1, 2 and 3?

	Test 1	*Test 2*	*Test 3*
(A)	Bright orange flame	No bubbles	White precipitate formed
(B)	Bright orange flame	Bubbles	No precipitate formed
(C)	Blue-green flame	No bubbles	No precipitate formed
(D)	Blue-green flame	Bubbles	White precipitate formed

14 Consider the following endothermic reaction taking place in a closed vessel.

$$N_2O_4(g) \rightleftharpoons 2NO_2(g)$$

Which of the following actions would cause more N_2O_4 to be produced?

(A) Adding a catalyst

(B) Decreasing the volume

(C) Decreasing the pressure

(D) Increasing the temperature

15 The table lists some properties of the straight-chained carbon compounds *W*, *X*, *Y* and *Z*.

Compound	*Reactivity in bromine water*	*Solubility in water*
W	Rapidly decolourises	Insoluble
X	Unreactive	Insoluble
Y	Unreactive	Soluble
Z	Unreactive	Partly soluble

Which row of the following table best identifies the compounds *W*, *X*, *Y* and *Z*?

	W	*X*	*Y*	*Z*
(A)	C_3H_6	C_3H_8	CH_3OH	C_4H_9OH
(B)	C_3H_8	C_3H_6	CH_3OH	C_4H_9OH
(C)	C_3H_6	C_3H_8	C_4H_9OH	CH_3OH
(D)	C_3H_8	C_3H_6	C_4H_9OH	CH_3OH

16 An electrochemical cell has the following structure.

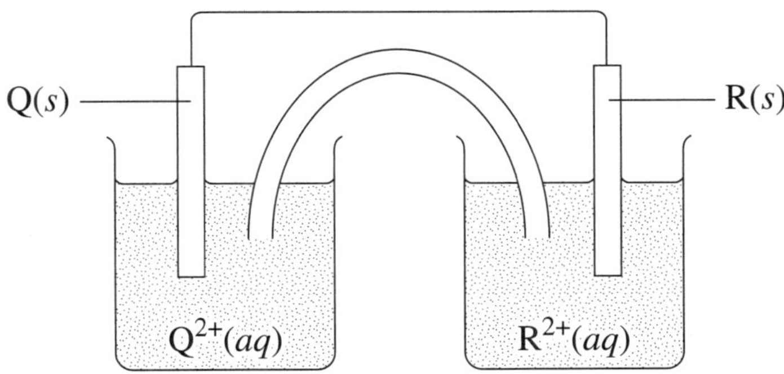

This particular cell can be represented as:

$$Q \mid Q^{2+} \parallel R^{2+} \mid R$$

Which of the following cells would produce the highest cell potential at standard conditions?

(A) $Mg \mid Mg^{2+} \parallel Fe^{2+} \mid Fe$

(B) $Al \mid Al^{3+} \parallel Cu^{2+} \mid Cu$

(C) $Zn \mid Zn^{2+} \parallel Pb^{2+} \mid Pb$

(D) $Ni \mid Ni^{2+} \parallel Ag^{+} \mid Ag$

17 A polymer has the following structure.

$$\left(\begin{array}{cccccccccc} CH_3 & H & H & H & CH_3 & H & H & H & CH_3 & H \\ | & | & | & | & | & | & | & | & | & | \\ C - & C - & C - & C - & C - & C - & C - & C - & C - & C \\ | & | & | & | & | & | & | & | & | & | \\ H & H & CH_3 & H & H & H & CH_3 & H & H & H \end{array}\right)_n$$

Which of the following represents the monomer from which this polymer can be produced?

(A) $\begin{array}{l} H_3C - \underset{\displaystyle H}{\underset{|}{C}} = \underset{\displaystyle H}{\underset{|}{C}} - CH_3 \end{array}$

(B) $H - \underset{\displaystyle H}{\underset{|}{C}} = \underset{\displaystyle H}{\underset{|}{C}} - \overset{\displaystyle H}{\overset{|}{\underset{\displaystyle H}{\underset{|}{C}}}} - CH_3$

(C) $HO - \overset{\displaystyle H}{\overset{|}{\underset{\displaystyle CH_3}{\underset{|}{C}}}} - \overset{\displaystyle H}{\overset{|}{\underset{\displaystyle H}{\underset{|}{C}}}} - OH$

(D) $H - \underset{\displaystyle H}{\underset{|}{C}} = \underset{\displaystyle H}{\underset{|}{C}} - CH_3$

18 40 mL of 0.10 mol L^{-1} NaOH is mixed with 60 mL of 0.10 mol L^{-1} HCl.

What is the pH of the resulting solution?

(A) 7.0

(B) 1.7

(C) 1.4

(D) 1.2

19 Excess barium nitrate solution is added to 200 mL of 0.200 mol L^{-1} sodium sulfate.

What is the mass of the solid formed?

(A) 4.65 g

(B) 8.69 g

(C) 9.33 g

(D) 31.5 g

20 A section of the emission spectrum of a mercury lamp is shown.

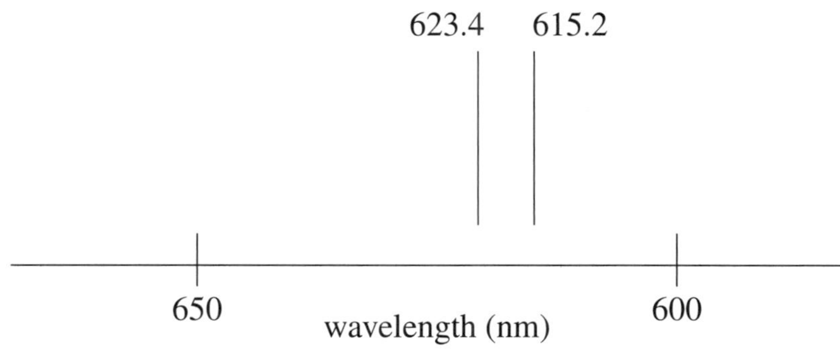

Light at 623.4 nm and 615.2 nm from the mercury lamp was passed through a sample of water containing mercury, and the intensities were then measured by a detector.

I (x nm) = Intensity of light at a wavelength of x nm from the lamp
I_d (x nm) = Intensity of light at a wavelength of x nm at the detector

Which of the following pairs of intensities can be used in the determination of the amount of mercury in the water sample using atomic absorption spectroscopy (AAS)?

(A) I (615.2 nm) and I_d (615.2 nm)

(B) I (615.2 nm) and I_d (623.4 nm)

(C) I (615.2 nm) and I (623.4 nm)

(D) I_d (615.2 nm) and I_d (623.4 nm)

2016 HIGHER SCHOOL CERTIFICATE EXAMINATION

Chemistry

Centre Number

Student Number

Section I (continued)

Part B – 55 marks
Attempt Questions 21–30
Allow about 1 hour and 40 minutes for this part

Answer the questions in the spaces provided. These spaces provide guidance for the expected length of response.

Show all relevant working in questions involving calculations.

Extra writing space is provided on pages 25–27. If you use this space, clearly indicate which question you are answering.

Write your Centre Number and Student Number at the top of this page.

Question 21 (5 marks)

A student set up the following galvanic cell.

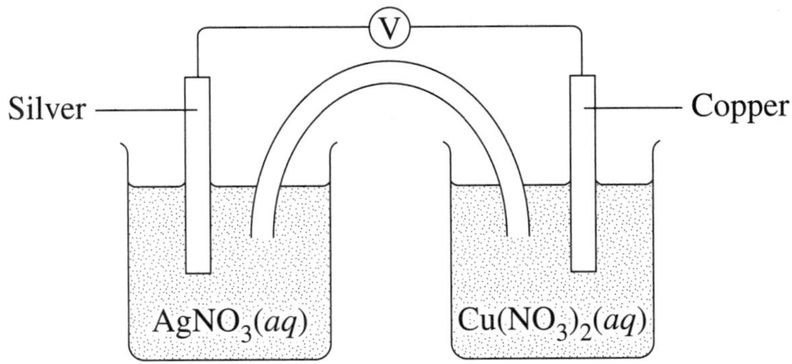

(a) On the diagram clearly indicate the direction of electron flow. **1**

(b) Complete the following table for this galvanic cell. **4**

Anode half equation	
Cathode half equation	
Overall cell equation	
Overall cell potential	

Question 22 (5 marks)

This apparatus was set up to produce methyl butanoate.

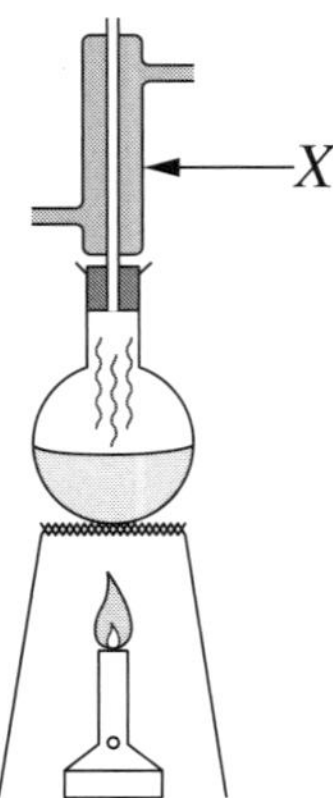

(a) Identify a safety issue in this experiment. **1**

....

(b) Using structural formulae, write the equation for the production of methyl butanoate. **2**

(c) Justify the use of apparatus *X* in this experiment. **2**

....

....

....

....

....

....

Question 23 (6 marks)

A spirit burner containing ethanol was used to heat water in a conical flask for three minutes to measure the molar heat of combustion of ethanol.

The results from the investigation are shown.

Time (min)	0	0.5	1.5	2.0	2.5	3.0	3.5	4.5	5.0
Temperature of water (°C)	18.5	20.5	25.0	27.0	29.5	31.0	30.5	28.5	27.5

(a) On the grid, draw a line graph to represent the data contained in the table. **3**

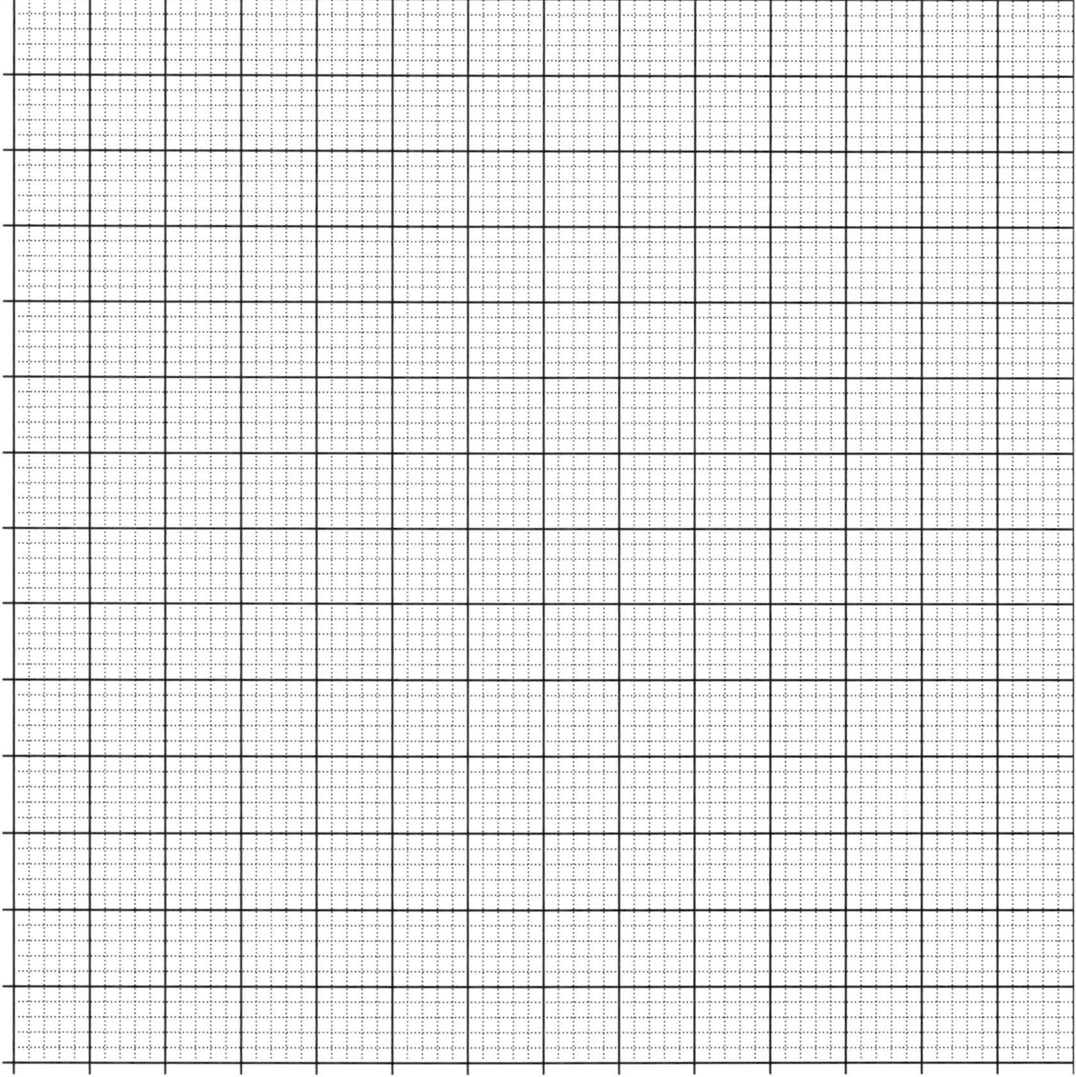

Question 23 continues

Question 23 (continued)

(b) The following values were also recorded during the investigation: **3**

Initial mass of spirit burner = 236.14 g

Final mass of spirit burner = 235.56 g

Calculated experimental molar heat of combustion of ethanol = $-827\ \text{kJ mol}^{-1}$.

Using information from the previous page and the above values, determine the mass of water that was in the conical flask.

....

....

....

....

....

....

....

End of Question 23

Question 24 (7 marks)

(a) Explain how microscopic membrane filters are used to purify contaminated water. Use a labelled diagram to support your answer. **4**

....

....

....

....

....

....

....

....

....

(b) Explain why dissolved oxygen levels can be used to measure the extent of eutrophication. **3**

....

....

....

....

....

....

....

....

Question 25 (4 marks)

An unattended car is stationary with its engine running in a closed workshop. The workshop is 5.0 m × 5.0 m × 4.0 m and its volume is 1.0×10^5 L. The engine of the car is producing carbon monoxide in an incomplete combustion according to the following chemical equation: **4**

$$C_8H_{18}(l) + \tfrac{17}{2}O_2(g) \rightarrow 8CO(g) + 9H_2O(l)$$

Exposure to carbon monoxide at levels greater than 0.100 g L^{-1} of air can be dangerous to human health.

6.0 kg of octane was combusted by the car in this workshop.

Using the equation provided, determine if the level of carbon monoxide produced in the workshop would be dangerous to human health. Support your answer with relevant calculations.

....

....

....

....

....

....

....

....

....

....

Question 26 (6 marks)

(a) Explain why cellulose is classified as a condensation polymer. **2**

....

....

....

....

....

(b) Justify the need for research into biopolymers. **4**

....

....

....

....

....

....

....

....

....

....

Question 27 (4 marks)

The volume of gas formed at 25°C and 100 kPa as hydrochloric acid was added to a pure sample of aluminium is shown in the graph. **4**

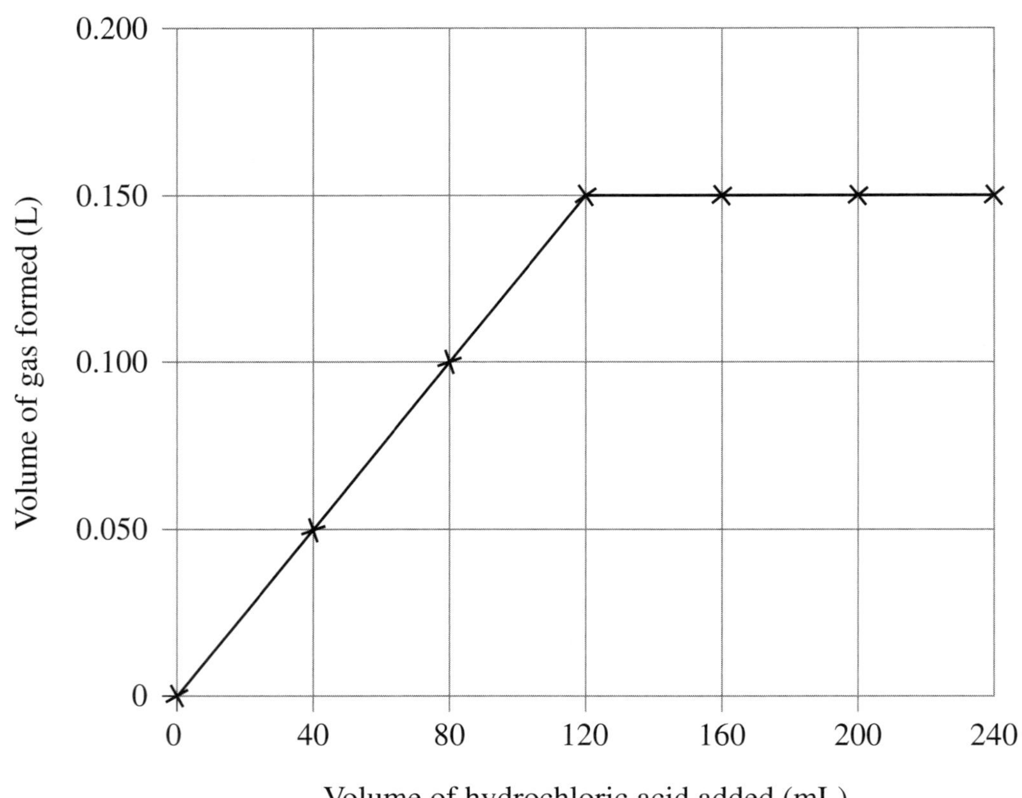

Calculate the original mass of the aluminium sample used in the reaction.

....

....

....

....

....

....

....

....

Question 28 (5 marks)

A mixture of carbon monoxide, chlorine and phosgene ($COCl_2$) gases was placed in a closed container. The concentrations of the gases were monitored over time.

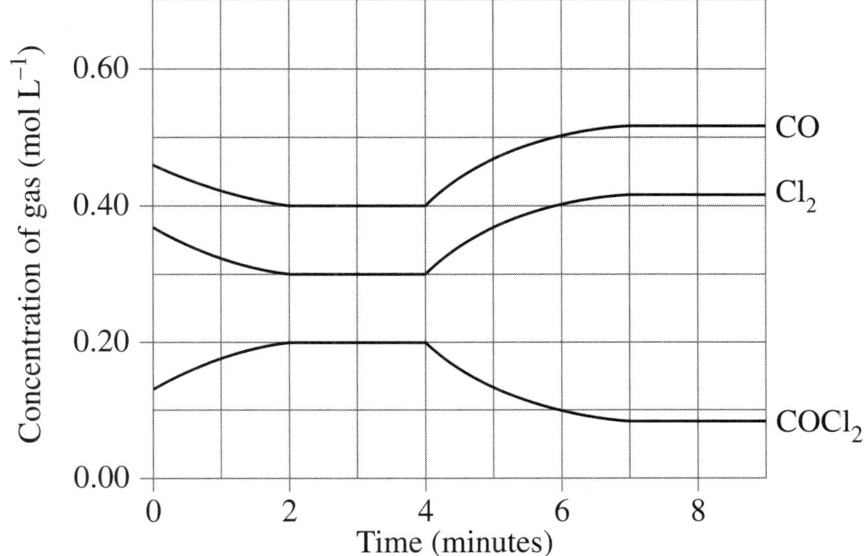

(a) At what time does the system first reach equilibrium? Justify your answer. **2**

....

....

(b) At four minutes, the temperature of the container was increased. **3**

Explain, with reference to the graph, whether the decomposition of $COCl_2$ into CO and Cl_2 is exothermic or endothermic.

....

....

....

....

....

....

....

....

....

Question 29 (6 marks)

A solution of hydrochloric acid was standardised by titration against a sodium carbonate solution using the following procedure.

- All glassware was rinsed correctly to remove possible contaminants.
- Hydrochloric acid was placed in the burette.
- 25.0 mL of sodium carbonate solution was pipetted into the conical flask.

The titration was performed and the hydrochloric acid was found to be 0.200 mol L^{-1}.

(a) Identify the substance used to rinse the conical flask and justify your answer. **2**

....

....

(b) Seashells contain a mixture of carbonate compounds. The standardised hydrochloric acid was used to determine the percentage by mass of carbonate in a seashell using the following procedure. **4**

- A 0.145 g sample of the seashell was placed in a conical flask.
- 50.0 mL of the standardised hydrochloric acid was added to the conical flask.
- At the completion of the reaction, the mixture in the conical flask was titrated with 0.250 mol L^{-1} sodium hydroxide.

The volume of sodium hydroxide used in the titration was 29.5 mL.

Calculate the percentage by mass of carbonate in the sample of the seashell.

....

....

....

....

....

....

....

....

....

....

....

....

Question 30 (7 marks)

The use of CFCs has caused ozone depletion in the stratosphere. **7**

Explain the steps that have been taken to reduce this problem. Include relevant chemical equations in your answer.

Section II

25 marks
Attempt ONE question from Questions 31–35
Allow about 45 minutes for this section

Answer parts (a)–(d) of one question in the Section II Writing Booklet. Extra writing booklets are available.

Show all relevant working in questions involving calculations.

Question 31 Industrial Chemistry

Question 32 Shipwrecks, Corrosion and Conservation

Question 33 The Biochemistry of Movement *(Not included in this reproduction)*

Question 34 The Chemistry of Art *(Not included in this reproduction)*

Question 35 Forensic Chemistry

Question 31 — Industrial Chemistry (25 marks)

Answer parts (a) and (b) of the question on pages 2–4 of the Section II Writing Booklet. Start each part of the question on a new page.

(a) The diagram shows a method used to extract sulfur from an underground sulfur deposit.

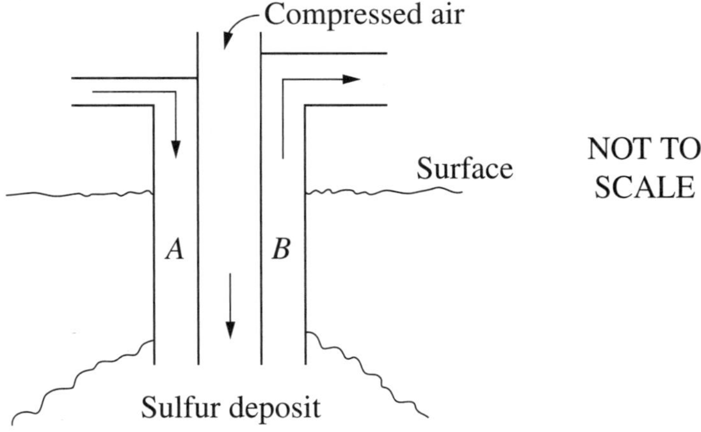

(i) Identify the substances travelling through pipes *A* and *B*. **2**

(ii) Explain how the properties of sulfur allow it to be extracted using this method. **3**

(b) A first-hand investigation to electrolyse a solution of sodium chloride is to be performed.

(i) Outline a procedure that is suitable for carrying out this investigation in a school laboratory. In your answer, address a safety issue. **3**

(ii) Describe how one of the products of the electrolysis of the sodium chloride solution can be identified. In your answer, refer to the chemistry occurring at each of the electrodes. **4**

Question 31 continues

Question 31 (continued)

Answer parts (c) and (d) of the question on pages 5–8 of the Section II Writing Booklet. Start each part of the question on a new page.

(c) (i) Methane and water vapour react to form carbon monoxide and hydrogen in a closed container as shown. **3**

$$CH_4(g) + H_2O(g) \rightleftharpoons CO(g) + 3H_2(g) \quad \Delta H = +206 \text{ kJ}$$

Compare the impact on the equilibrium system of a decrease in volume of the container to the impact of a decrease in temperature. Refer to the equilibrium constant in your answer.

(ii) Solid ammonium hydrogen sulfide (NH_4HS) decomposes to form ammonia gas and hydrogen sulfide gas (H_2S). **4**

2.00 moles of ammonium hydrogen sulfide were placed in a sealed 3.00 L container and the system was allowed to reach equilibrium. At equilibrium, there were 0.0328 moles of ammonia gas.

Calculate the equilibrium constant for this reaction.

(d) Compare the process of saponification in a school laboratory with the industrial preparation of soap and justify any differences in the methods used. Include a relevant chemical equation in your answer. **6**

End of Question 31

Question 32 — Shipwrecks, Corrosion and Conservation (25 marks)

Answer parts (a) and (b) of the question on pages 2–4 of the Section II Writing Booklet. Start each part of the question on a new page.

(a) The diagram shows an electrochemical cell with graphite electrodes.

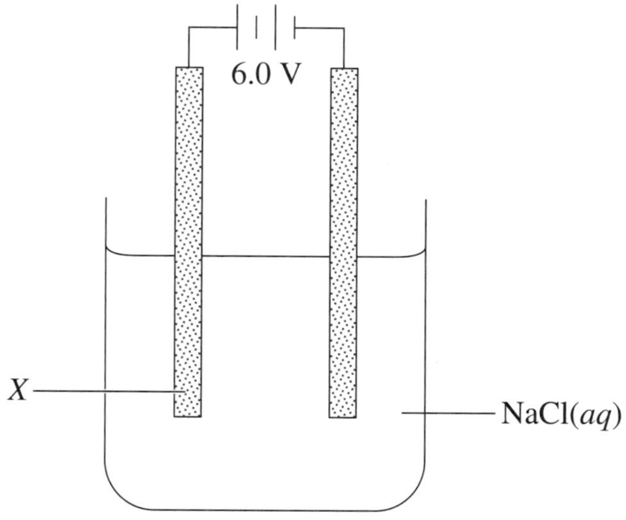

(i) Identify the type of electrochemical cell shown in the diagram, giving a reason for your answer. **2**

(ii) Describe a chemical process that could occur at the electrode labelled *X* in the electrochemical cell in terms of electron transfer. Include a relevant chemical equation in your answer. **3**

(b) A first-hand investigation to compare the rate of corrosion of materials at different temperatures is to be performed.

(i) Describe a procedure that can be used to carry out this investigation safely and reliably. **3**

(ii) Explain how the results of this investigation AND one other factor can be used to predict the rate of corrosion of a wreck at great depth in the ocean. **4**

Question 32 continues

Question 32 (continued)

Answer parts (c) and (d) of the question on pages 5–8 of the Section II Writing Booklet. Start each part of the question on a new page.

(c) (i) Using an example, explain how *passivating metals* resist corrosion. **3**

(ii) Using an example, explain how a sacrificial electrode can prevent corrosion of an iron ocean-going vessel. **4**

(d) Iron cannons have been found in a wreck on a coral reef. It is estimated that the wreck occurred about 250 years ago. **6**

Explain the processes involved in restoring the cannons after they have been salvaged from the wreck.

End of Question 32

Question 35 — Forensic Chemistry (25 marks)

Answer parts (a) and (b) of the question on pages 2–4 of the Section II Writing Booklet. Start each part of the question on a new page.

(a) (i) The photo shows a shoe print in soil obtained at a crime scene. **2**

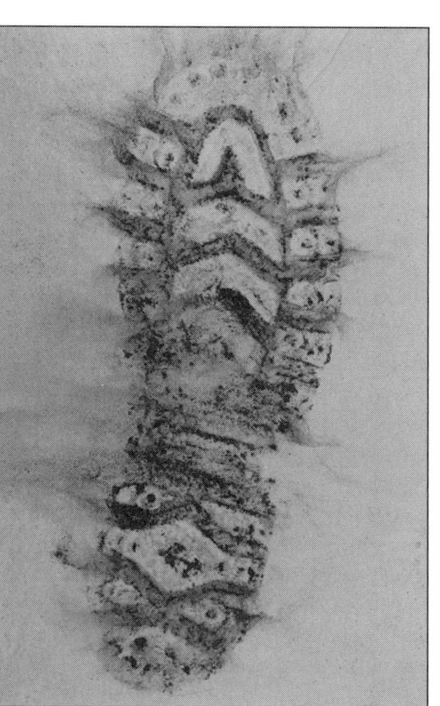

Describe how chemical analysis of material on the shoe that made this print could be used to link a suspect to the crime scene.

(ii) Describe ways of collecting samples from a crime scene that improve the accuracy of evidence presented in court. **3**

Question 35 continues

Question 35 (continued)

(b) (i) A first-hand investigation to separate a mixture using electrophoresis is to be performed in a school laboratory. **3**

Describe a procedure that is suitable for use in this investigation. In your answer, address a safety issue.

(ii) Electrophoresis was performed on a mixture of three amino acids, *X*, *Y* and *Z*, at a pH of 6.0. **4**

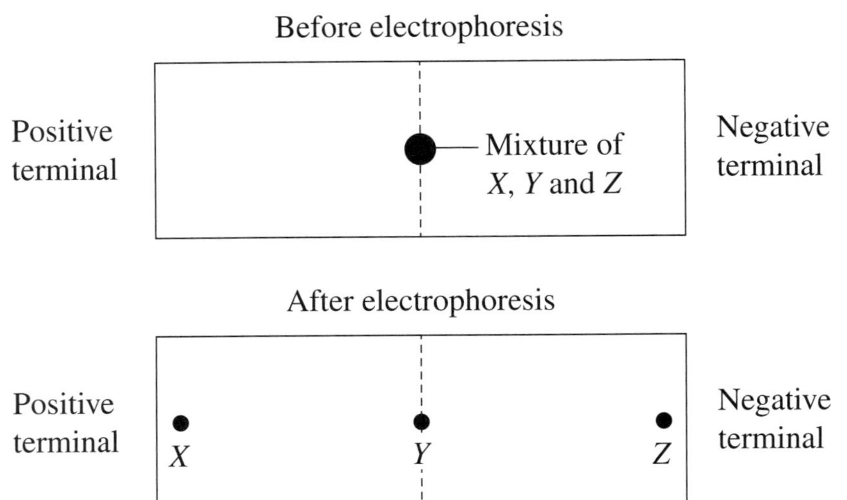

Explain how electrophoresis has separated the amino acids *X*, *Y* and *Z*.

Question 35 continues

Question 35 (continued)

Answer parts (c) and (d) of the question on pages 5–8 of the Section II Writing Booklet. Start each part of the question on a new page.

(c) The diagrams below show schematic representations of atomic emission spectra from a range of metals and an unknown mixture from a paint sample.

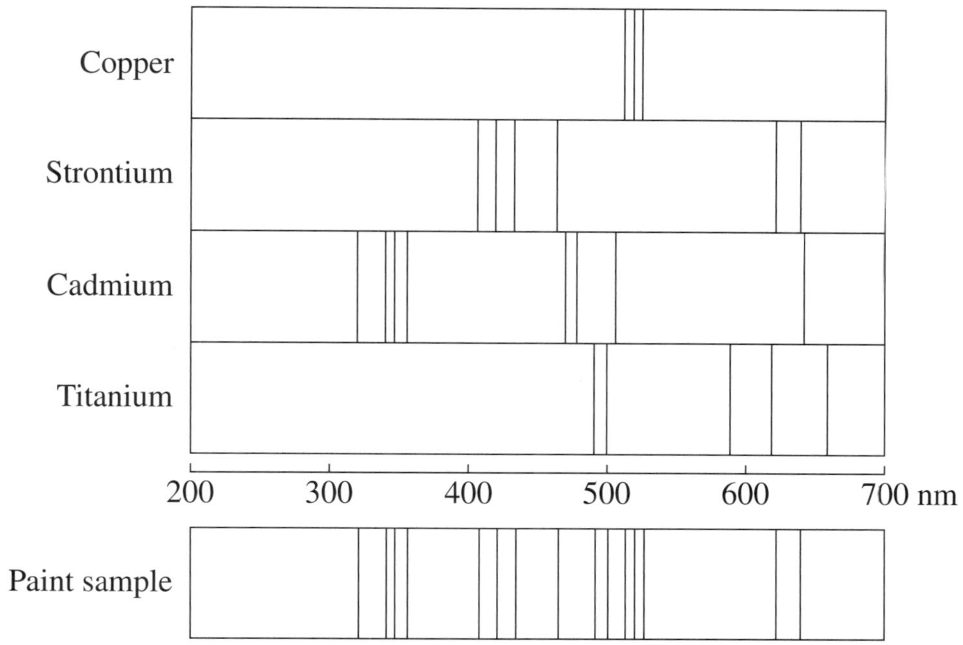

(i) Identify TWO of the metal ions present in the paint sample and justify your answer. **3**

(ii) Describe how atomic emission spectra are produced and used in forensic analysis. **4**

(d) Explain how technology allows forensic scientists to use the features of DNA to improve the accuracy of evidence presented in criminal cases. **6**

End of paper

2016 HSC Examination Paper

Sample Answers

Section I Part A

(Total 20 marks)

1 D

2 D

3 C

4 C

5 B

6 A

7 B

8 D

9 D

10 C

11 D

12 C Diluted concentration of HCl $= 1 \times 10^{-4}$ mol.L^{-1}; pH $= -\log(1 \times 10^{-4}) = 4$

13 A

14 B

15 A

16 B

17 D

18 B Excess HCl = 0.0020 mol in 100 mL of solution

$$c(\text{HCl}) = \frac{0.0020}{0.100} = 0.0200 \text{ mol.L}^{-1}$$

pH $= -\log(0.0200) = 1.7$

19 C $n(BaSO_4) = 0.0400$ mol; $m(BaSO_4) = nM = (0.0400)(233.37) = 9.33$ g

20 A

Section I Part B

21 (a) Electron flow is from the copper anode to the silver cathode.

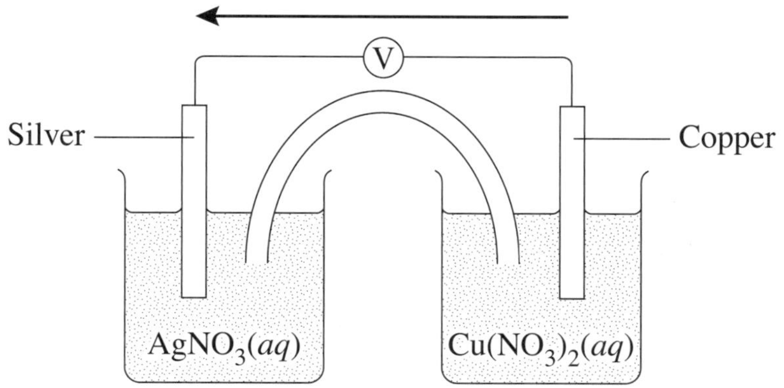

(*1 mark*)

(b)

Anode half equation	$Cu(s) \rightarrow Cu^{2+}(aq) + 2e^-$
Cathode half equation	$Ag^+(aq) + e^- \rightarrow Ag(s)$
Overall cell equation	$Cu(s) + 2Ag^+(aq) \rightarrow Cu^{2+}(aq) + 2Ag(s)$
Overall cell potential	$0.8 + (-0.34) = +0.46$ V

(*4 marks*)

22 (a) The reaction requires the addition of concentrated sulfuric acid as a catalyst. This acid is corrosive to skin and damaging to eyes. (*1 mark*)

(b)

```
    H  H  H    O              H                       H  H  H    O                  
    |  |  |   //              |                       |  |  |   //     H         O—H
 H—C—C—C—C           +   H—C—O—H   ⇆   H—C—C—C—C          |     +    |
    |  |  |   \               |                       |  |  |   \O—C—H        H
    H  H  H    O—H            H                       H  H  H          |
                                                                       H
```

(*2 marks*)

(c) The reactants in this reaction are volatile so as they are heated they readily turn to gas and will escape from the reaction mixture if they are not condensed back to liquids. The apparatus X is a cooling condenser. It facilitates the cooling of the reactant gases condensing them to liquids so that continued heating can take place. (*2 marks*)

23 (a)

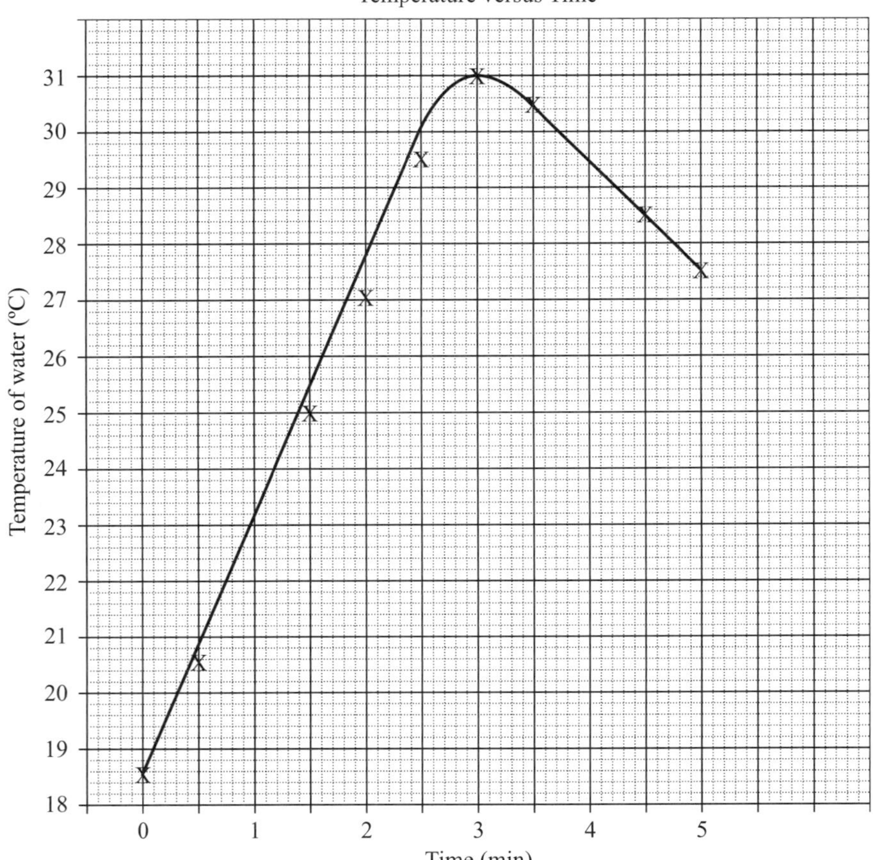

(3 marks)

(b) $\Delta T = 31.0 - 18.5 = 12.5°$

$C = 4.18 \times 10^3 \text{ J kg}^{-1} \text{ K}^{-1}$

$\Delta_c H = -827 \text{ kJ mol}^{-1}$

Δm ethanol = 236.14 – 235.56 = 0.58 g

$n \text{ (ethanol reacting)} = \dfrac{0.58}{46.068} = 0.01259 \text{ mol}$

$\Delta H \text{ (reaction)} = -0.01259 \times 827000 = -10412 \text{ J}$

$\Delta H = -mC\Delta T$

$$m \text{ (water)} = -\frac{\Delta H}{C\Delta T}$$

$$= \frac{-(-10\,412)}{(4.18 \times 10^3 \times 12.5)}$$

$$= 0.199 \text{ kg water}$$

(3 marks)

24 (a)

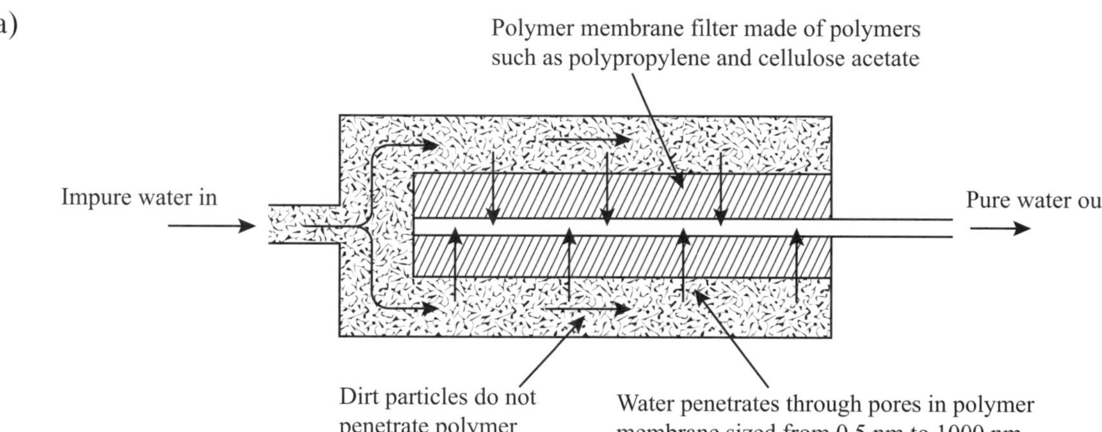

Microscopic membrane filters are composed of a variety of polymers such as polypropylene and cellulose acetate and contain tiny pores through which water can pass but colloidal material, microorganisms and some organic molecules cannot. The pore size can vary depending on what is being filtered. Microfiltration involves a pore size of 100 to 1000 nanometres and will filter silt and microbes such as bacteria and protozoans. Ultrafiltration involves pore sizes varying from 5 to 50 nanometres and will trap smaller microbes such as viruses and larger organic molecules. Nanofiltration uses a pore size between 0.5 and 5 nanometres and will filter most organic molecules and some larger ions. These filters are designed so that water flows across the surface of the filter rather than at 90° to its surface to avoid the filters clogging. *(4 marks)*

(b) Eutrophication describes the enrichment of waterways by nutrients such as phosphate and nitrate. It results in the excessive growth of photosynthetic organisms such as algae and cyanobacteria. This growth blocks the penetration of sunlight into waterways and results in the death of the organisms that initially flourished. Decomposer bacteria then flourish as they feed on the dead photosynthetic organisms and use all the available dissolved oxygen in the waterway. Therefore, measuring levels of dissolved oxygen gives an indirect measure of the extent of eutrophication. *(3 marks)*

25 $n\ (C_8H_{18}) = \frac{6000\ \text{g}}{114.224} = 52.53\ \text{mol}$

$n\ (CO)$ produced $= 52.53 \times 8 = 420.2$ mol

$m\ (CO)$ produced $= 420.2 \times 28.01 = 11\ 770$ g

$c\ (CO)$ in the workshop $= \frac{11\ 770}{1.0} \times 10^5\ \text{L} = 0.11\ 770\ \text{g L}^{-1}$

$= 0.12\ \text{g L}^{-1}$ (2 sig. fig.)

To two significant figures the concentration of carbon monoxide is greater than 0.100 g L^{-1}. Therefore, at this level the concentration of carbon monoxide is dangerous to human health. *(4 marks)*

26 (a) Condensation polymers are those that are formed due to the elimination of small molecules, such as water, from the monomers, allowing them to link up and form a long chain.

In the case of cellulose, the monomers are beta glucose. In order for monomers to link together, water is removed at the junction of monomers by one monomer losing an OH group and the adjacent one losing an H. *(2 marks)*

(b) Biopolymers are long chain molecules formed when organic monomers link together. The organic monomers are produced by living organisms. As they are made by living things, biopolymers are renewable and biodegradable. Research into these polymers is important as most polymers produced in the world today are synthetic and made from crude oil. Crude oil is a non-renewable resource and so will run out in time. As well as plastics being non-renewable they are also non-biodegradable and add to landfill. For these two reasons and the fact that plastic products are so important in the world for packaging, containers, furniture, clothing, etc., research is needed to find replacements that are renewable and biodegradable. *(4 marks)*

27 $2\ Al(s) + 6\ HCl(aq) \rightarrow 2\ AlCl_3(aq) + 3\ H_2(g)$

From the graph it can be determined that all of the aluminium is reacting when no more gas is being produced. At this point 0.150 L of hydrogen gas is formed.

$$n\ (H_2) \text{ formed at this point} = \frac{0.1500}{24.79} = 0.006\,05 \text{ mol}$$

$$n\ (Al) \text{ reacting} = 0.006\,05 \times \frac{2}{3} = 0.004\,03 \text{ mol}$$

$$m\ (Al) = 0.004\,03 \times 26.98 = 0.109 \text{ g}$$

(4 marks)

28 (a) Equilibrium is first reached at 2 minutes. At this time the concentration of all gases remains constant. *(2 marks)*

(b) $CO(g) + Cl_2(g) \rightleftharpoons COCl_2(g)$

The graph shows that the concentrations of CO and Cl_2 increase and the concentration of $COCl_2$ decreases when the temperature is increased. This indicates that the reaction (given by the equation above) shifts in reverse. Therefore the reverse reaction must be acting to partially counteract the increase in temperature, by absorbing heat. The reverse reaction, which is the decomposition of $COCl_2$, is therefore endothermic. *(3 marks)*

29 (a) The conical flask would have been rinsed with distilled water because the 25.0 mL aliquot of sodium carbonate pipetted into the flask was the standard used in the titration and as such had a known concentration and number of moles. Rinsing with distilled water cleans the flask without influencing the number of moles of the standard. *(2 marks)*

(b) n (NaOH) at the end point of the titration = 0.250 × 0.0295 = 0.00738 mol

$NaOH(aq) + HCl(aq) \rightarrow NaCl(aq) + H_2O(l)$

n (excess HCl) = 0.00738 mol

n (HCl initial) = 0.200 × 0.0500 = 0.0100 mol

n (HCl reacting) = initial moles – excess moles

= 0.0100 – 0.00738 = 0.00262 mol

$2H^+(aq) + CO_3^{2-}(aq) \rightarrow CO_2(g) + H_2O(l)$

$n(CO_3^{2-}) = n\frac{(H^+)}{2} = \frac{0.00262}{2} = 0.00131$ mol

$m(CO_3^{2-}) = 0.00131 \times 60.01 = 0.00786$ g

$\%$ mass $CO_3^{2-} = \frac{0.00786}{0.145} \times 100 = 54.2\%$ *(4 marks)*

30 CFCs cause ozone depletion because these synthetic molecules decompose in the stratosphere to chlorine radicals which react with and decompose ozone molecules.

For example, $CFCl_3(g) \rightarrow CFCl_2\bullet(g) + Cl\bullet(g)$

Chlorine radicals react with ozone molecules decomposing them and leading to ozone depletion.

$O_3(g) + Cl\bullet(g) \rightarrow O_2(g) + ClO(g)$

The chlorine radicals are recycled and a single CFC can lead to the decomposition of thousands of ozone molecules.

$ClO(g) + O\bullet(g) \rightarrow O_2(g) + Cl\bullet(g)$

The steps taken to address this problem have involved phasing out CFCs and replacing them with compounds that can be used as refrigerants and propellants but which cause less damage to ozone.

The first step in this process was to obtain accurate measurements of ozone levels in the stratosphere. These measurements were taken using Total Ozone Mapping Spectrometers (TOMS) on satellites. Only when it was proven that ozone levels had been dramatically reduced were follow-up steps taken to address the problem.

The second step was for the international community to meet and formulate a plan to reduce ozone destruction. This occurred in 1987 at a climate conference in Montreal. The Montreal Protocol was formulated where it was mandated that countries that could afford to would phase out the use of CFCs and replace them with hydrochlorofluorocarbons (HCFCs). These compounds are far more reactive than CFCs and tend to decompose before they reach the stratosphere. A phase-out date of 1995 for first-world countries and 2010 for developing countries was agreed upon.

The third step was to continue to measure ozone levels and to hold follow-up meetings to modify the Montreal Protocol in light of current levels. Subsequent meetings were held in

London (1990), Copenhagen (1992), Vienna (1995), Montreal (1997) and Beijing (1999). The initial protocol has seen recommendations to phase out HCFCs and replace them with HFCs and eventually with hydrocarbons so that no halogens, the highly destructive component of refrigerants, are being used.

These steps have proven to be most effective as ozone levels have improved in recent years and are on track to be back at 1970 levels by 2050. *(7 marks)*

Section II—Options

Question 31—Industrial Chemistry

(a) (i) Pipe *A* contains superheated and pressurised liquid water. Pipe *B* contains a sulfur emulsion (liquid sulfur droplets suspended in liquid water). *(2 marks)*

(ii) The following properties of sulfur allow it to be extracted using the method shown.

- Sulfur has a melting point of 113 °C which means that it can be melted using superheated water at 160 °C. The sulfur emulsion can then be forced under pressure to the surface.
- Sulfur has a boiling point of 445 °C which means that it will not boil under the conditions used in the extraction.
- Sulfur is not water-soluble, which means that at no stage does an aqueous solution of sulfur form. At the surface the pressure drops, the emulsion cools and the sulfur emulsion separates into solid sulfur and water. As a result, high-purity solid sulfur is obtained. *(3 marks)*

(b) (i) To observe the electrolysis of 0.1 mol L^{-1} $NaCl(aq)$ use Hoffman's voltameter. Fill the voltameter with 0.1 mol L^{-1} $NaCl(aq)$ to which a few drops of universal indicator have been added. Connect the voltameter electrodes to a power pack set to 6V, and turn the power on. A small amount of chlorine gas may be produced in this reaction. Chlorine is a toxic gas and to prevent inhalation the experiment should be performed in a fume cupboard. The electricity can be turned off and the experiment stopped when one of the tubes of the voltameter is half full of gas. *(3 marks)*

(ii) The electrolysis of 0.1 mol L^{-1} $NaCl(aq)$ produces hydrogen gas at the cathode, and oxygen and chlorine gas at the anode (the proportions of oxygen and chlorine depend on the salt concentration), according to the following equations:

Anode: $2H_2O(l) \rightarrow O_2(g) + 4e^- + 4H^+(aq)$, and

$2Cl^-(aq) \rightarrow Cl_2(g) + 2e^-$

Cathode: $2H_2O(l) + 2e^- \rightarrow H_2(g) + 2OH^-(aq)$

The hydrogen gas produced at the cathode can be identified using the 'pop test'. Some of the gas is collected in an inverted test tube, and a lit match is placed into the mouth of the test tube. An audible 'pop' confirms that the gas is hydrogen. This is also supported by the fact that the universal indicator in this electrolysis compartment of the voltameter turns purple because of the basic OH^- ions also produced. *(4 marks)*

(c) (i) For this reaction $K = \frac{[H_2]^3[CO]}{[CH_4][H_2O]}$

Decreasing the volume of the system will increase the pressure and Le Chatelier's Principle predicts that the equilibrium position will shift to minimise this change, that is, it will produce fewer moles of gas. Thus the reverse reaction will be favoured, and the concentrations of methane and water vapour will increase while the concentrations of carbon monoxide and hydrogen gas will decrease. The value of K will not change.

According to LCP, a decrease in temperature will favour the exothermic reaction, which is also the reverse reaction. Thus the equilibrium position will shift to the left, and in this case the value of K will get smaller as the concentrations of CH_4 and H_2O increase and those of CO and H_2 decrease. *(3 marks)*

(ii) $NH_4HS(s) \leftrightarrows NH_3(g) + H_2S(g)$ and $K = [H_2S][NH_3]$

(Note that solids do not appear in the equilibrium-constant expression.)

If 0.0328 mol of NH_3 is formed, then 0.0328 mol of H_2S is also formed since there is a 1:1 mole ratio of products in the forward reaction.

Thus at equilibrium the concentration of both NH_3 and $H_2S = \frac{0.0328}{3.00} = 0.0109\ mol.L^{-1}$.

$K = 0.0109 \times 0.0109 = 1.19 \times 10^{-4}$ *(4 marks)*

(d) Saponification is the hydrolysis under basic conditions of a triglyceride into glycerol and the sodium salt of fatty acids, for example:

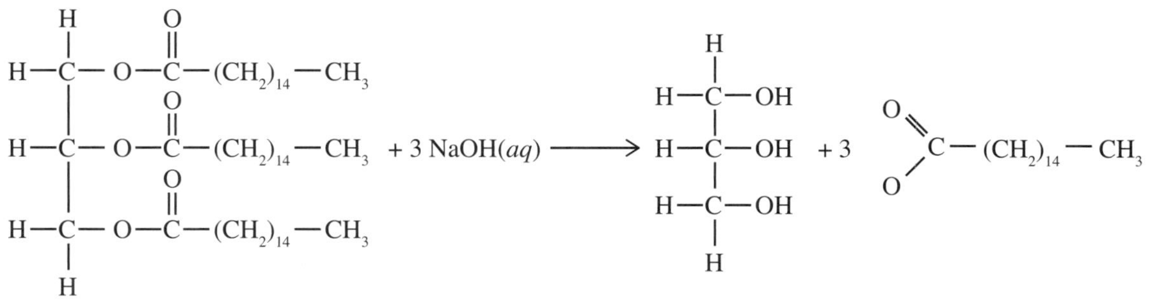

$$\begin{array}{l} H-C-O-C(=O)-(CH_2)_{14}-CH_3 \\ H-C-O-C(=O)-(CH_2)_{14}-CH_3 \\ H-C-O-C(=O)-(CH_2)_{14}-CH_3 \end{array} + 3\,NaOH(aq) \longrightarrow \begin{array}{l} H-C-OH \\ H-C-OH \\ H-C-OH \end{array} + 3\ O=C(-O)-(CH_2)_{14}-CH_3$$

Comparing industrial production with the school laboratory:

Differences	School laboratory	Industrially
Scale	Grams—for practicality and safety (concentrated NaOH is highly caustic)	Kilograms or tonnes: economies of scale for commercial viability
Reactants	Cooking oil—for ease of handling and long-term storage of the starting material	Beef and pig fat—they are cheap by-products of the meat industry
Heating methods	Bunsen burner or hotplate	High-pressure steam
Purity of product	Not completely purified—still very alkaline—insufficient time to neutralise and purify	Highly purified before being released from the factory because an alkaline product will damage skin
Glycerol	Not recovered	Glycerol recovered, purified and used to manufacture other products
Similarities		
Reactants	Both use a strong base such as NaOH or KOH to saponify the fat/oil.	
Products	Both produce the salt of fatty acids and glycerol.	
Process	Both use NaCl to saturate the aqueous phase, making the sodium salt of the fatty acid less soluble, causing it to precipitate so that it can be obtained by filtration.	

(*6 marks*)

Question 32—Shipwrecks, Corrosion and Conservation

(a) (i) This is an electrolytic cell as energy is supplied to bring about a chemical change. *(2 marks)*

(ii) Oxidation occurs at *X*, the positive electrode in an electrolytic cell. At this electrode, water is oxidised, losing electrons and becoming oxygen gas and hydrogen ions.

$$H_2O(l) \rightleftharpoons \frac{1}{2} O_2(g) + 2H^+(aq) + 2e^-$$

The chloride ions in the saltwater can also be oxidised to form chlorine.

$$2Cl^-(aq) \rightleftharpoons Cl_2(g) + 2e^-$$

(3 marks)

(b) (i) 1. Set up the following equipment.

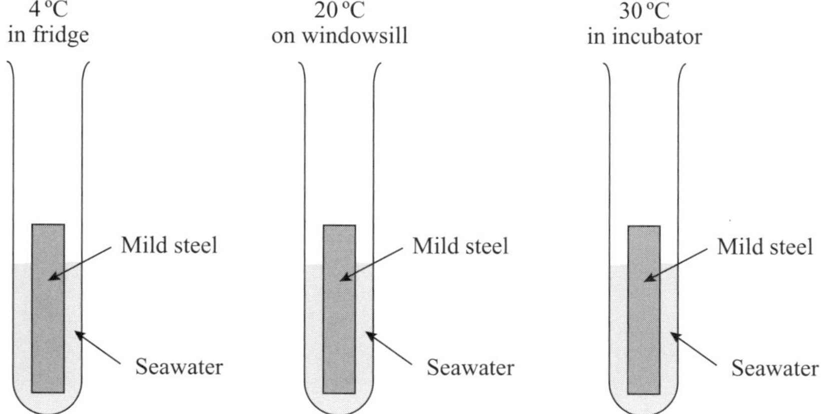

2. Each test tube contains identical 5-cm strips of mild steel half submerged in seawater. The first is placed in a fridge at 4 °C to simulate the temperature in the deep ocean. The second is placed on a windowsill at 20 °C to simulate the average surface ocean temperature and the third is placed in an incubator at 30 °C simulating hot tropical surface waters.

3. Qualitative results are recorded in each test tube every day for 7 days. Levels of corrosion are graded on a 1 to 5 scale, with 1 being no signs of corrosion and 5 being extensive evidence of rusty-brown precipitate on the metal and in the water around it.

4. All variables other than temperature are held constant, such as the volume of seawater and the surface area of the steel strips.

5. The investigation is repeated in exactly the same way but with aluminium strips and then copper strips in order to compare the corrosion rate of steel with other materials. In the case of aluminium, corrosion is graded by the extent of white precipitate in the water and for copper a blue-green precipitate is observed.

6. To improve reliability of results the class is divided into five groups which all carry out the same investigation and pool the results.

7. In order to carry out the investigation safely, care is taken to keep water away from electrical devices, such as the incubator, by placing all test tubes in a rack in a water-resistant tray. *(3 marks)*

(ii) The expected results of this investigation are that the steel will corrode more slowly as the temperature gets colder and the amount of corrosion at 4 °C in the fridge would be little to none. Steel will corrode more rapidly than both aluminium and copper but these materials will also corrode more slowly in the colder water.

You would, therefore, predict that corrosion of metal wrecks at great depth in the ocean would be slow as the temperature is lower than in the shallower water. The temperature at the bottom of the ocean is around 4 °C and is close to what was set up in the investigation in the fridge. The results of the investigation in the fridge would show very little evidence of corrosion at this temperature and you would predict the same for a wreck in the ocean.

Another factor that can be used to predict the rate of corrosion of a wreck at great depth is dissolved oxygen. Corrosion usually requires oxygen to act as the oxidant in the corrosion process. The higher the levels of dissolved oxygen, the greater the rate of corrosion. Oxygen enters the ocean at the surface and the concentration gradually reduces with depth as organisms consume it. Levels of DO at depth are very low and you would therefore predict corrosion rates to be low. *(4 marks)*

(c) (i) Passivating metals resist corrosion because they form stable, impervious oxide layers on their surface. Aluminium is one such metal. When it corrodes, it forms a thin layer of aluminium oxide on its surface. This layer is impervious to water and oxygen so that these oxidants do not touch the underlying metal and thus do not remove electrons from it which would cause it to corrode. *(3 marks)*

(ii) Sacrificial electrodes are those that oxidise in preference to another metal, preventing the other metal from corroding by forcing it to become cathodic. In the case of iron ocean-going vessels, the common sacrificial electrode used is zinc. Zinc blocks are attached periodically to the hull of ships. The oxidation potential of zinc is higher than that of iron, so that the zinc becomes the anode and corrodes.

$Fe(s) \rightleftharpoons Fe^{2+}(aq) + 2e^- \quad E° = 0.44\ V$

$Zn(s) \rightleftharpoons Zn^{2+}(aq) + 2e^- \quad E° = 0.76\ V$

The surrounding iron becomes cathodic and, although reduction occurs there, the iron itself is protected. *(4 marks)*

(d) When artefacts, such as iron cannons, are discovered submerged in the ocean they are not moved immediately as this might accelerate corrosion. When the conditions surrounding the artefacts have been studied and analysed they may be moved and taken through a number of steps to restore them.

The first step is to stabilise the metal cannons by placing them in dilute sodium hydroxide solution. This ensures that the solution does not become acidic which would accelerate corrosion.

The cannons can then be X-rayed so that the extent of the damage can be assessed.

Many animals and plants grow a calcium carbonate shell or skeleton. These cling to the artefacts and must be removed. The majority of calcareous deposits are chiselled off before a dental drill is used to carefully remove the remainder. The X-ray will reveal the extent of concretions and must be used in conjunction with the drill. When most calcium carbonate is removed by physical means, the artefacts can be placed in dilute weak acid to dissolve the remaining encrustaceans.

$$CaCO_3(s) + 2CH_3COOH(aq) \rightarrow CO_2(g) + H_2O(l) + 2CH_3COO^-(aq) + Ca^{2+}(aq)$$

Desalination is the next step in the cleaning process. Chloride and sulfate ions are removed from the structure by electrolysis. The cannons are connected to the negative terminal of a power source and become the cathode and a stainless steel anode is usually used. This is called electrolytic reduction cleaning. A dilute sodium hydroxide electrolyte is used. The Fe^{3+} oxide layer (rust) is reduced to black magnetite (Fe_3O_4). This stabilises the iron surface. Some Fe^{2+} is also reduced to metallic iron. The stainless steel anode is used as a mesh wrapped around the artefact.

The iron artefact can then be heat-treated in a reducing atmosphere of hydrogen gas. This converts iron(III) oxides to metallic iron.

$$Fe_2O_3(s) + 3H_2(g) \rightarrow 2Fe(s) + 3H_2O(l)$$

The cannons are then dried for 48 hours at 120 °C.

To prevent the cannons corroding further following the cleaning treatment, they are coated with a special wax to prevent oxidants reaching the iron and to give it an attractive lustre.

They can be immersed in molten microcrystalline wax which is kept at 135 °C for 5 days until no gas bubbles are evident. Allowing the wax to cool to just above melting point before removing the cannon from the wax ensures maximum penetration of wax.

If the original cannons had wooden stocks, these can be replaced to recreate the original appearance. (*6 marks*)

Question 35—Forensic Chemistry

(a) (i) If the shoe is discovered in a suspect's possession, material on the shoe can be analysed to determine if it matches material found in the soil at the crime scene. For example, the soil at the crime scene may have a particular profile of organic matter that is also found on the shoe, which would be determined by gas chromatography coupled with mass spectroscopy (GCMS). The mineral composition of the soil found at the crime scene can be compared with that found on the shoe using atomic emission spectroscopy. Furthermore, the size and brand of the shoe that made the print would be compared to those of the shoe found in the suspect's possession. *(2 marks)*

(ii) In order for forensic evidence to be deemed accurate in court, it must be collected, transported, sorted and transferred from one person/entity to another following strict guidelines. For example, the crime scene must not be contaminated with investigators' biological material, and investigators must wear gloves, shoe covers and possibly hairnets to prevent this. Where possible, large enough samples must be collected so that multiple analyses can be carried out, in case ambiguous results are obtained or results are called into question. Samples must be meticulously photographed and labelled as they are collected, and strict chain-of-custody procedures must be adhered to so that interference with samples cannot be claimed. *(3 marks)*

(b) (i) A mixture of unknown amino acids can be separated using electrophoresis. A cellulose sheet containing a fluorescent indicator is used as the solid matrix, marked at one end with a '+' sign, and at the other with a '–' sign. A pencil line is drawn across the middle of the sheet between the positive and negative signs. Two crosses are drawn on the dividing line to indicate where samples will be placed; they should not be closer than 1 cm, nor closer to the edge than 1 cm. Using a glass capillary tube, a solution of amino-acid standards is spotted onto one of the crosses. The mixture of unknowns is spotted onto the other cross using a clean glass capillary tube. A phosphate buffer solution (pH = 6) is applied to the plate, which is then placed into the electrophoresis apparatus so that the positive and negative markings on the plate align correctly with the positive and negative terminals of the apparatus. The apparatus is turned on and allowed to run for 40 minutes, after which the plate is removed and sprayed with ninhydrin spray. This spray is an eye and skin irritant and is harmful if inhaled. Therefore, safety glasses and gloves must be worn and the spray applied in a fume cupboard. After the plate has been sprayed it is placed on a hotplate on a low temperature until purple spots appear. The unknown amino acids can be identified by comparing their migration with that of the known amino acids in the mixture of standards. *(3 marks)*

(ii) Electrophoresis is a separation technique that works by using an electric field to cause species in a mixture to move at different rates towards the positive or negative terminal of the field, on the basis of their overall charge and size. Because of the presence of the amine and the carboxylic acid functional group in amino acids, their overall charge will depend on the pH of the buffer solution used, since the amino group can become the positively charged ammonium analogue, and the acid group can become negatively charged. If the overall charge on the amino acid is positive it will migrate towards the negative terminal, such as amino acid Z in the experiment shown, and amino acids with an overall negative charge will migrate to the positive terminal, such as amino acid X in the experiment shown. The rate of movement is determined by the amino acid's overall charge and its mass; the larger the size and the smaller the overall charge, the slower the amino acid will move through the solid matrix. Both amino acids X and Z have moved the same distance from the centre of the plate in the experiment shown, so their overall mass/charge ratio must be the same or very similar. After separation the amino acids must be stained so that they can be visualised and compared with amino acid standards and their rate of migration in the field can be calculated. In the experiment shown, the amino acid Y must exist in its neutral zwitter ion form at pH = 6; that is, it has no overall charge and therefore has not moved during the electrophoresis. *(4 marks)*

(c) (i) Copper ions are in the sample, because all of the spectral lines from the copper reference spectrum (between 500 and 550 nm) appear in the spectrum of the unknown sample, and these three closely spaced spectral lines do not appear in the spectra of the other metal references. Strontium ions are also in the sample, for the same reasons as those given for copper. Cadmium and titanium can be excluded from the sample because, while some of their spectral lines appear in the spectrum of the unknown, not all of their spectral lines appear; for example, neither the Cd spectral line at approximately 480 nm nor the Ti spectral line at approximately 590 nm appear in the spectrum of the sample. The presence of other spectral lines in the spectrum of the sample indicates that there are other elements present in addition to Cu and Sr. *(3 marks)*

(ii) An atomic emission spectrum is a series of coloured lines on a black background, at wavelengths characteristic of each element. Comparison of the spectrum with the reference spectra of different elements allows the elements present in the same to be identified.

Atomic emission spectra are used in forensic analysis to identify elements, particularly metals, present in samples found at crime scenes. For example, the elements present in a sample of paint found at a crime scene may help investigators to identify the car used; a heavy metal present in poisoned food could be identified; the origin of a soil sample found on a suspect's shoe could be linked to a particular location with the same elemental profile.

Atomic emission spectra are produced by vaporising the sample and exciting the electrons in atoms by irradiating the same with electromagnetic radiation of an appropriate frequency range. Those frequencies that correspond exactly to the electrons' transitions to higher energy levels are absorbed and, when the electrons relax back to the ground state, exactly the same frequencies of energy are re-emitted, and observed as a series of coloured lines on a black background. Because the number of protons, and the energy of electron shells are different for different elements, the energy required to excite electrons, and therefore the energy released when electrons relax back to the ground state, is different for each element. Therefore each element's atomic emission spectrum is unique, making this a useful tool in forensic analysis. *(4 marks)*

(d) DNA has characteristics that make it extremely useful in forensic analysis, thanks to the application of modern technology. DNA contains an individual's genetic code and is found in almost all biological samples. DNA is present in the cell as chromosomes, which are long, double-helical strands of DNA. Chromosomes contain many genes, and they also contain regions which do not function as genes. These regions are known as 'fingerprint regions' (or introns) because they are as specific to an individual as their fingerprint. The process of analysing these segments of chromosomes is known as 'DNA fingerprinting'.

Technology is used extensively in DNA analysis. First, samples found at a crime scene must be treated so that the cell membranes break down, which allows the DNA to be extracted from within the nucleus, and separated from the other cell contents. One of the most important characteristics of DNA is the complementary base-pairing found in the double helix structure. This complementary base pairing is what allows multiple copies of the DNA to be made, for analysis. The polymerise chain reaction (PCR) is used to do this (that is, to amplify the DNA). The sample is heated so that the double-stranded helix separates, and then enzymes are added which catalyse the amplification (copying) of the DNA from individual nucleotides in solution. Because of complementary base pairing, the original DNA strands are used as templates in the construction of the complementary DNA strands. The PCR is repeated many times so that large quantities of the DNA are available for analysis. The amplified DNA is then cut at specific locations in the base sequence by the addition of restriction enzymes, and the resulting short fragments of DNA are separated using another example of technology, electrophoresis. In this technique, the sample of DNA fragments is applied to a solid matrix (the electrophoresis gel), and a voltage is applied, causing the fragments to migrate through the gel at different rates depending on their overall charge/mass ratio. Thus, the fragments are separated and, after electrophoresis, stained to make them visible. In this way the DNA from a suspect can be compared with the DNA in a crime-scene sample.

DNA analysis is the most useful of crime scene analysis techniques when biological samples are involved, because it is found in almost all biological samples, which means that different sample types can be compared. For example, the DNA in a blood sample found at a crime scene can be compared with the DNA in a suspect's skin cells, in order to identify or eliminate a suspect. The accuracy of the PCR means that even tiny samples can be successfully analysed. Caution must be exercised with the use of DNA analysis because the presence of a suspect's biological material at a crime scene does not necessarily mean that they committed the crime. Other evidence must also be gathered to establish a person's guilt or innocence. *(6 marks)*

2017 | HIGHER SCHOOL CERTIFICATE EXAMINATION

Chemistry

General Instructions

- Reading time – 5 minutes
- Working time – 3 hours
- Write using black pen
- Draw diagrams using pencil
- NESA approved calculators may be used
- A data sheet and Periodic Table are provided at the back of this paper

Total marks: 100

Section I – 75 marks

This section has two parts, Part A and Part B

Part A – 20 marks

- Attempt Questions 1–20
- Allow about 35 minutes for this part

Part B – 55 marks

- Attempt Questions 21–30
- Allow about 1 hour and 40 minutes for this part

Section II – 25 marks

- Attempt ONE question from Questions 31–35
- Allow about 45 minutes for this section

Section I
75 marks

Part A – 20 marks
Attempt Questions 1–20
Allow about 35 minutes for this part

Use the multiple-choice answer sheet for Questions 1–20.

---

1 In an experiment, 30 mL of water is to be transferred into a conical flask.

Which piece of equipment would deliver the volume with the greatest accuracy?

A. Burette

B. Beaker

C. Test tube

D. Measuring cylinder

2 Which row of the table correctly matches an ion with its flame colour during a flame test?

	Ion	*Flame colour*
A.	Barium	Orange-red
B.	Calcium	Blue-green
C.	Carbonate	Orange-red
D.	Copper	Blue-green

3 What is the name of this compound?

```
     H   H   Cl  H
     |   |   |   |
 H — C — C — C — C — H
     |   |   |   |
     H   H   H   F
```

A. 2-chloro-1-fluorobutane

B. 3-chloro-4-fluorobutane

C. 1-fluoro-2-chlorobutane

D. 4-fluoro-3-chlorobutane

4 Esterification can be carried out in a school laboratory using the equipment shown.

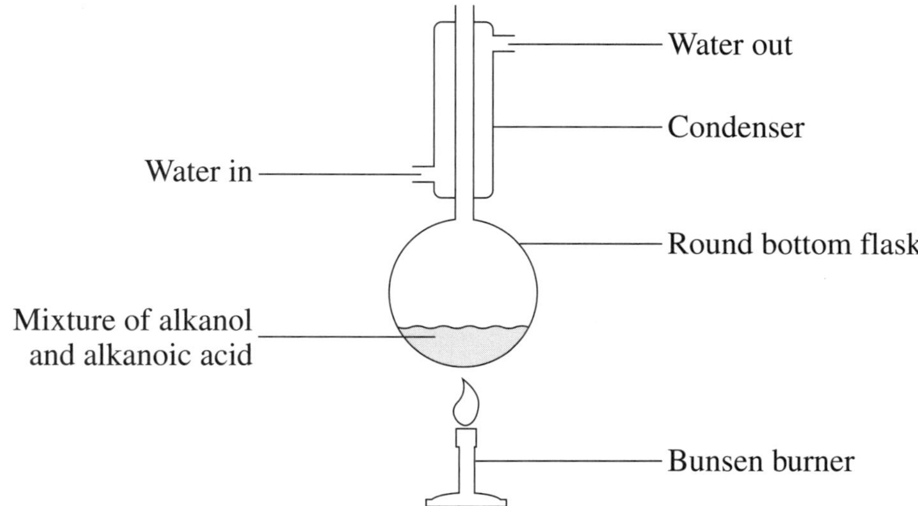

How could the safety of the process shown be improved?

A. Place a stopper on top of the condenser.

B. Add concentrated sulfuric acid to the flask.

C. Change the direction of water flow through the condenser.

D. Replace the Bunsen burner with an electric heating mantle.

5 Which of the following substances is amphiprotic in nature?

A. HSO_4^-

B. H_2SO_4

C. SO_4^{2-}

D. H_2SO_3

6 Which of the following is a transuranic element that is most likely to have been produced in a nuclear reactor?

A. Co-60

B. Np-239

C. U-238

D. Hs-265

7 Three test tubes were set up as shown.

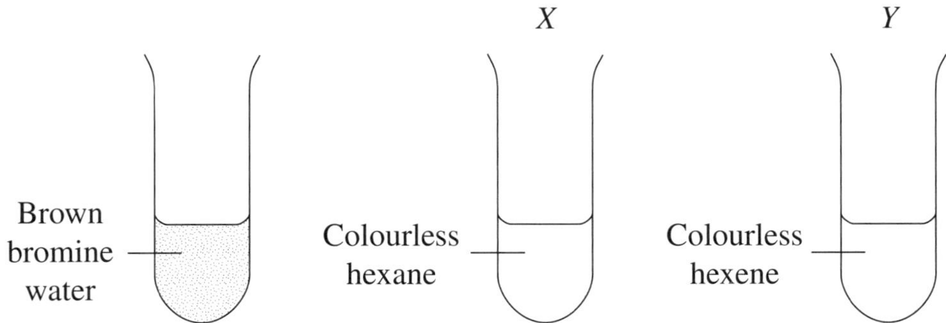

Bromine water was added to *X* and *Y* in the absence of UV light.

Which of the following best represents the changes in test tubes *X* and *Y*?

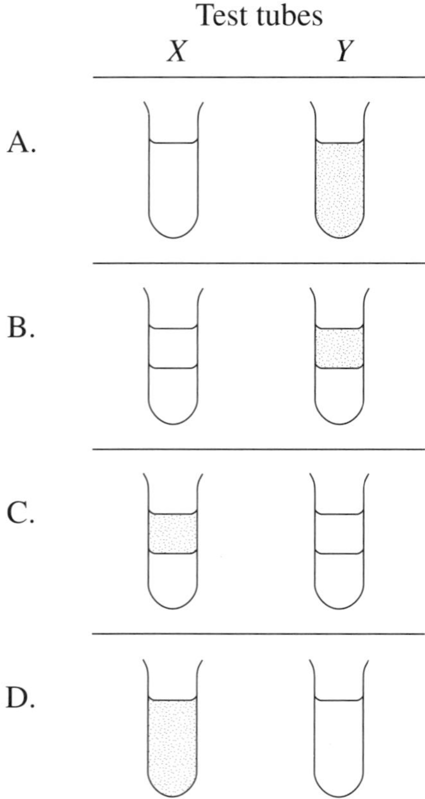

8 There are two unlabelled solutions. One is barium nitrate and the other lead nitrate.

Which of the following could be added to the two unlabelled solutions to distinguish between them?

A. Sodium sulfate

B. Sodium nitrate

C. Sodium chloride

D. Sodium carbonate

9 The following equipment was set up to measure the heat of combustion of an alkanol.

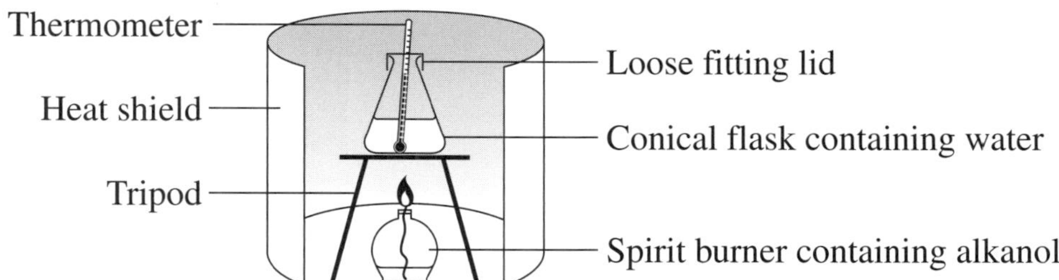

Black deposits were observed on the bottom of the conical flask and the heat of combustion measured was lower than the theoretical value.

Which of the following equations could account for these observations?

A. $2C_2H_6(g) + 7O_2(g) \longrightarrow 4CO_2(g) + 6H_2O(g)$

B. $C_3H_8O(g) + 4O_2(g) \longrightarrow CO_2(g) + CO(g) + 4H_2O(g)$

C. $2C_4H_{10}O(g) + 3O_2(g) \longrightarrow 8C(s) + 2H_2(g) + 8H_2O(g)$

D. $2C_2H_6O(g) + 4O_2(g) \longrightarrow 2CO_2(g) + 2C(s) + 6H_2O(g)$

10 The diagrams show a dry cell and a lead-acid cell. The electrodes are labelled (X) and (Y).

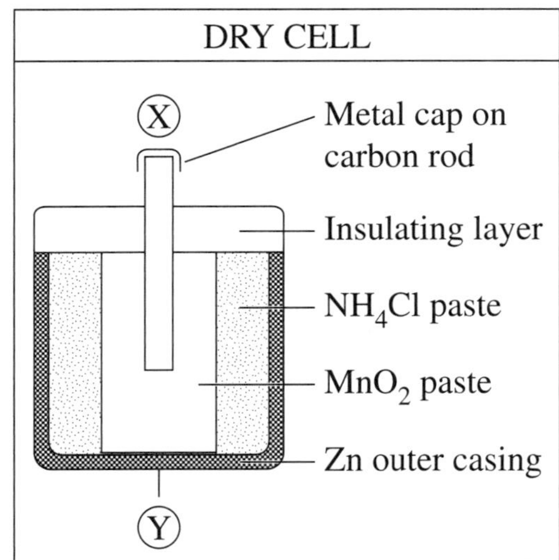

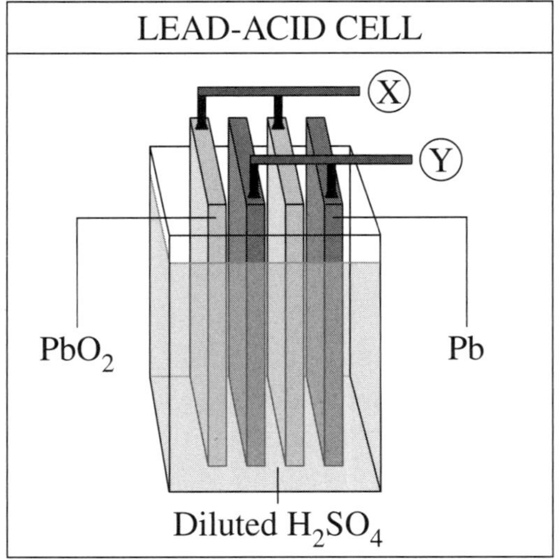

When either cell is connected to supply energy, which row of the table correctly describes the electrodes and their charges?

	Electrode (X)		*Electrode* (Y)	
A.	Anode	Positive	Cathode	Negative
B.	Cathode	Positive	Anode	Negative
C.	Anode	Negative	Cathode	Positive
D.	Cathode	Negative	Anode	Positive

11 Consider the following redox reaction.

$$2K_2Cr_2O_7(aq) + 2H_2O(l) + 3S(s) \longrightarrow 2Cr_2O_3(aq) + 4KOH(aq) + 3SO_2(g)$$

Which species is being oxidised?

A. Cr^{6+}

B. K^+

C. O^{2-}

D. S

12 What is the product when propene undergoes addition polymerisation?

A.

$$\begin{array}{ccccccccccccc} & \mathrm{H} & & \mathrm{H} & & \mathrm{H} & & \mathrm{H} & & \mathrm{H} & & \mathrm{H} & \\ & | & & | & & | & & | & & | & & | & \\ - & \mathrm{C} & = & \mathrm{C} & - & \mathrm{C} & - & \mathrm{C} & = & \mathrm{C} & - & \mathrm{C} & - \\ & & & & & | & & & & & & | & \\ & & & & & \mathrm{H} & & & & & & \mathrm{H} & \end{array}$$

B.

$$\begin{array}{ccccccccccccc} & \mathrm{H} & & \mathrm{H} & & \mathrm{H} & & \mathrm{H} & & \mathrm{H} & & \mathrm{H} & \\ & | & & | & & | & & | & & | & & | & \\ - & \mathrm{C} & - & \mathrm{C} & - & \mathrm{C} & - & \mathrm{C} & - & \mathrm{C} & - & \mathrm{C} & - \\ & | & & | & & | & & | & & | & & | & \\ & \mathrm{H} & & \mathrm{H} & & \mathrm{H} & & \mathrm{H} & & \mathrm{H} & & \mathrm{H} & \end{array}$$

C.

$$\begin{array}{ccccccccccccc} & \mathrm{H} & & \mathrm{CH_3} & & \mathrm{H} & & \mathrm{CH_3} & & \mathrm{H} & & \mathrm{CH_3} & \\ & | & & | & & | & & | & & | & & | & \\ - & \mathrm{C} & = & \mathrm{C} & - & \mathrm{C} & = & \mathrm{C} & - & \mathrm{C} & = & \mathrm{C} & - \\ & & & | & & & & | & & & & & \\ & & & \mathrm{H} & & & & \mathrm{H} & & & & & \end{array}$$

D.

$$\begin{array}{ccccccccccccc} & \mathrm{H} & & \mathrm{CH_3} & & \mathrm{H} & & \mathrm{CH_3} & & \mathrm{H} & & \mathrm{CH_3} & \\ & | & & | & & | & & | & & | & & | & \\ - & \mathrm{C} & - & \mathrm{C} & - & \mathrm{C} & - & \mathrm{C} & - & \mathrm{C} & - & \mathrm{C} & - \\ & | & & | & & | & & | & & | & & | & \\ & \mathrm{H} & & \mathrm{H} & & \mathrm{H} & & \mathrm{H} & & \mathrm{H} & & \mathrm{H} & \end{array}$$

13 25.0 mL of a 0.100 mol L^{-1} acid is to be titrated against a sodium hydroxide solution until final equivalence is reached.

Which of the following acids, if used in the titration, would require the greatest volume of sodium hydroxide?

A. Acetic

B. Citric

C. Hydrochloric

D. Sulfuric

14 One litre of an aqueous solution is formed from mixing equal volumes of 0.2 mol L^{-1} hydrochloric acid (HCl) and 0.2 mol L^{-1} sodium chloride (NaCl).

How effective as a buffer is the aqueous solution formed?

A. Ineffective, because HCl is a strong acid

B. Effective, because Cl^- is the conjugate base of HCl

C. Ineffective, because NaCl forms a neutral salt solution

D. Effective, because the pH would change when a solution of NaOH is added

15 Dinitrogen oxide (N_2O) contains a coordinate covalent bond.

Which Lewis electron dot structure correctly represents N_2O?

A. $:N\vdots\vdots N::\ddot{\underset{\cdot\cdot}{O}}:$

B. $:N\vdots\vdots N:\ddot{\underset{\cdot\cdot}{O}}:$

C. $\ddot{\underset{\cdot\cdot}{N}}\cdot\cdot\ddot{\underset{\cdot\cdot}{N}}::\dot{\underset{\cdot}{O}}:$

D. $:N\vdots\vdots N::\dot{\underset{\cdot}{O}}:$

16 The following equilibrium is established in a closed system.

$$CO_2(g) + H_2O(l) \rightleftharpoons H_2CO_3(aq) \qquad \Delta H = -19.4 \text{ kJ mol}^{-1}$$

How can the gas pressure in the system be decreased?

A. Add more $CO_2(g)$

B. Add hydroxide ions to the solution

C. Decrease the volume of the container

D. Increase the temperature of the system

17 What is the density of ozone at 25°C and 100 kPa?

A. 1.291 g L^{-1}

B. 1.500 g L^{-1}

C. 1.936 g L^{-1}

D. 2.114 g L^{-1}

18 Three gases X, Y and Z were mixed in a closed container and allowed to reach equilibrium. A change was imposed at time *T* and the equilibrium was re-established. The concentration of each gas is plotted against time.

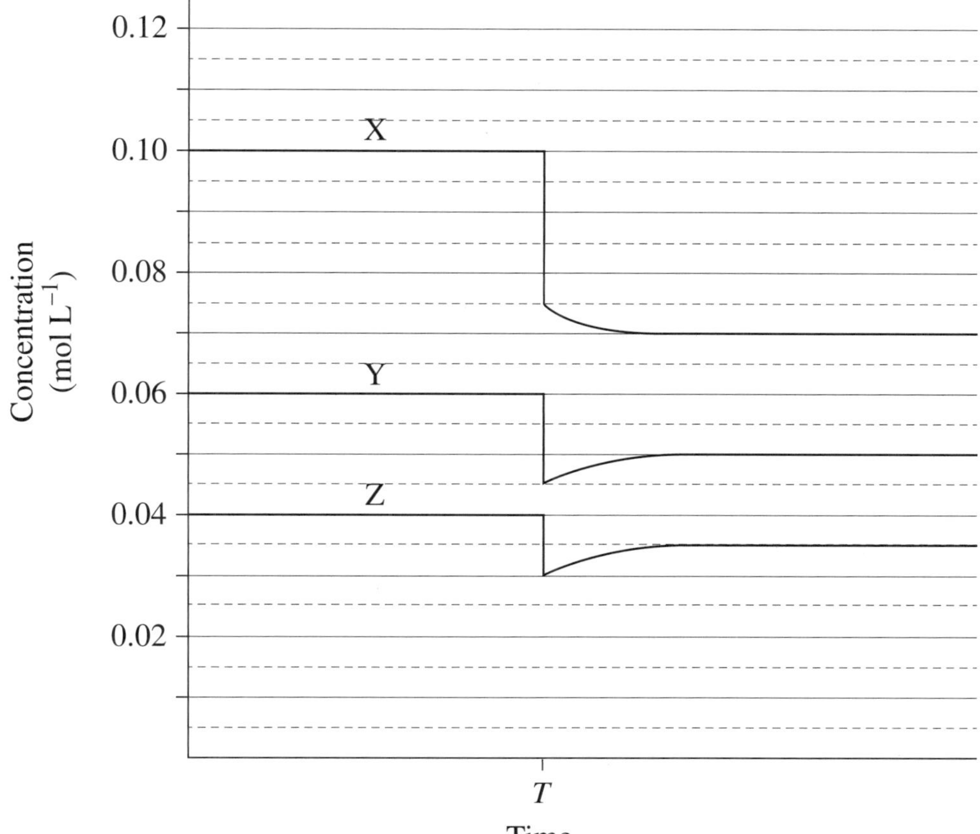

Which reaction is represented by the graph?

A. $X(g) + Y(g) \rightleftharpoons 2Z(g)$

B. $2X(g) \rightleftharpoons Y(g) + Z(g)$

C. $2X(g) \rightleftharpoons Y(g) + 3Z(g)$

D. $X(g) \rightleftharpoons Y(g) + Z(g)$

19 The sulfate content of a fertiliser is 48% by mass. 1.20 g of this fertiliser is completely dissolved in water and an excess of $Ba(NO_3)_2(aq)$ is added.

What mass of precipitate would be formed?

A. 0.006 g

B. 0.58 g

C. 1.40 g

D. 1.57 g

20 20.0 mL of 0.020 mol L^{-1} barium hydroxide solution is added to 50.0 mL of 0.040 mol L^{-1} hydrochloric acid solution.

What is the pH of the final solution?

A. 0.2

B. 1.6

C. 1.8

D. 2.9

2017 | HIGHER SCHOOL CERTIFICATE EXAMINATION

Centre Number

Student Number

Chemistry

Section I Part B Answer Booklet

55 marks
Attempt Questions 21–30
Allow about 1 hour and 40 minutes for this part

Instructions

- Write your Centre Number and Student Number at the top of this page.
- Answer the questions in the spaces provided. These spaces provide guidance for the expected length of response.
- Extra writing space is provided at the back of this booklet. If you use this space, clearly indicate which question you are answering.
- Show all relevant working in questions involving calculations.

Question 21 (5 marks)

(a) Outline ONE effect of ozone in the troposphere and ONE in the stratosphere. **2**

....

....

....

....

(b) Qualitatively compare TWO properties of oxygen (O_2) and ozone (O_3). **2**

....

....

....

(c) Using ONE chemical equation, show how a chlorine radical (Cl•) reacts with ozone. **1**

....

Question 22 (5 marks)

Atomic absorption spectroscopy was used to determine the concentration of zinc in a water sample. The absorbance of a series of standard solutions of known concentration of zinc was measured. The results are shown in the table.

Zinc concentration (ppm)	0.00	1.00	2.00	3.00	4.00	5.00
Absorbance	0.00	0.17	0.34	0.48	0.65	0.83

(a) Plot the data on the grid and draw a line of best fit. **3**

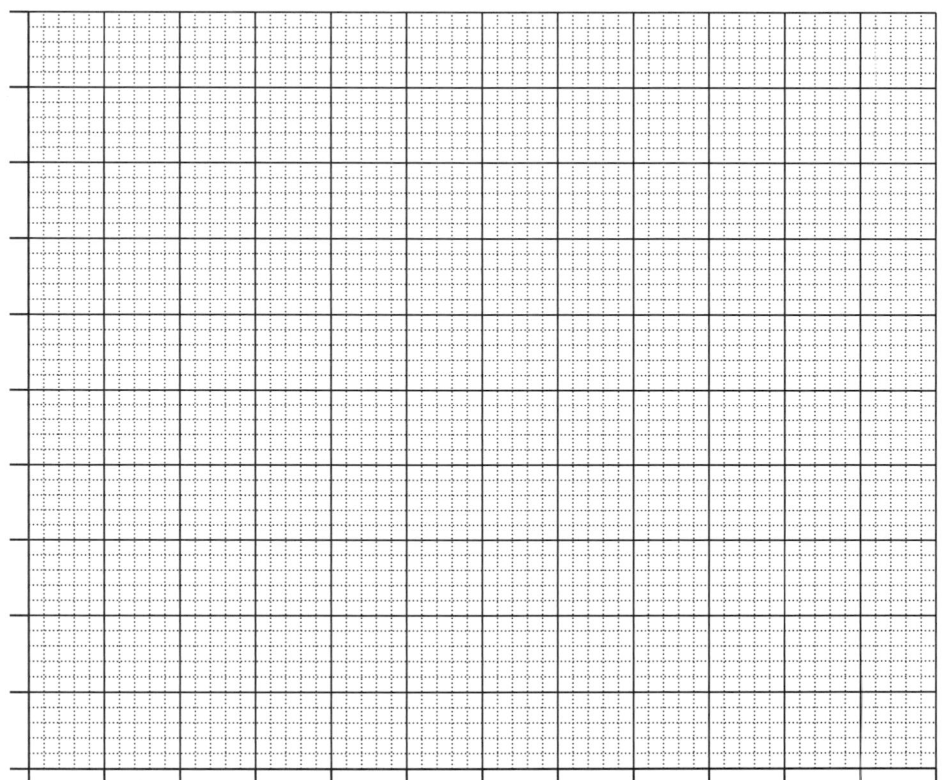

(b) In order for water to be considered safe for drinking, the concentration of zinc must be less than 2.80 ppm. **2**

The absorbance of the water sample was 0.58. Explain whether this water is safe for drinking.

....

....

....

....

Question 23 (6 marks)

The diagram shows a galvanic cell.

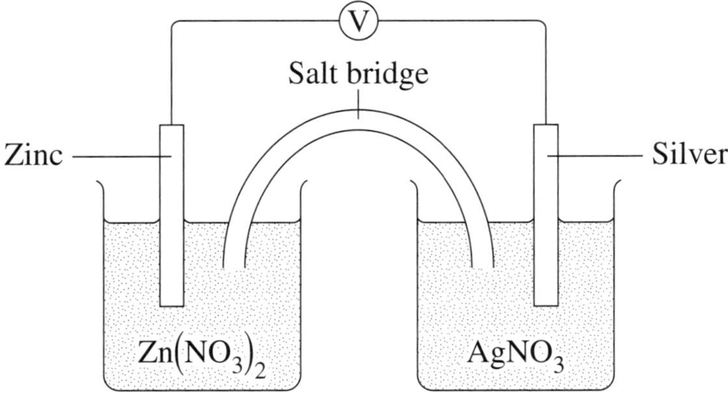

(a) Explain why a salt bridge is required. **2**

....

....

....

....

(b) The initial mass of each electrode was 10.0 g. After some time, the electrodes were removed from the solutions, dried and reweighed. The mass of the zinc electrode had changed by 1.00 g. **4**

Calculate the new mass of the silver electrode.

....

....

....

....

....

....

....

....

....

....

....

....

Question 24 (5 marks)

A solution of sodium hydroxide was titrated against a standardised solution of acetic acid which had a concentration of 0.5020 mol L^{-1}.

(a) The end point was reached when 19.30 mL of sodium hydroxide solution had been added to 25.00 mL of the acetic acid solution. **3**

Calculate the concentration of the sodium hydroxide solution.

....

....

....

....

....

....

....

....

....

(b) Explain why the pH of the resulting salt solution was not 7. Include a relevant chemical equation in your answer. **2**

....

....

....

....

....

....

Question 25 (4 marks)

Outline the steps that can be used to convert cellulose to polyethylene. Include relevant chemical equations in your answer. **4**

....

....

....

....

....

....

....

....

....

....

....

....

Question 26 (7 marks)

(a) Outline TWO reasons why the release of sulfur dioxide into the atmosphere is a concern. **3**

....

....

....

....

....

....

....

....

....

Question 26 continues

Question 26 (continued)

(b) The diagram shows the concentrations of sulfur dioxide (SO_2) in the lower atmosphere above a landmass. High concentrations of sulfur dioxide were recorded above a metal smelter and coal fired power stations. **4**

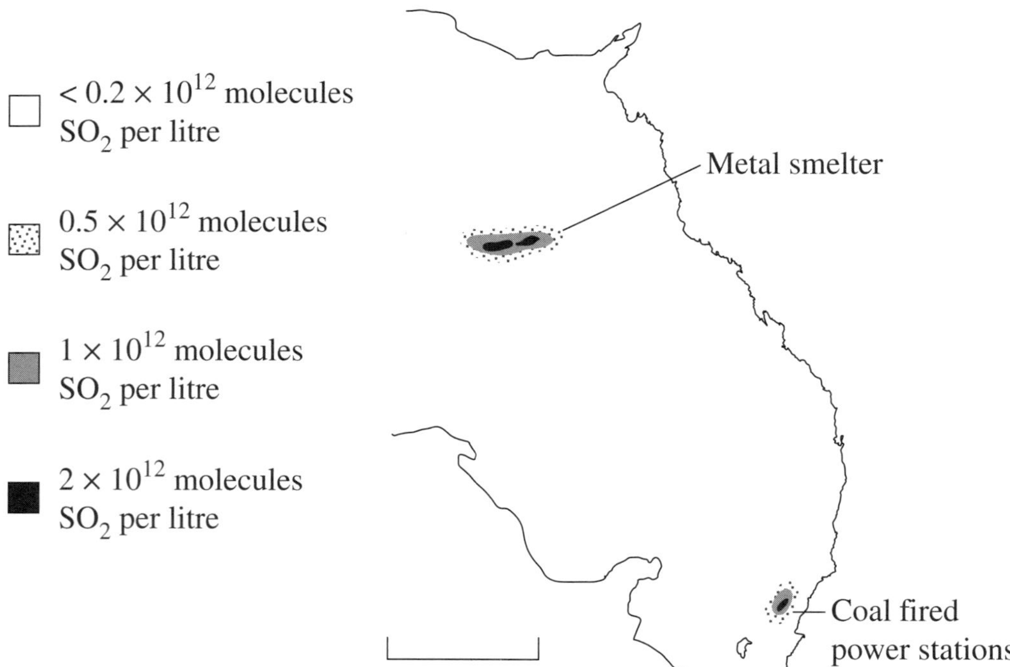

Explain the varying concentrations of sulfur dioxide shown on the map. Include relevant chemical equations in your answer.

....

....

....

....

....

....

....

....

....

....

....

....

End of Question 26

Question 27 (5 marks)

The boiling points and molar masses of three compounds are shown in the table. **5**

Compound	*Boiling point* (°C)	*Molar mass* (g mol^{-1})
Acetic acid	118	60
Butan-1-ol	117	74
Butyl acetate	116	116

Acetic acid, butan-1-ol and butyl acetate have very different molar masses but similar boiling points. Explain why in terms of the structure and bonding of the three compounds.

....

....

....

....

....

....

....

....

....

....

....

....

....

....

....

Question 28 (7 marks)

(a) Outline TWO advantages and TWO disadvantages of using ethanol as an alternative fuel for motor vehicles. **4**

....

....

....

....

....

....

....

....

....

....

....

....

(b) The molar heat of combustion (ΔH_c) for ethanol is 1360 kJ mol^{-1}. **3**

Calculate the energy generated per kg of CO_2 released by the combustion of ethanol.

....

....

....

....

....

....

....

....

....

Question 29 (4 marks)

Rivertown sits at the junction of two rivers. **4**

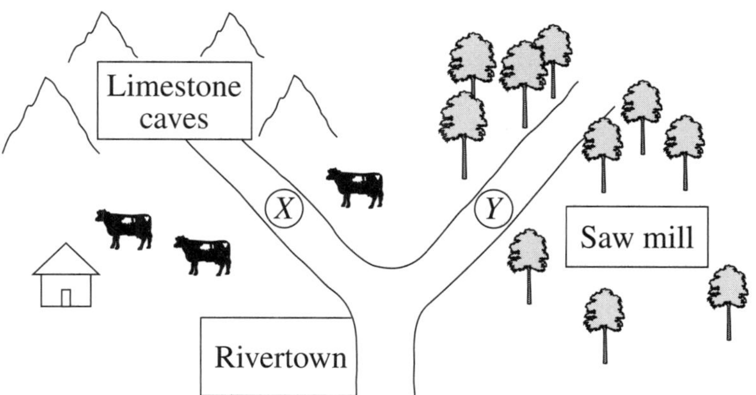

A simple water purification system has been purchased for the town water supply. It consists ONLY of a sedimentation tank, pH control, sand filters and a chlorination facility.

The system is to draw water either from Site *X* or Site *Y*.

A water chemist has obtained the following data from each site.

Factor	*X*	*Y*
Turbidity (NTU)	2	400
pH	7.3	6.0
Calcium (ppm)	120	5
Phosphate (ppm)	1.00	0.0001

Question 29 continues

Question 29 (continued)

With reference to the information provided, justify which of the sites, *X* or *Y*, would be the preferred water source for the town water supply.

....

....

....

....

....

....

....

....

....

....

....

....

....

....

....

....

End of Question 29

Question 30 (7 marks)

Analyse the conditions required to optimise the production of ammonia using the Haber process. **7**

2017 | HIGHER SCHOOL CERTIFICATE EXAMINATION

Chemistry

Section II

25 marks
Attempt ONE question from Questions 31–35
Allow about 45 minutes for this section

Answer parts (a)–(d) of one question in the Section II Writing Booklet. Extra writing booklets are available.

Show all relevant working in questions involving calculations.

Question 31 Industrial Chemistry

Question 32 Shipwrecks, Corrosion and Conservation

Question 33 The Biochemistry of Movement *(Not included in this reproduction)*

Question 34 The Chemistry of Art *(Not included in this reproduction)*

Question 35 Forensic Chemistry

Question 31 — Industrial Chemistry (25 marks)

Answer parts (a) and (b) of the question on pages 2–4 of the Section II Writing Booklet. Start each part of the question on a new page.

(a) (i) Write an equation using structural formulae to describe saponification. Use ethyl propanoate as one of the reactants. **2**

(ii) Describe a procedure that can be used to carry out saponification and test the product in a school laboratory. **3**

(b) The Contact process for the production of sulfuric acid includes a step whereby sulfur dioxide is converted to sulfur trioxide in an equilibrium reaction:

$$SO_2(g) + \frac{1}{2}O_2(g) \rightleftharpoons SO_3(g) \quad \Delta H = -99 \text{ kJ mol}^{-1}$$

SO_2 and O_2 were added to a closed container. The production of SO_3 over time is shown on the graph below.

(i) Copy the graph below into your answer booklet and sketch a second curve on the same axes to demonstrate the production of SO_3 over time when the reaction is repeated at a higher temperature. Clearly label the two curves. **2**

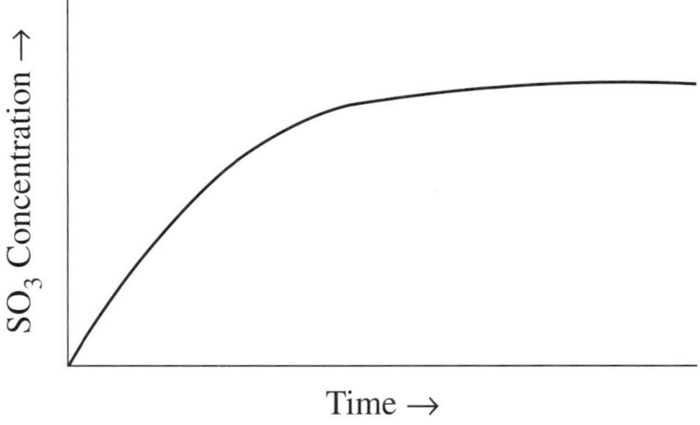

(ii) At a certain temperature, the equilibrium constant, *K*, is 12.1 for this reaction as written in the equation above. **4**

At the same temperature, 1.0 mol SO_2 and 1.0 mol O_2 were added to a 1.0 litre closed container. At a point in time, the concentration of SO_3 in the container was measured as 0.70 mol L^{-1}.

Had equilibrium been reached in the container at this point? Use calculations to justify your answer.

Question 31 continues

Question 31 (continued)

Answer parts (c) and (d) of the question on pages 5–8 of the Section II Writing Booklet. Start each part of the question on a new page.

(c) The raw materials required to produce sodium carbonate by the Solvay process are limestone and brine. Part of the process is represented by the flow chart.

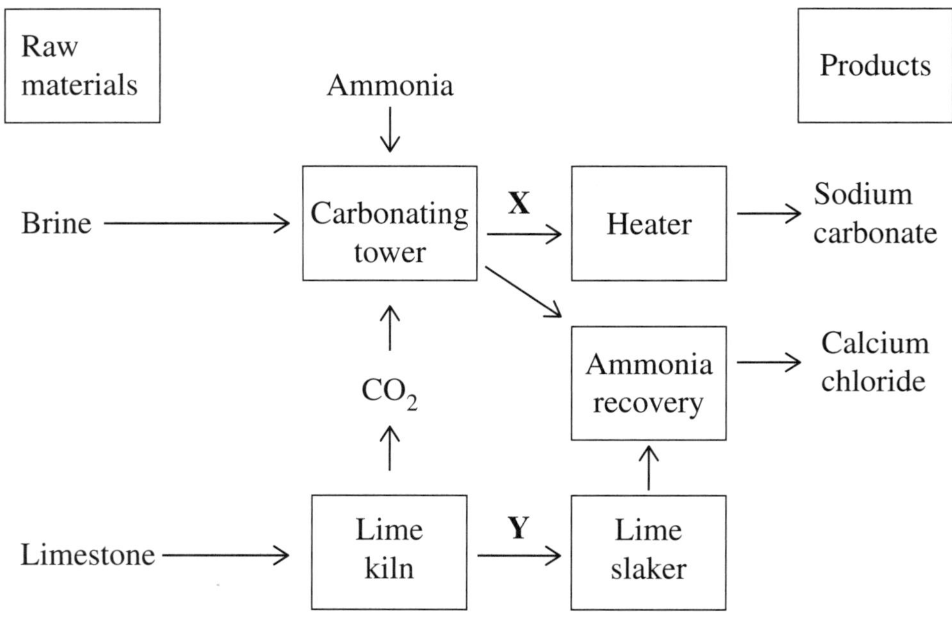

(i) Name the products labelled **X** and **Y**. **2**

(ii) How is the brine purified? Include a relevant chemical equation in your answer. **2**

(iii) Explain the role of ammonia in the manufacture of sodium carbonate. Include relevant chemical equation(s) in your answer. **3**

(d) Assess the extent to which technological advances have overcome the technical and environmental issues associated with the industrial production of sodium hydroxide. **7**

End of Question 31

Question 32 — Shipwrecks, Corrosion and Conservation (25 marks)

Answer parts (a) and (b) of the question on pages 2–4 of the Section II Writing Booklet. Start each part of the question on a new page.

(a) A first-hand investigation is to be carried out to compare the rate of corrosion of metals in acidic and neutral solutions.

(i) Outline a suitable procedure for this investigation. **2**

(ii) Explain the expected result. Include relevant half equations in your answer. **3**

(b) (i) Outline TWO different ways to increase the rate of an electrolysis reaction. **2**

(ii) When two carbon electrodes were placed in a blue, 3 mol L^{-1} solution of $CuCl_2$ and connected to a 6 V DC power supply, bubbles started appearing at the anode. After some time, the mass of the cathode had increased and the intensity of the blue colour had decreased. **4**

Account for these observations. Include relevant chemical equations in your answer.

Answer parts (c) and (d) of the question on pages 5–8 of the Section II Writing Booklet. Start each part of the question on a new page.

(c) In 1912, the ship *Titanic* sank in water over 3 km deep. The wreck was discovered in 1985.

(i) Why did scientists predict that the rate of corrosion would have been slow at this depth? **2**

(ii) Explain why extra precautions need to be taken when retrieving wooden artefacts from sea water compared to fresh water. **2**

(iii) Describe how electrolysis can be used to remove chloride ions from a metallic artefact. Support your answer with a labelled diagram. **3**

(d) Analyse the effect that the work of Volta and Davy had in reducing corrosion of ocean-going vessels. **7**

Question 35 — Forensic Chemistry (25 marks)

Answer parts (a) and (b) of the question on pages 2–4 of the Section II Writing Booklet. Start each part of the question on a new page.

(a) (i) Outline a test that could be carried out in a school laboratory to distinguish between sodium carbonate and starch. **2**

(ii) The emission spectrum of sodium is shown. **3**

Wavelength (nm)

400 450 500 550 600 650 700

467 515 588 616

With reference to atomic structure, explain how the lines on the emission spectrum are produced.

(b) (i) A group of students tried to use chromatography to separate the pigments extracted from a plant leaf. They put a small spot of the pigments on a strip of paper and drew a line across the paper to show the starting point. They then allowed solvent to run up the paper. **2**

Explain why a pencil rather than a pen should be used to draw the line.

(ii) The students then tried to use chromatography to separate an alkanol and an alkanoic acid. They found that the spot of the alkanol and alkanoic acid mixture did not move when an alkane was used as the solvent. **4**

Justify a modification to the experiment that would lead to the separation of the alkanol and the alkanoic acid.

Question 35 continues

Question 35 (continued)

Answer parts (c) and (d) of the question on pages 5–8 of the Section II Writing Booklet. Start each part of the question on a new page.

(c) Electrophoresis of proteins can be used to test for performance enhancing drugs.

(i) Using structural formulae, draw a chemical equation to show how a protein fragment can become positively charged in a low pH solution. **2**

(ii) Explain how electrophoresis is used to separate and identify proteins. Include a labelled diagram in your answer. **3**

(iii) Mass spectrometry can be used to identify other drugs in a blood sample. Below is a spectrum obtained from the analysis of a blood sample. **2**

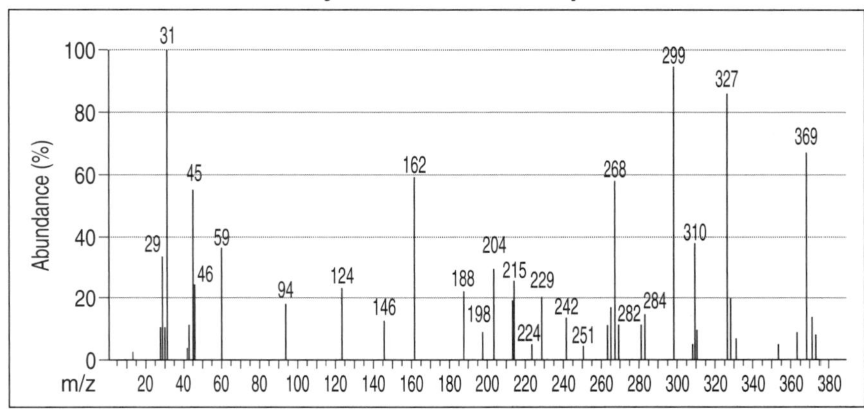

The mass spectra of four substances (Diagrams 1 to 4) are shown on the next page.

Using the diagrams, deduce which of the substances are present in the blood sample. Justify your answer.

(d) Discuss the use of DNA analysis in finding lost relatives. Include the relevant underlying chemistry in your answer. **7**

Question 35 continues

Question 35 (continued)

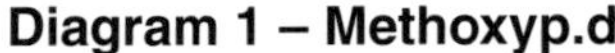

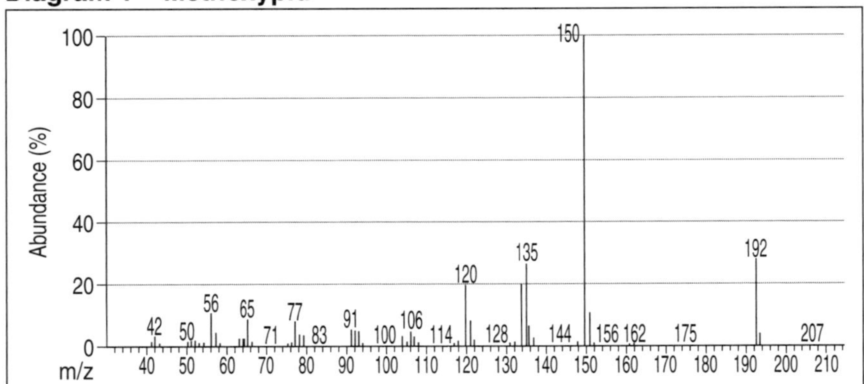

Diagram 2 – Trifluor.d

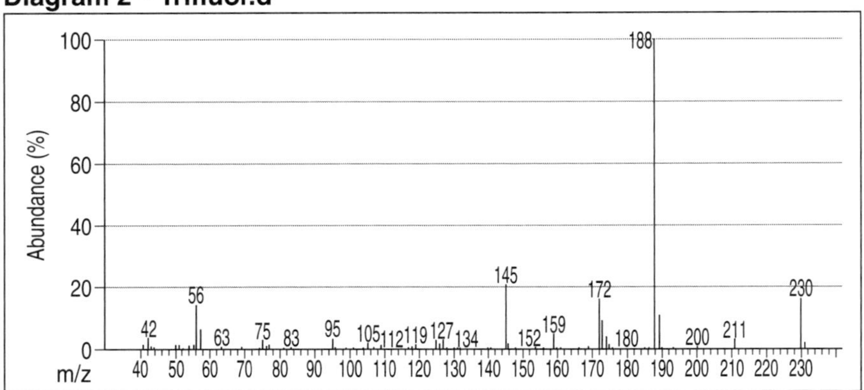

Diagram 3 – Heroin

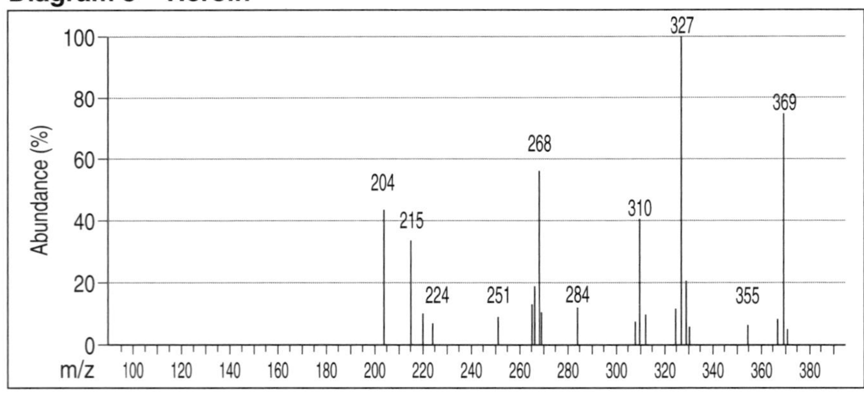

Diagram 4 – Ethanol

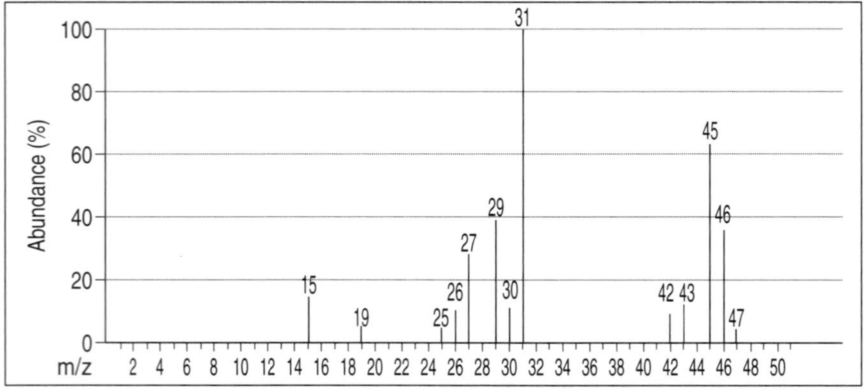

End of paper

2017 HSC Examination Paper

Sample Answers

Section I Part A

(Total 20 marks)

1 A Burettes deliver volumes to a 0.05 mL accuracy. B and D are incorrect as the scales are only approximate. C is incorrect as test tubes have no scales and are not used for measuring.

2 D Copper (II) ions produce blue-green flames. A is incorrect as barium ions produce yellow-green flames. B is incorrect as calcium ions produce orange-red flames. C is incorrect as carbonate ions do not produce a flame colour.

3 A 2-chloro-1-fluorobutane is correct as the hydrocarbon chain is numbered from right to left to give the lowest locant set (1, 2) and then the functional groups are named alphabetically. B is incorrect as the locant set is not the lowest set. C and D are incorrect as the functional groups are not named alphabetically.

4 D The use of a Bunsen burner flame is dangerous as the vapours produced during reflux are flammable and could catch fire if a leak of these vapours occurred. A heating mantle minimises this problem. A is incorrect as a stopper will lead to dangerous pressure build-up inside the apparatus. B is incorrect as the reaction mixture already contains concentrated sulfuric acid as a catalyst. C is incorrect as cooling water enters at the base of the reflux condenser where the vapours are hottest and greatest cooling is achieved.

5 A The hydrogen sulfate ion can donate a proton and form the sulfate ion as its conjugate base or it can accept a proton and form H_2SO_4 as its conjugate acid. B is incorrect as H_2SO_4 cannot accept a proton to form $H_3SO_4^+$. C is incorrect as sulfate ions do not have a proton to donate. D is incorrect as sulfurous acid cannot accept a proton to form $H_3SO_3^+$.

6 B Neptunium-239 (a transuranic element) is formed using a nuclear reactor to fire neutrons into a U-238 nucleus. A is incorrect as Co-60 is not a transuranic element although it is formed by neutron bombardment in a nuclear reactor. C is incorrect as U-238 is a natural radioactive element. D is incorrect as hassium-265 is produced using a particle accelerator by the collision of iron-58 ions with bismuth-208 nuclei.

7 C No decolourisation occurs with hexane whereas decolourisation occurs with hexene. The hydrocarbons do not mix with the bromine water and will float on top as they are less dense. A and D are incorrect as the liquids are immiscible. B is incorrect as the bromine water does not decolourise with alkanes but does decolourise with alkenes.

8 C Barium chloride is soluble and so no precipitate forms on the addition of NaCl solution to the barium nitrate solution whereas a white precipitate will form with the lead nitrate solution as lead chloride is insoluble. A is incorrect as both solutions will form white precipitates as barium sulfate and lead sulfate are insoluble. B is incorrect as all nitrates are soluble and no precipitates can form. D is incorrect as carbonate ions will precipitate both barium and lead ions.

9 D The black deposit is soot (carbon). A is incorrect as the ethane is not an alkanol and no soot is formed. B is incorrect as no soot is formed. C is incorrect as hydrogen gas is not a product of the combustion.

10 B In the dry cell and the lead-acid battery the cathode (X) is positive and the anode (Y) is negative. A and C are incorrect as X is not the anode. D is incorrect as the cathode in either cell is not negatively charged.

11 D The sulfur (oxidation state = 0) is oxidised to sulfur dioxide in which the oxidation state of the sulfur is +4. A is incorrect as the chromium atoms (not ions) in the dichromate ion exist in a +6 oxidation state and it is reduced to a +3 oxidation state in Cr_2O_3. B is incorrect as potassium ions (oxidation state = +1) do not change their oxidation state. C is incorrect as no oxide ions exist in either reactant—the oxygen atoms in the reactants and products do not change their oxidation state of –2.

12 D Methyl functional groups are produced along the polymer chain as addition polymerisation occurs. A and C are incorrect as double covalent bonds are still present. B is incorrect as this structure is polyethylene and not polypropene.

13 B Citric acid is triprotic and requires 3 moles of NaOH to neutralise 1 mole of citric acid. A and C are incorrect as both acids are monoprotic and the mole ratio of NaOH:acid = 1:1. D is incorrect as sulfuric acid is diprotic and requires 2 moles of NaOH to neutralise 1 mole of the sulfuric acid.

14 A Buffer solutions contain a weak acid and its conjugate base or a weak base and its conjugate acid. HCl is a strong acid and is completely dissociated to form hydronium ions and chloride ions. Hydronium ions and water will not produce a buffer solution. B is incorrect as the solution formed is not an effective buffer as a strong acid and the weak conjugate base (Cl^-) cannot be an effective buffer as no dynamic equilibrium exists because no HCl molecules exist in solution. C is incorrect as the neutrality of the NaCl is irrelevant to buffering; buffer solutions may be acidic or alkaline. D is incorrect as the change in pH would be very large and pH changes should be very small when small amounts of strong acids or bases are added. Buffering requires only small pH changes.

15 B This structure shows a coordinate bond between the N and O atoms. A, C and D do not show a coordinate bond and too many electrons are located around the central N atom.

16 B The NaOH will neutralise the carbonic acid and cause the equilibrium to shift to the right. Therefore the pressure of carbon dioxide gas will decrease as it reacts to replace some of the carbonic acid that was neutralised. A is incorrect as more carbon dioxide gas will raise the total gas pressure. C is incorrect as decreasing the volume will lead to a gas pressure increase (Boyle's Law). D is incorrect as an increase in temperature for this exothermic equilibrium will cause the equilibrium to shift to the left and raise the carbon dioxide pressure.

17 C $M(O_3) = 48.00$ g mol^{-1}; Molar volume (V_M) = 24.79 L mol^{-1};

$$d = \frac{M}{V_M} = \frac{48.00}{24.79} = 1.936 \text{ g L}^{-1}$$

Thus A, B and D are incorrect.

18 D At *T* the volume of the system has suddenly increased by 25% and the concentrations of X, Y and Z decreased by 25%. After *T* a new equilibrium is established by the following changes: X decreases by 0.005 mol L^{-1}; Y increases by 0.005 mol L^{-1}; and Z increases by 0.005 mol L^{-1}. Thus D is the correct response as the mole ratio is 1:1:1 in that equilibrium equation. The change at *T* has caused the equilibrium to shift to the right. A, B and C are incorrect as the mole ratio of reactants and products is incorrect.

19 C $m(SO_4^{2-}) = \left(\frac{48}{100}\right)(1.20) = 0.576$ g

$$n(SO_4^{2-}) = \frac{m}{M} = \frac{(0.576)}{(32.07 + 4 \times 16.00)} = \frac{0.576}{96.07} = 5.996 \times 10^{-3} \text{ mol}$$

$Ba^{2+}(aq) + SO_4^{2-}(aq) \rightarrow BaSO_4(s)$

$n(BaSO_4) = n(SO_4^{2-}) = 5.996 \times 10^{-3}$ mol

$m(BaSO_4) = nM = (5.996 \times 10^{-3})(233.37) = 1.40$ g

Thus A, B and D are incorrect.

20 C $n(Ba(OH)_2) = cV = (0.020)(0.0200) = 4.00 \times 10^{-4}$ mol

$n(HCl) = cV = (0.040)(0.0500) = 2.00 \times 10^{-3}$ mol

$Ba(OH)_2(aq) + 2HCl(aq) \rightarrow BaCl_2(aq) + 2H_2O(l)$

$n(Ba(OH)_2):n(HCl) = 1:2$

HCl is in excess. $n(HCl)$excess $= 2.00 \times 10^{-3} - 2(4.00 \times 10^{-4}) = 1.2 \times 10^{-3}$ mol

Excess H^+:$n(H^+) = 1.20 \times 10^{-3}$ mol

$$[H^+] = \frac{n}{V} = \frac{(1.20 \times 10^{-3})}{(0.020 + 0.050)} = 1.714 \times 10^{-2} \text{ mol/L}$$

$pH = -\log_{10}[H''] = -\log_{10}(1.714 \times 10^{-2}) = 1.8$

Thus A, B and D are incorrect.

Section I Part B

21 (a) *Troposphere*: Ozone occurs in very low concentrations (0.02 ppm) in the troposphere. At higher concentrations it is a lower atmosphere pollutant. It is very poisonous (> 20ppm) as it rapidly oxidises living tissue.

Stratosphere: Ozone absorbs UV radiation (UV-B and UV-C) and converts the radiation to heat energy which warms the stratosphere. A warm stratosphere overlying a cooler troposphere reduces the upward movement of gases from the troposphere. It also prevents the downward movement of ozone into the troposphere. The ozone layer acts as an ozone shield and reduces the UV level at the Earth's surface.

(2 marks)

(b) Ozone is a much more powerful oxidant than oxygen. Ozone can oxidise and damage living tissue. Ozone very rapidly oxidises metals to form metal oxides with the release of oxygen gas whereas oxygen oxidises reactive metals less rapidly.

$Mg(s) + O_3(g) \rightarrow MgO(s) + O_2(g)$

$2Mg(s) + O_2(g) \rightarrow 2MgO(s)$

Ozone has higher melting and boiling points than oxygen. Its density is also greater. Ozone is a polar molecule with a higher molecular weight than non-polar oxygen. Thus the intermolecular forces are greater between ozone than oxygen.

(2 marks)

(c) $O_3(g) + Cl\cdot(g) \rightarrow O_2(g) + ClO\cdot(g)$

(1 mark)

22 (a)

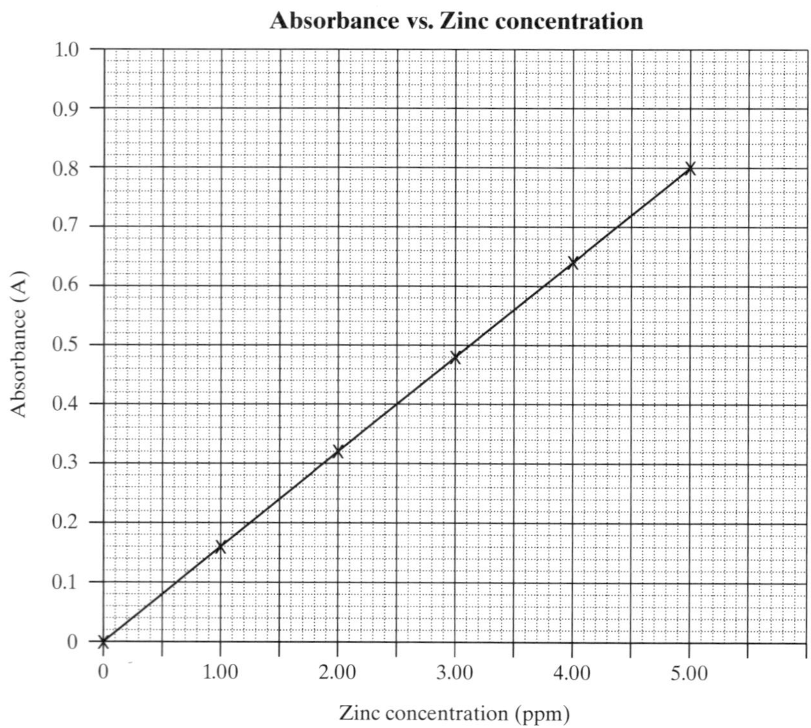

(3 marks)

(b) Interpolation from graph: $[Zn^{2+}] = 3.52$ ppm > 2.80 ppm. Therefore the water is not fit to drink.

(2 marks)

23 (a) A salt bridge completes the electric circuit by allowing the movement of ions between the two half-cells. Electrons travel through the copper wires in the external circuit and the charge transfer is completed by ion movement though the salt bridge. Positive ions move towards the cathode and negative ions move towards the anode.

(2 marks)

(b) The zinc anode oxidises: $Zn(s) \rightarrow Zn^{2+}(aq) + 2e^-$

The silver ion is reduced: $Ag^+(aq) + e^- \rightarrow Ag(s)$

Net equation: $Zn(s) + 2Ag^+(aq) \rightarrow Zn^{2+}(aq) + 2Ag(s)$

$$n(Zn) = \frac{m}{M} = \frac{1.00}{65.38} = 0.0153 \text{ mol}$$

$\therefore n(Ag)$ formed = 2(0.015 30) = 0.0306 mol

$m(Ag) = nM = (0.0306)(107.9) = 3.30$ g

Final mass of silver electrode = 10.0 + 3.30 = 13.3 g

(4 marks)

24 (a) $CH_3COOH(aq) + NaOH(aq) \rightarrow NaCH_3COO(aq) + H_2O(l)$

$n(CH_3COOH) = cV = (0.5020)(0.025\,00) = 0.012\,55$ mol

$n(NaOH) = n(CH_3COOH) = 0.012\,55$ mol

$$c(NaOH) = \frac{n}{V} = \frac{(0.012\,55)}{(0.019\,30)} = 0.6503 \text{ mol L}^{-1}$$

(3 marks)

(b) The acetate ion of the sodium acetate salt is a Brønsted–Lowry base in water.

$CH_3COO^-(aq) + H_2O(l) \rightarrow CH_3COOH(aq) + OH^-(aq)$

The acetic acid formed is a weak acid whereas the hydroxide ion is a strong base and the pH of the salt solution will be greater than 7 as the final solution is alkaline due to the OH^- ions.

(2 marks)

25 Cellulose is broken down into glucose monomers using cellulase enzymes or heating with dilute hydrochloric acid.

The glucose is extracted and anaerobically fermented using yeast to produce ethanol.

$C_6H_{12}O_6(aq) \rightarrow 2C_2H_5OH(aq) + 2CO_2(g)$

The ethanol solution is distilled to extract hydrous ethanol which is dehydrated using zeolite crystals to form anhydrous ethanol.

The ethanol is converted by dehydration to ethylene using concentrated sulfuric acid.

$C_2H_5OH(l) \rightarrow C_2H_4(g) + H_2O(l)$

The ethylene undergoes addition polymerisation to form polyethylene using an initiator or a catalyst.

(4 marks)

26 (a) Sulfur dioxide dissolves in rainwater to form acid rain. Sulfurous acid is formed. This acid can be oxidised by atmospheric oxygen to form sulfuric acid.

$SO_2(g) + H_2O(l) \rightarrow H_2SO_3(aq)$

$2H_2SO_3(aq) + O_2(g) \rightarrow 2H_2SO_4(aq)$

Acid rain has a pH <5 and causes serious environmental damage such as destruction of forests and chemical weathering of structures made of metals/alloys and marble in urban environments.

$CaCO_3(s) + H_2SO_4(aq) \rightarrow CaSO_4(s) + H_2O(l) + CO_2(g)$

Sulfur dioxide is poisonous. It irritates the nose, eyes and lungs. Sulfur dioxide can be produced naturally by volcanoes or by industries such as smelting of ores. Air quality standards in Australia state that sulfur dioxide levels should be less than 0.08 ppm over a 24-hour period.

(3 marks)

(b) Sulfide ore smelters produce sulfur dioxide when ores such as chalcopyrite are smelted to produce copper.

$2CuFeS_2(s) + 5O_2(g) \rightarrow 2Cu(l) + 2FeO(s) + 4SO_2(g)$

Sulfur dioxide is also produced when coal is burnt to generate electricity in power stations. Coal contains small amounts of sulfide minerals which are oxidised in the combustion process. For example, pyrite (FeS_2) is oxidised to form sulfur dioxide.

$4FeS_2(s) + 11O_2(g) \rightarrow 2Fe_2O_3(s) + 8SO_2(g)$

In the map the coal-fired power station is near the coast where sea breezes will assist in the dispersal and lowering of the sulfur dioxide levels. The metal smelter, however, is inland where wind levels are lower on average and the sulfur dioxide concentrations are higher than areas closer to the coast. The rest of the map shows very low sulfur dioxide levels in the atmosphere due to various factors including wind dispersal, dissolution in local waterways and precipitation in rain leading to acid rain production.

(4 marks)

27 Acetic acid is a highly polar molecule due to the presence of the carboxyl group (COOH). The dipole-dipole attractions and hydrogen bonding between acetic acid molecules leads to dimerisation and the boiling point is quite high even though the molar mass is relatively low.

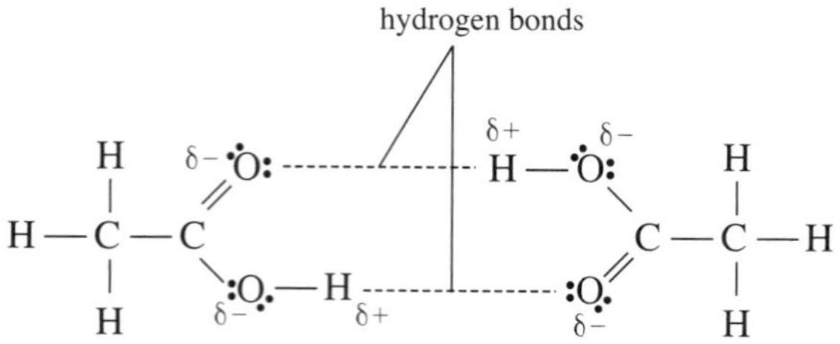

hydrogen bonds

The three organic molecules have similar strengths of dispersion forces as they have similar molar masses. The ester (butyl acetate) is the least polar of the three molecules and it does not hydrogen bond as no hydrogen atoms are bonded to the oxygen atoms. Consequently, its boiling point is reduced even though it has a much greater molar mass. Butan-1-ol is intermediate in molar mass and less polar than acetic acid but more polar than the ester. Therefore the intermolecular forces are intermediate. In butan-1-ol the dispersion forces between the hydrocarbon chains are the main intermolecular force apart from the dipole-dipole attractions.

(5 marks)

28 (a) *Advantages of ethanol as a fuel*

1. Ethanol burns more completely and efficiently than petrol or diesel. It reduces the net increase in atmospheric carbon dioxide compared with fossil fuels. Less CO and soot is also produced.
2. Ethanol is a renewable fuel derived from biomass whereas petrol or diesel are fossil fuels and non-renewable.

Disadvantages of ethanol as a fuel

1. Petrol or diesel release more energy (on a mole or kg basis) on complete combustion than ethanol. Therefore the driving range on a tank of ethanol is less than using petrol or diesel.
2. Vast areas of well-watered arable land are required to produce sufficient ethanol to meet demand. This quantity of arable land is not available. Arable land is used to produce food instead. The cost of production compared with petroleum fuels is also much greater.

(4 marks)

(b) $C_2H_5OH(l) + 3O_2(g) \rightarrow 2CO_2(g) + 3H_2O(l)$; $\Delta_c H = -1360 \text{ kJ mol}^{-1}$

1 mole of ethanol releases 2 moles of carbon dioxide on complete combustion

$M(CO_2) = (12.01 + (2 \times 16.00)) = 44.01 \text{ g mol}^{-1} = 0.04401 \text{ kg mol}^{-1}$

2 moles of carbon dioxide = 0.088 02 kg

Energy generated per kg of carbon dioxide produced

$$= \frac{1360}{0.088\,02}$$

$$= 1.55 \times 10^4 \text{ kJ}$$

(3 marks)

29 The turbidity of river Y is due to sawdust washed into the river from the sawmill.

The slightly higher pH of river X is due to the dissolution of limestone (a base) in water flowing out of the limestone caves. The slight acidity of water in river Y can be caused by acidic compounds present in the sawdust.

The significantly higher concentration of calcium ions in river X is also due to the dissolution of limestone in the caves.

The higher concentration of phosphate ions in river X is likely due to the farm on either side of river X.

The proposed water purification plant has some limited abilities to purify the water.

The desirable pH range for potable water of 6.5 to 8.5 can be achieved by the plant as it has a pH control system.

The sedimentation tank is used to allow suspended materials such as silt and sawdust to settle out prior to filtration.

The sand filter will remove any remaining silt and sawdust from the water leaving the sedimentation tank. Sand filters are often used to produce high-quality water without the use of chemicals, especially in rural areas.

The chlorination facility will kill microbes from either water source.

The water from river X will be hard as the calcium ion concentration is > 75ppm. Drinking water guidelines state that calcium ion concentrations should be less than 200 ppm to avoid scale build up in water heaters. The levels of calcium ions in both rivers is less than 200 ppm.

Phosphate levels in potable water should be less than 5 ppm and the levels in rivers X and Y are lower than 5 ppm.

Conclusions

The simple water purification system would not soften the water from river X but it would produce potable water for the town.

The purification system would remove sawdust from river Y using the settling tank and sand filter. The water produced would also be soft. The hardness of water from river X will prevent the lathering of soap during washing.

The pH of the water from river Y is readily adjusted to the desirable range.

High sawdust levels would require regular maintenance of the sand filter. Potable water can therefore be produced from river Y.

Consequently, sourcing water from river Y is a good strategy as the sawdust is removed and the water is soft.

(4 marks)

30 The manufacture of ammonia by the Haber process illustrates the industrial requirements of rapid production rates and high yields.

Increasing the reaction rate

Kinetic studies show that the rate of the ammonia synthesis reaction increases as the temperature increases. Hot molecules have greater kinetic energies to overcome the activation energy barrier between reactants and products.

Using a suitable catalyst

The porous iron catalyst lowers the activation energy for the reaction and so the reaction rate increases. The catalyst also allows lower temperatures to be used.

Equilibrium yield

The ammonia synthesis reaction is an exothermic equilibrium.

$N_2(g) + 3H_2(g) \leftrightharpoons 2NH_3(g)$ + heat

As the temperature increases, the equilibrium shifts back to the reactants to counteract the increase in temperature (Le Chatelier's principle). Thus an increase in temperature decreases the yield of ammonia. If the temperature is too high (> 700 °C), the yield of ammonia drops to less than 5% at 20 MPa.

The stoichiometry of the ammonia synthesis equilibrium shows that an increase in total pressure will cause the equilibrium to shift to the product side to counteract the change in pressure. The number of particles is reduced (4 to 2) by a shift to the ammonia side of the equilibrium. At 500 °C an increase in pressure from 20 MPa to 40 MPa causes the yield to increase from about 18% to about 32%. In engineering terms, however, the cost and maintenance of high-pressure equipment is very high.

The iron catalyst has no effect on the equilibrium yield. The catalyst reduces the time to reach equilibrium. Carbon monoxide must be removed from the feedstocks as it will poison the catalyst.

Compromise conditions

Because the kinetic and equilibrium effects are conflicting, a compromise set of conditions is used in the Haber process. The conditions used are:

- 1:3 mole ratio of nitrogen to hydrogen in reaction mixture
- high pressures (~18 MPa in Australia)
- moderate temperatures (450–500 °C in Australia)
- iron catalyst, which allows the reaction rate to remain high despite the moderate temperatures
- constant removal of ammonia (by liquefaction under pressure) as it forms to drive the equilibrium to the product side
- although the yield under the above condition is less than 20% per run, recycling of unused reactants occurs so that after 5 to 6 cycles nearly 100% of the reactants have been converted to ammonia.

Heat released by the exothermic process is not wasted. It is recycled back to heat up the reactant mixture as it enters the vessel.

(7 marks)

Section II—Options

Question 31—Industrial Chemistry

(a) (i) Saponification is the hydrolysis in basic conditions of a triacyl glyceride to give glycerol and the sodium salt of fatty acids. The reaction can be illustrated using ethyl propanoate as a model:

$$CH_3CH_2COOCH_2CH_3 + NaOH(aq) \longrightarrow CH_3CH_2COO^- Na^+ + HO{-}CH_2CH_3$$

(Note: No soap will form in the alkaline hydrolysis as propanoic acid is not a fatty acid.)

(2 marks)

(ii) Saponification can be modelled in the school laboratory using the following method:

1. Wear safety glasses and rubber gloves because both 4M NaOH (corrosive) and saturated NaCl solution (irritant) are harmful to eyes and skin.
2. Use a measuring cylinder to transfer 10 mL of 4M $NaOH(aq)$ to a large evaporating basin.
3. Add 10 mL of olive oil using a measuring cylinder.
4. Boil the mixture gently for 30 minutes using a Bunsen burner.
5. Carefully add water as necessary from a wash bottle to maintain approximately 20 mL.
6. When white solids have been formed allow the mixture to cool.
7. Once cool, add 20 mL of saturated $NaCl(aq)$ and gently stir.
8. Separate the white solid by filtration.
9. Wash the residue with cold, deionised water and allow it to dry. The residue is the desired product: soap.

Testing the product:

1. Add 20 mL of water to a 100 mL conical flask.
2. Add 5 mL of olive oil.
3. Stopper and shake, then leave to settle into two distinct layers.
4. Add one spatula-full of the soap product to the flask, stopper again and shake. Observe that an emulsion has formed, demonstrating that the product of the saponification reaction is able to emulsify oil and water.

(3 marks)

(b) (i)

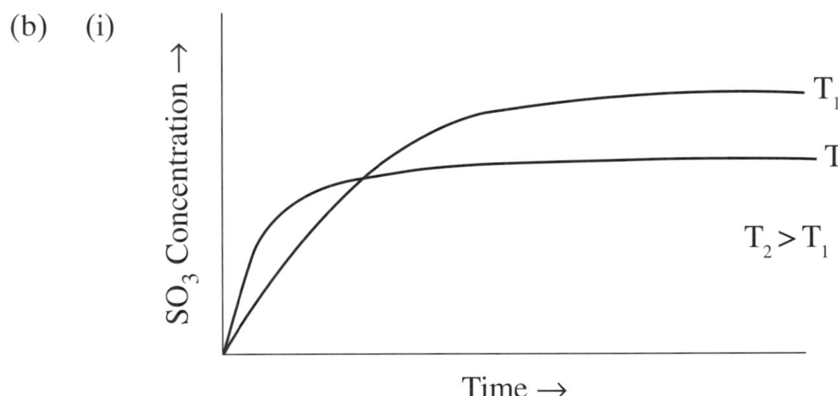

At T_2, higher than T_1 the rate is faster; therefore the equilibrium position is reached sooner.

However, because ΔH is negative, a higher temperature favours the reverse reaction so there will be a lower concentration of SO_3 in the equilibrium mixture at T_2.

(2 marks)

(ii)

	$SO_2(g)$	$+\frac{1}{2}O_2(g)$	$\rightleftharpoons$	$SO_3(g)$
Initial conc.	1.0	1.0		0
Change	–0.7	–0.35		+0.7
Conc. at time *t*	1.0 – 0.7 = **0.3**	1.0 – 0.35 = **0.65**		0.7

$$Q = \frac{[SO_3]}{[SO_2][O_2]^{\frac{1}{2}}}$$
$$= \frac{0.7}{0.3 \times 0.65^{\frac{1}{2}}}$$
$$= 2.89$$

Because *Q* does not equal *K* (12.1) the reaction had not reached equilibrium when the mixture was sampled.

(4 marks)

(c) (i) X is solid sodium hydrogen carbonate, $NaHCO_3(s)$.

Y is solid calcium oxide, $CaO(s)$.

(2 marks)

(ii) Brine (a concentrated solution of NaCl) may contain impurities of Ca^{2+} and Mg^{2+}. These are removed by precipitation followed by either filtration or flocculation. $Na_2CO_3(aq)$ is added to precipitate the Ca^{2+} ions:

$Ca^{2+}(aq) + CO_3^{2-}(aq) \rightarrow CaCO_3(s)$ and $NaOH(aq)$ is added to precipitate out the Mg^{2+} ions:

$Mg^{2+}(aq) + 2OH^-(aq) \rightarrow Mg(OH)_2(s)$.

(2 marks)

(iii) Ammonia is vital to the success of the Solvay process, even though it does not appear in the overall equation: $CaCO_3(s) + 2NaCl(aq) \rightarrow Na_2CO_3(s) + CaCl_2(s)$. Ammonia is added to the 'carbonating' (Solvay) tower, where it mixes with the brine and $CO_2(g)$. Carbon dioxide reacts with water to produce the weak acid, carbonic acid, which reacts with ammonia (a weak base) to produce ammonium ions and hydrogencarbonate ions: $NH_3(aq) + H_2CO_3(aq) \rightarrow NH_4^+(aq) + HCO_3^-(aq)$. Sodium and chloride ions are present in the brine, and when this mixture is cooled, $NaHCO_3(s)$ precipitates, leaving ammonium chloride in the solution. Thus ammonia is important because it facilitates the production of large amounts of hydrogencarbonate ions, which is necessary for the formation of sodium hydrogencarbonate in the 'carbonating' (Solvay) tower. [Note: There are other ways of representing the chemistry that results in the formation of Na^+, Cl^-, NH_4^+ and HCO_3^- ions in this mixture, using different equations.] *(3 marks)*

(d) Three methods of NaOH production have been used by the chemical industry, all of which use saturated NaCl solution (brine) as the feedstock. These methods are the mercury method, the diaphragm method and the membrane method (the most recent).

The mercury method of NaOH production uses an inert electrode as the anode, oxidising chloride ions to chlorine gas: $2Cl^- \rightarrow Cl_2(g) + 2e^-$. The cathode is liquid mercury, which is pumped around the electrolytic cell, into and out of the cathode compartment. In the cathode compartment, Na^+ ions are reduced to sodium metal which dissolves in the liquid mercury to form an amalgam: $Na^+ + e^- \rightarrow Na(Hg)$. The amalgam is pumped into a separate chamber where it is sprayed into water, where NaOH(aq) is formed: $2Na + 2H_2O(l) \rightarrow 2NaOH(aq) + H_2(g)$. The most significant technical issue in mercury plants is the leaking of mercury from the plant, which also has devastating environmental impacts because mercury is a neurotoxin and bio-accumulates in the environment. The process has large energy requirements, causing the production and release of large volumes of carbon dioxide from power stations, contributing to global warming.

The next method of NaOH production, the diaphragm method, represents a technological advance, with the incorporation of an asbestos membrane separating anode and cathode compartments in the electrolytic cell. In this method chloride ions are oxidised at an inert anode (steel mesh) to give chlorine gas and the reduction of water occurs directly at the cathode, producing hydrogen gas and hydroxide ions: $2H_2O(l) \rightarrow 2OH^-(aq) + H_2(g) + 2e^-$. The asbestos diaphragm separates the anode and cathode compartments, and allows sodium ions to pass through, producing NaOH in the cathode compartment. The inclusion of an asbestos membrane allows the elimination of mercury, thus eliminating the technical issue of mercury leaks and the associated environmental devastation. As a result the asbestos process represents a technological advance but it also has serious technical and environmental limitations of its own. Some sodium chloride also passes through the diaphragm. This introduces a different technical issue (reduced NaOH purity) and necessitates additional processes to remove the NaCl from the desired product. On the other hand this process has lower energy requirements, reducing CO_2 emissions. Most importantly though, the process uses asbestos. This has a huge impact on the working environment; workers exposed to asbestos develop asbestosis and lung cancer.

The most recent process is the membrane process which has relatively few technical and environmental issues associated with it compared to earlier processes. In the membrane process chloride ions are oxidised and water reduced at inert electrodes, with the two compartments separated by a highly selective Teflon-based membrane. Membrane technology is relatively new and ion-selective membranes represent an important technological advance. It allows only Na^+ ions to pass through, producing $NaOH(aq)$ in the cathode compartment and eliminating the technical issue of impure $NaOH$. It uses neither mercury nor asbestos and its energy requirements are relatively low.

Of the three methods, the membrane method—based on technological advancements in membrane chemistry—overcomes the technical issues associated with $NaOH$ production and eliminates most of the associated environmental issues; it is a great example of how technological developments can overcome issues in industrial chemistry.

(7 marks)

Question 32—Shipwrecks, Corrosion and Conservation

(a) (i) Metals to test: magnesium, zinc, iron, copper (use metal strips of similar size)

Solutions to test: water, 0.001 mol L^{-1} HCl

1. Ensure that all metals are clean and free of corrosion.
2. Place eight test tubes in test tube racks. Place 5 mL of deionised water in the first four tubes and 5 mL of dilute HCl in the last four tubes.
3. Place a sample of each metal in the water and the dilute acid.
4. Observe and record any changes to the surfaces of each metal during the first lesson as well as each lesson over the next week.

(2 marks)

(ii) Corrosion will occur faster in dilute HCl than in water. The magnesium is the most active of the four metals and will react rapidly. The magnesium dissolves in excess acid and hydrogen gas is released.

Zinc reacts less rapidly and iron relatively slowly.

$Mg(s) + 2HCl(aq) \rightarrow MgCl_2(aq) + H_2(g)$

In water the magnesium slowly reacts and small bubbles of hydrogen gas are seen on the surface.

$Mg(s) + H_2O(l) \rightarrow MgO(s) + H_2(g)$

Iron will rust in water forming hydrated iron (III) oxide.

$4Fe(s) + 2H_2O(l) + 3O_2(g) \rightarrow 2Fe_2O_3{\cdot}H_2O(s)$

In HCl the iron will slowly dissolve to produce a pale green solution of iron (II) ions which turn to yellow iron (III) ions due to oxidation by the dissolved oxygen.

The copper will not corrode in water or acid as it has a low activity (weak reductant).

(3 marks)

(b) (i) The rate of electrolysis can be increased by any two of the following:

- increasing the applied voltage
- increasing the concentration of the electrolyte being electrolysed
- increasing the surface area of the electrodes
- reducing the separation between the two electrodes. *(2 marks)*

(ii) Electrolysis of 3 mol L^{-1} $CuCl_2$

At the anode, water or chloride ions can be oxidised. Due to the high chloride ion concentration, oxidation of chloride ions most likely occurs and gaseous chlorine forms. (Note: Some oxygen may form if water is oxidised.)

Anode: $2Cl^-(aq) \rightarrow Cl_2(g) + 2e^-$

The cathode increases in mass as Cu is deposited. The blue colour fades in the solution as copper (II) ions are reduced.

Cathode: $Cu^{2+} + 2e^- \rightarrow Cu(s)$ *(4 marks)*

(c) (i) At great ocean depths, the following conditions exist:

- temperatures are very low (0 °C – 4 °C)
- oxygen levels are very low (1–3 ppm). In the ooze at the bottom, oxygen levels may be zero or almost zero due to the activity of various decomposer organisms.

Scientists made predictions about the rate of metal corrosion in the ocean depths:

- low temperatures should slow down the rate of corrosion as the interacting particles have lower kinetic energies
- low oxygen levels should slow down corrosion as oxygen is necessary for the formation of metal oxides.

Thus the wreck of the *Titanic* was expected to show little or no corrosion.

(2 marks)

(ii) Wooden artefacts collected from shipwrecks are saturated with mineral salts. Sulfate and chloride salts are the major contaminants. The cellular contents of wooden artefacts are leached out and replaced by salty solutions that eventually fill the spaces between the cellulose and lignin fibres. Eventually the cellulose also breaks down due to hydrolysis and the action of bacteria. When wooden artefacts are brought to the surface they are kept wet in sea water to prevent the destruction which occurs when they dry out. When a salt-saturated artefact is allowed to dry out, the salt crystals grow and cause weakening and distortion of the artefact. *(2 marks)*

(iii) The removal of chloride salts can be achieved using electrolysis. The metal artefact becomes the cathode in the electrolytic cell. A stainless steel mesh anode is used and the electrolyte consists of a dilute (~5%) sodium hydroxide solution. The half-reactions are:

Anode: $4OH^-(aq) \rightarrow 2H_2O(l) + O_2(g) + 4e^-$

Cathode: $2H_2O(l) + 2e^- \rightarrow H_2(g) + 2OH^-(aq)$.

Chloride ions continue to leach out of the artefact as they are attracted to the positive anode. The alkaline electrolyte helps to passivate any iron in the artefact. Hydrogen released at the cathode also provides a reducing environment for an iron artefact. The electrolysis may continue to remove the chloride ions for months or years.

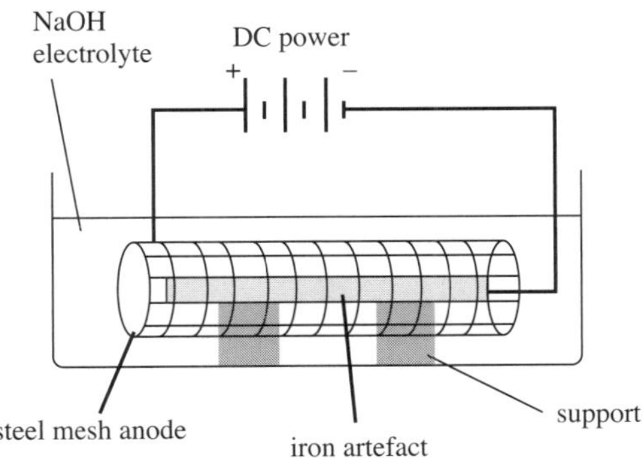

(3 marks)

(d) The reduction of corrosion in ocean-going vessels became very important as wooden ships were replaced by metal-hulled ships.

In 1792 Alessandro Volta investigated the effect of the contact between two metals in the presence of a conducting solution (electrolyte). He discovered that sparks jumped between the two metals when they were almost touching. He concluded that a force was generated in which one metal gave up the 'electric fluid' and the other metal received it. He determined that some metals were more active than others. For example, iron was more active than copper and reacted more rapidly in electrolyte solutions.

Humphry Davy continued Volta's experiments in relation to the prevention of corrosion in ships. He established that two metals in contact would not corrode if water was excluded. In 1824 he conducted experiments at the request of the British Admiralty to investigate why copper sheets attached to the wooden hulls of ships with iron nails led to corrosion of the nails and the detachment of the copper sheets. Davy's experiments showed that the problem was caused by a reaction between the two metals in contact with the sea water electrolyte. This was an example of galvanic corrosion in which the copper accelerated the corrosion of the iron nails. He proposed the attachment of pieces of zinc to the copper plates and the painting of the copper to reduce its surface area. Even though the Admiralty abandoned this solution as too impractical, Davy's conclusions were correct and were applied in later years when steel ships were built.

Davy's experiments demonstrated the principle of a sacrificial anode which protects the steel from corrosion. The zinc oxidises in preference to the iron. The iron becomes cathodically protected and does not oxidise.

Zinc anode: $Zn(s) \rightarrow Zn^{2+}(aq) + 2e^-$
Iron cathode: $2H_2O(l) + O_2(aq) + 4e^- \rightarrow 4OH^-(aq)$

Galvanic protection continues to be used on the hulls of modern steel ships.

(7 marks)

Question 35—Forensic Chemistry

(a) (i) Sodium carbonate can be distinguished from starch by the addition of a few drops of dilute acid to either a solid or aqueous sample. If the sample is sodium carbonate the addition of acid will cause gas bubbles to appear, as CO_2 is formed:

$Na_2CO_3(s) + 2HCl(aq) \rightarrow 2NaCl(aq) + H_2O(l) + CO_2(g)$.

To confirm that the other substance is starch a few drops of iodine solution can be added to an aqueous solution of the substance. The mixture will turn blue-black if the substance is starch.

(2 marks)

(ii) Emission spectra are produced when excited electrons relax back to their ground state and release energy. Atoms of each element have a unique number of protons and electrons, and thus electron shells and sub-shells are of different energies. When electrons are excited from the ground state to a higher energy level they absorb a quantum of energy exactly equivalent to the energy difference between the two energy levels. This quantum of energy is released when they relax back to the ground state and this gives rise to the discrete lines (at specific energies) in the emission spectrum. For example, in the case of the sodium emission spectrum the two very closely spaced lines at 588 nm correspond to transitions from the 3p to the 3s orbitals.

(3 marks)

(b) (i) Pen ink consists of pigments dissolved in a volatile carrier. These pigments will interact with the solvent and move up the paper, potentially interfering with the separation of the pigments in the plant leaf. The line left by a pencil is made of graphite which is insoluble in the solvent and will not move up the paper.

(2 marks)

(ii) Alkanols and alkanoic acids both contain highly polar hydroxyl groups (-OH). The hydrogen bonding between the molecules in the sample is too strong to be overcome by interactions with the non-polar alkane molecules, and so they do not move with the solvent through the stationary phase. Replacing the solvent with a polar solvent would allow their separation. The experiment should be repeated a number of times with different solvents or solvent combinations, such as water, water+ethanol or ethanol, to determine which mobile phase gives the best separation. Note that the spots will not be visible after the separation and another reagent will need to be sprayed on the dried chromatogram in order to make the spots visible and determine whether separation has been achieved.

(4 marks)

(c) (i)

$$HOOC{-}CH(R_1){-}NH{-}CO{-}CH(R_2){-}NH{-}CO{-}CH(R_3){-}NH_2 + 3H^+ \rightarrow HOOC{-}CH(R_1){-}NH_2^+{-}CO{-}CH(R_2){-}NH_2^+{-}CO{-}CH(R_3){-}NH_3^+$$

(Note: There are other correct examples, such as using a zwitter ion.)

(2 marks)

(ii) Proteins are biopolymers made of repeating amino acids. There are 22 naturally occuring amino acids, and thus there is a vast variation in protein primary structure. Mixtures of proteins can be separated and analysed using electrophoresis, a separation technique based on the relative movement of ions of different mass in an electric field. The 22 amino acids that make up proteins have different side chains and these ionise in solution to give protein molecules different overall charges, depending on the pH of the buffer used. The sample is placed in the middle of the supporting matrix (e.g. paper), which is then immersed in the buffer solution and connected to the terminals of a DC power source. The proteins with an overall negative charge will migrate towards the positive electrode and those that are positively charged overall to the negative electrode. Proteins are separated because of their different rates of migration towards the electrodes. Reference protein mixtures can be run alongside the sample so that after the separation and subsequent staining, identification of proteins in the mixture can be made.

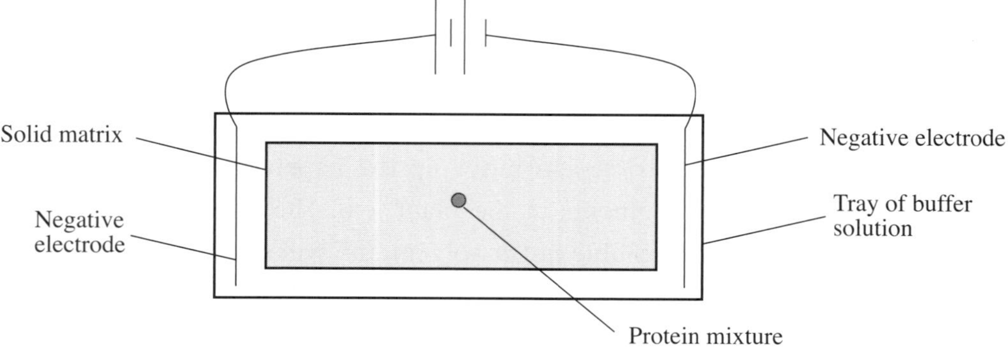

(3 marks)

(iii) The sample contains heroin, evidenced by the presence of the molecular ion at 369 and strong peaks at 327, 310 and 268.

It also contains ethanol, based on the presence of strong peaks at 46 (the molecular ion), 45, 31 and 29.

The lack of peaks for the molecular ions of methoxyp (at 192) and trifluor (at 230) exclude their presence from the sample.

(2 marks)

(d) DNA is a biopolymer made from repeating nucleotides, which consist of a phosphate group, the sugar deoxyribose and one of four nitrogenous bases (nucleobases): adenosine, thymine, guanine and cytosine. These nucleotides bond together via the phosphate and sugar groups. The bases on one strand of DNA bond in a complementary fashion with those on a second strand (A with T, G with C), resulting in a double-stranded DNA molecule with a helical overall shape (the DNA double-helix). Certain specific sequences of bases on the DNA strand (a chromosome) are known as genes, and provide the molecular code for the construction of proteins. The majority of DNA does not consist of genes, and the base sequences do not code for proteins.

Although all humans have the majority of their nucleobase sequence in common, there are sections of the DNA called exons that are unique to each person. These *exons* have small differences in the base sequence that lead to differences in appearance and bodily characteristics. *Introns* are other nucleobase sequences between the exons. Family members have similarities in their introns. Children inherit half their introns from each parent. Fewer introns are shared when people are more distantly related. DNA profiling (also known as 'DNA fingerprinting') involves the comparison of repeated nucleobase sequences on these introns.

DNA is first treated using the polymerase chain reaction, which involves breaking the cell membranes in a biological sample such as saliva, and separating the DNA from the rest of the cell by centrifugation. The DNA sample is heated to separate the strands, and enzymes are added which build a new strand of DNA based on the original strand, using a sequence of nucleotides with complementary bases. Repeating this process many times produces sufficient DNA for analysis. Analysis consists of cutting the DNA into fragments at specific locations using restriction enzymes and separating these fragments using electrophoresis. Once separated the DNA fragments are stained and compared to reference fragments electrophoresed under identical conditions, such as a sample taken from a crime scene or from a potential relative. Closely related individuals such as family members have more similarities in their DNA than unrelated or distantly related people and, as a result, DNA analysis can be used to identify lost relatives.

For people who do not know their biological parent(s) the psychological benefit of their discovery is often enormous, as is the discovery of other unknown close relatives such as lost siblings. There can also be positive outcomes if it allows illnesses to be avoided or treated more effectively. On the other hand the unexpected discovery of lost relatives could cause emotional stress in an adoptive family, financial disputes could arise from the appearance of an unknown relative at the time a will comes into effect and biological parents who have chosen to be anonymous at the time of conception may still want and expect their anonymity to be respected. As a result DNA analysis to find lost relatives cannot be assumed to be a welcome thing for all individuals.

(7 marks)

Acknowledgements

Graph Q10, 2003 HSC, *Chemistry of the Atmosphere*, 3rd ed. (2000), Richard P Wayne, Oxford University Press, Cornwell, UK

Diagram Q34(d), 2003 HSC, University of Virginia

Graph Q33(a)(ii), 2003 HSC, *Chemistry* 2nd ed. (2002), CE Housecroft and EC Constable, Prentice Hall

Image Q30, 2013 HSC, NASA

2017 | HIGHER SCHOOL CERTIFICATE EXAMINATION

Chemistry

DATA SHEET

Avogadro constant, N_A .... 6.022×10^{23} mol^{-1}

Volume of 1 mole ideal gas: at 100 kPa and
at 0°C (273.15 K) .... 22.71 L
at 25°C (298.15 K) .... 24.79 L

Ionisation constant for water at 25°C (298.15 K), K_w .... 1.0×10^{-14}

Specific heat capacity of water .... 4.18×10^3 J kg^{-1} K^{-1}

Some useful formulae

$\mathrm{pH} = -\log_{10}[\mathrm{H}^+]$ $\Delta H = -mC\Delta T$

Some standard potentials

$K^+ + e^-$	$\rightleftharpoons$	$K(s)$	–2.94 V
$Ba^{2+} + 2e^-$	$\rightleftharpoons$	$Ba(s)$	–2.91 V
$Ca^{2+} + 2e^-$	$\rightleftharpoons$	$Ca(s)$	–2.87 V
$Na^+ + e^-$	$\rightleftharpoons$	$Na(s)$	–2.71 V
$Mg^{2+} + 2e^-$	$\rightleftharpoons$	$Mg(s)$	–2.36 V
$Al^{3+} + 3e^-$	$\rightleftharpoons$	$Al(s)$	–1.68 V
$Mn^{2+} + 2e^-$	$\rightleftharpoons$	$Mn(s)$	–1.18 V
$H_2O + e^-$	$\rightleftharpoons$	$\frac{1}{2}H_2(g) + OH^-$	–0.83 V
$Zn^{2+} + 2e^-$	$\rightleftharpoons$	$Zn(s)$	–0.76 V
$Fe^{2+} + 2e^-$	$\rightleftharpoons$	$Fe(s)$	–0.44 V
$Ni^{2+} + 2e^-$	$\rightleftharpoons$	$Ni(s)$	–0.24 V
$Sn^{2+} + 2e^-$	$\rightleftharpoons$	$Sn(s)$	–0.14 V
$Pb^{2+} + 2e^-$	$\rightleftharpoons$	$Pb(s)$	–0.13 V
$H^+ + e^-$	$\rightleftharpoons$	$\frac{1}{2}H_2(g)$	0.00 V
$SO_4^{2-} + 4H^+ + 2e^-$	$\rightleftharpoons$	$SO_2(aq) + 2H_2O$	0.16 V
$Cu^{2+} + 2e^-$	$\rightleftharpoons$	$Cu(s)$	0.34 V
$\frac{1}{2}O_2(g) + H_2O + 2e^-$	$\rightleftharpoons$	$2OH^-$	0.40 V
$Cu^+ + e^-$	$\rightleftharpoons$	$Cu(s)$	0.52 V
$\frac{1}{2}I_2(s) + e^-$	$\rightleftharpoons$	I^-	0.54 V
$\frac{1}{2}I_2(aq) + e^-$	$\rightleftharpoons$	I^-	0.62 V
$Fe^{3+} + e^-$	$\rightleftharpoons$	Fe^{2+}	0.77 V
$Ag^+ + e^-$	$\rightleftharpoons$	$Ag(s)$	0.80 V
$\frac{1}{2}Br_2(l) + e^-$	$\rightleftharpoons$	Br^-	1.08 V
$\frac{1}{2}Br_2(aq) + e^-$	$\rightleftharpoons$	Br^-	1.10 V
$\frac{1}{2}O_2(g) + 2H^+ + 2e^-$	$\rightleftharpoons$	H_2O	1.23 V
$\frac{1}{2}Cl_2(g) + e^-$	$\rightleftharpoons$	Cl^-	1.36 V
$\frac{1}{2}Cr_2O_7^{2-} + 7H^+ + 3e^-$	$\rightleftharpoons$	$Cr^{3+} + \frac{7}{2}H_2O$	1.36 V
$\frac{1}{2}Cl_2(aq) + e^-$	$\rightleftharpoons$	Cl^-	1.40 V
$MnO_4^- + 8H^+ + 5e^-$	$\rightleftharpoons$	$Mn^{2+} + 4H_2O$	1.51 V
$\frac{1}{2}F_2(g) + e^-$	$\rightleftharpoons$	F^-	2.89 V

Aylward and Findlay, *SI Chemical Data* (5th Edition) is the principal source of data for this examination paper. Some data may have been modified for examination purposes.

PERIODIC TABLE OF THE ELEMENTS

KEY

Atomic Number	79
Symbol	Au
Standard Atomic Weight	197.0
Name	Gold

1 H 1.008 Hydrogen																	2 He 4.003 Helium
3 Li 6.941 Lithium	4 Be 9.012 Beryllium											5 B 10.81 Boron	6 C 12.01 Carbon	7 N 14.01 Nitrogen	8 O 16.00 Oxygen	9 F 19.00 Fluorine	10 Ne 20.18 Neon
11 Na 22.99 Sodium	12 Mg 24.31 Magnesium											13 Al 26.98 Aluminium	14 Si 28.09 Silicon	15 P 30.97 Phosphorus	16 S 32.07 Sulfur	17 Cl 35.45 Chlorine	18 Ar 39.95 Argon
19 K 39.10 Potassium	20 Ca 40.08 Calcium	21 Sc 44.96 Scandium	22 Ti 47.87 Titanium	23 V 50.94 Vanadium	24 Cr 52.00 Chromium	25 Mn 54.94 Manganese	26 Fe 55.85 Iron	27 Co 58.93 Cobalt	28 Ni 58.69 Nickel	29 Cu 63.55 Copper	30 Zn 65.38 Zinc	31 Ga 69.72 Gallium	32 Ge 72.64 Germanium	33 As 74.92 Arsenic	34 Se 78.96 Selenium	35 Br 79.90 Bromine	36 Kr 83.80 Krypton
37 Rb 85.47 Rubidium	38 Sr 87.61 Strontium	39 Y 88.91 Yttrium	40 Zr 91.22 Zirconium	41 Nb 92.91 Niobium	42 Mo 95.96 Molybdenum	43 Tc Technetium	44 Ru 101.1 Ruthenium	45 Rh 102.9 Rhodium	46 Pd 106.4 Palladium	47 Ag 107.9 Silver	48 Cd 112.4 Cadmium	49 In 114.8 Indium	50 Sn 118.7 Tin	51 Sb 121.8 Antimony	52 Te 127.6 Tellurium	53 I 126.9 Iodine	54 Xe 131.3 Xenon
55 Cs 132.9 Caesium	56 Ba 137.3 Barium	57–71 Lanthanoids	72 Hf 178.5 Hafnium	73 Ta 180.9 Tantalum	74 W 183.9 Tungsten	75 Re 186.2 Rhenium	76 Os 190.2 Osmium	77 Ir 192.2 Iridium	78 Pt 195.1 Platinum	79 Au 197.0 Gold	80 Hg 200.6 Mercury	81 Tl 204.4 Thallium	82 Pb 207.2 Lead	83 Bi 209.0 Bismuth	84 Po Polonium	85 At Astatine	86 Rn Radon
87 Fr Francium	88 Ra Radium	89–103 Actinoids	104 Rf Rutherfordium	105 Db Dubnium	106 Sg Seaborgium	107 Bh Bohrium	108 Hs Hassium	109 Mt Meitnerium	110 Ds Darmstadtium	111 Rg Roentgenium	112 Cn Copernicium	113 Nh Nihonium	114 Fl Flerovium	115 Mc Moscovium	116 Lv Livermorium	117 Ts Tennessine	118 Og Oganesson

Lanthanoids

57 La 138.9 Lanthanum	58 Ce 140.1 Cerium	59 Pr 140.9 Praseodymium	60 Nd 144.2 Neodymium	61 Pm Promethium	62 Sm 150.4 Samarium	63 Eu 152.0 Europium	64 Gd 157.3 Gadolinium	65 Tb 158.9 Terbium	66 Dy 162.5 Dysprosium	67 Ho 164.9 Holmium	68 Er 167.3 Erbium	69 Tm 168.9 Thulium	70 Yb 173.1 Ytterbium	71 Lu 175.0 Lutetium

Actinoids

89 Ac Actinium	90 Th 232.0 Thorium	91 Pa 231.0 Protactinium	92 U 238.0 Uranium	93 Np Neptunium	94 Pu Plutonium	95 Am Americium	96 Cm Curium	97 Bk Berkelium	98 Cf Californium	99 Es Einsteinium	100 Fm Fermium	101 Md Mendelevium	102 No Nobelium	103 Lr Lawrencium

Standard atomic weights are abridged to four significant figures.

Elements with no reported values in the table have no stable nuclides.

Information on elements with atomic numbers 113 and above is sourced from the International Union of Pure and Applied Chemistry Periodic Table of the Elements (November 2016 version).

The International Union of Pure and Applied Chemistry Periodic Table of the Elements (February 2010 version) is the principal source of all other data. Some data may have been modified.